# INTRODUCTION TO
# IMAGE
# PROCESSING
## and
# ANALYSIS

The software mentioned in this book is now available for download on our Web site
at: http://www.crcpress.com/e_products/downloads/default.asp

# INTRODUCTION TO
# IMAGE
# PROCESSING
# and
# ANALYSIS

JOHN C. RUSS

J. CHRISTIAN RUSS

## CRC Press
Taylor & Francis Group
Boca Raton   London   New York

CRC Press is an imprint of the
Taylor & Francis Group, an informa business

CRC Press
Taylor & Francis Group
6000 Broken Sound Parkway NW, Suite 300
Boca Raton, FL 33487-2742

© 2008 by Taylor & Francis Group, LLC
CRC Press is an imprint of Taylor & Francis Group, an Informa business

No claim to original U.S. Government works
Printed in the United States of America on acid-free paper
10 9 8 7 6 5 4 3 2 1

International Standard Book Number-13: 978-0-8493-7073-1 (Hardcover)

**Library of Congress Cataloging-in-Publication Data**

Russ, John C.
    Introduction to image processing and analysis / John C. Russ and J. Christian Russ.
        p. cm.
    Includes bibliographical references and index.
    ISBN 978-0-8493-7073-1 (hardback : alk. paper) 1. Image processing. I. Russ, Christian. II. Title.

TA1637.R86 2008
621.36'7--dc22                                                                     2007016992

**Visit the Taylor & Francis Web site at**
**http://www.taylorandfrancis.com**

**and the CRC Press Web site at**
**http://www.crcpress.com**

*For Helen, Jenn, and Colette*

# Contents

# Introduction

## Assumptions

This text is intended to introduce students to the programming involved in image processing and analysis. The student is assumed to have some previous programming experience, and to be able to use a C compiler and programming environment. Either a Windows (Win 2K or later) or Macintosh (OS X) computer may be used. The concentration of topics here is on the processing and measurement of images, not on peripheral subjects such as the wide variety of file formats in which images are stored, nor the display and printing of images, nor the statistical techniques for interpreting measurement data. This text is not intended to be an encyclopedia of image processing (see the References for appropriate choices), but is instead focused on the implementation of many of the most widely used and most important image processing and analysis algorithms, which requires also an understanding of their results and purpose.

Unlike many texts that deal with image processing algorithms, this book does not concentrate on the mathematical underpinnings of the field (this is particularly true in the section on Fourier transforms). Rather, it introduces just enough of the math to explain the workings of the algorithms while emphasizing the practical reasons for the use of the methods, their effects on images, and their appropriate applications. Also, the intent in the chapters that follow is to combine image processing with image analysis. As indicated in Figure 0.1, **_image processing_** comprises a broad variety of methods that operate on images to produce another image. The changes that are introduced are generally intended to improve the visibility of features and detail, or to improve the images for printing or transmission, or to facilitate subsequent analysis. Chapters 1, 2 and 3 deal with the algorithms used in these processing steps. **_Image analysis_** is the process of obtaining numerical data from images. This is usually accomplished by a combination of measurement and processing operations, as described in Chapters 4 and 5. The data obtained by measurement may subsequently be analyzed statistically, or used to generate graphs or other visualizations.

## The Program Environment

The selection of a host program for running the routines to be developed has an important consequence for the choice of a working environment to be used for writing them. We wanted to eliminate the need to create an entire working program that can acquire images from cameras or scanners, read various image file formats, and display and print images before any image processing and analysis routines could be introduced. Accordingly, the decision was taken to make use of a widely used program as a host. Adobe Photoshop® handles all of those tasks, presents a consistent user appearance on both Mac and Windows computers, and has a well-documented application program interface (API, the relevant

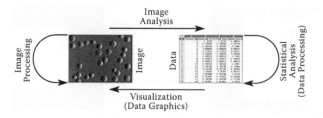

**FIGURE 0.1**
The relationships between image processing, image analysis, data processing, and data visualization.

portions of which are described in the Appendix) for "plug-ins" that can access the image data for processing and measurement.

The plug-ins themselves are dynamically linked libraries (DLLs) that can be separately compiled and placed in a folder where the program accesses them. Each individual plug-in can be dedicated to a specific function, which is listed as a separate menu item. This makes it possible for the student to easily and quickly compile and test routines, and to build a library of functions that can be applied in sequences and combinations to accomplish more complex tasks.

Furthermore, the API defined by Adobe engineers for Photoshop-compatible plug-ins has been adopted by many other programs, ranging from very inexpensive ones (e.g., Adobe Photoshop Elements®, Corel Paint Shop Pro®, Lemke Graphic Converter®) to other graphic arts packages (e.g., Deneba Canvas®, Corel Painter®) and even professional image analysis packages (e.g., Media Cybernetics Image-Pro Plus®). Several of these programs (including Photoshop) are available at reduced cost for educational and student use, or may be available on a campuswide license. The widespread use of Photoshop and the many compatible programs means that students gaining experience with programming image processing routines using this text will be directly able to use the same knowledge in many practical settings.

The use of the standard C programming language ensures a consistent style and knowledge base with other programming courses, and allows students to take advantage of the skills they have already acquired. Unlike image processing texts that use languages like Basic or Java, the use of C is consistent with professional applications, and it ensures clarity, simplicity, flexibility, and good performance. The use of C++ would have added unnecessary overhead and complexity. The interface code (often referred to as the "glue" that implements the API and binds the plug-ins to the host program) and the utilities provided on the companion CD handle the details of converting various image modes to a common one and provide the necessary supporting routines, which allows the student to concentrate on the central topics of image processing and analysis, rather than spending time on other programming tasks.

Specialized image processing texts that use Matlab®, Mathematica®, Mathcad®, and other mathematics and visualization packages make use of highly optimized and preprogrammed subroutines that can be invoked, typically by a complex command line syntax that does little or nothing to help the student understand the actual procedures or to create new ones. Indeed, there is no effort made toward optimization or efficiency in the code examples used in this text. Instead, relying on the fact that computers have become quite fast in recent years, the emphasis is entirely on clarity and consistency. Once the principles are understood and the basic routines are implemented, the student who is interested in optimization of either speed or memory usage (for example) is welcome to explore those possibilities.

The plug-in interface used for the examples and projects covered by this text can be used with all versions of Photoshop from 5.0 and later (the current version as of this writing is Photoshop CS3, aka Photoshop 10). Some of the routines that are programmed in the examples shown here duplicate basic image processing routines that are built into Photoshop (and the other programs named above). For example, just about every image processing program has a Gaussian smoothing function. However, it is still important for the student to understand how the function can be implemented as well as what effect it has on the image.

Other functions covered in this text extend standard routines in novel and very useful ways. An example is the commonly used median filter, and the conditional versions introduced in this text. Still other routines that are developed in the text go beyond the basics to provide advanced processing and measurement capabilities.

This text does not try to present an encyclopedic compendium of image processing algorithms, nor to overwhelm the reader with their mathematical complexities, notations, or derivations. The emphasis is on clear explanations of the most commonly used and most universally useful techniques, with examples. The examples include both code fragments and illustrations of results, as well as appropriate diagrams and necessary equations. The student who masters these basic tools will be able to build upon that knowledge base to handle other techniques he or she may encounter later on, in the extensive literature covering this dynamic and rapidly expanding field.

## Image Values

Images may be acquired from many sources, including digital cameras and desktop scanners as well as scientific instruments, reconnaissance satellites and, of course, via the Internet. Each image is an array (generally a rectangular array) of *__pixels__* (short for picture elements), and each pixel contains information. The address of each pixel within the image is usually specified as an $(x,y)$ pair, with $x$ indicating the distance from the left edge and $y$ indicating the distance down from the top, both values starting at zero and increasing to (width − 1) and (height − 1), respectively. Measuring $y$ downward from the top rather than up from the bottom is a historical convention that arose because of the way that computer displays (and television sets) are scanned.

*Note:* Key words that are frequently encountered in image processing and analysis are highlighted in the text, as is *__pixels__* in the preceding paragraph. It is recommended that the student become familiar with them. The index also highlights the pages where these key words are defined.

In some cases, the information associated with each pixel may be a single value, representing the grayscale (monochrome) brightness of that point in a scene. The most common type of image encountered will contain color information, which the pixel may represent as a combination of red, green, and blue (RGB) values. Other combinations of values are possible and, indeed, very useful for processing as will be seen, but the RGB triplet is the most common format and is used in Photoshop and most image processing programs. It also corresponds to the colors used in cathode ray tube (CRT) and liquid crystal (LCD) displays used to present images to the viewer, and in most cameras or scanners that are used to acquire images.

Some image sources may provide more than three values at each pixel. For example, many satellite images consist of the visible red, green, and blue channels as well as several infrared values. Also, scientific instruments may have dozens or hundreds of channels, representing different signals or elemental composition, for instance. These are not dealt with explicitly in this text, but represent straightforward generalizations of three-channel RGB color images.

In many traditional image processing programs, the pixel values themselves are treated as integers that can range from 0 (black) to 255 (white). That artificial limitation was introduced historically because it corresponded to a single 8-bit byte of computer memory, was easily handled by integer math routines in slow CPUs, and was adequate to represent the dynamic range of signals from sources such as a video camera (which actually has fewer than 256 distinguishable levels). With the advent of digital cameras and scanners (not to mention scientific devices) with greater precision, and the appreciation that faster modern computers do not impose a significant penalty for using floating point arithmetic, it seems to us better to move beyond the "8-bit integer" limitation while still preserving a degree of consistency with the past.

Accordingly, the convention used for the image handling routines throughout this book is that the values can range from 0 (black) to 255 (maximum), but are not restricted to integer values. All values are inherently treated as floating point numbers. Furthermore, all images received from and returned to the host program are considered to be RGB color images. That means images which Photoshop considers to be 8 or 16 bits per channel, and either grayscale (monochrome) or RGB color, are all converted in the interface software to consist of three values per pixel, representing red, green, and blue, respectively, and with a floating point range from 0.0 to 255.0.

For an image that is actually monochrome, the red, green, and blue values will be identical. To confirm that this is true, simply examine the computer display with a magnifying glass. When a monochrome (grayscale) image is displayed, the signals sent to the red, green, and blue dots on the screen are equal in magnitude and visually blend to produce the impression of neutral grays. When red, green, and blue are all set to 255, the result is perceived as white. For an image with a precision greater than 8 bits, the pixel values in images will be represented as having fractional parts (e.g., 143.627).

Some of the plug-in routines will perform image processing by reading the original image from the host program, manipulating the pixel values, and writing the resulting image back to the host program. The conversion of values from whatever mode the image uses in the host program to the three-value floating point format, and back again, is handled by the interface software (which is fully documented in the Appendix).

As an example of a (trivial) program that reads, alters, and rewrites values in an image, Code Fragment 0.1 performs what most programs call "inverting" an image. This is not the inverse in the mathematical sense of 1.0 divided by the value, but rather replaces each value by 255 minus the value, and should more properly be called complementing or reversing the image values. For a monochrome (grayscale) image the result is like a photographic negative (Figure 0.2). For a color image, reds are replaced by cyans, blues by yellows and greens by magentas (Figure 0.3). The reasons for these complementary colors will become more apparent in the next chapter.

**Code Fragment 0.1 — Complementing/reversing an image**

```c
// this struct is defined in the PhotoshopShell.h file
// it holds three floating point values for each pixel
typedef struct
{
   float red, green, blue;
} RGBPixel;

// the following is the UserCode.c file for the plug-in
#include "PhotoshopShell.h"// include the interface 'glue' and utilities

ErrType  MainUserEntry()
{
   ErrType  result = noErr; // accumulate any errors that happen
   long     height, width;  // pixel dimensions of image
   long     x, y;           // loop counters to traverse image
   RGBPixel *line = NULL;    // memory buffer

   GetOriginalDimensions(&width, &height); // find out how big the image is
   // Allocate memory for one row of pixels
   line = CreateAPointer(width, sizeof(RGBPixel));
   //Image reversal example
   for (y = 0; y < height; y++)
   {  // test if user canceled (ESC/Cancel) & advance the progress bar
      if (DoProgressBarTestAbort(float)y/(float)height))// 0.0 .. 1.0
      {
         result = userCanceledErr;
         goto mainexit;
      }
      ReadOriginalLine(y, line);   // read a line of pixel values
      for (x = 0; x < width; x++)  // for each pixel on the line
      {  //reverse the values
         line[x].red =   255.0 - line[x].red;
         line[x].green = 255.0 - line[x].green;
         line[x].blue =  255.0 - line[x].blue;
      }//for x
      WriteResultLine(y, line);    //write changed line back
   }// for y

mainexit:                         // clean up everything we allocated
   if (line)  // test to see if the pointer is still null (not allocated)
      DisposeAPointer(line);
   return result;                 // pass any errors back
}// MainUserEntry
```

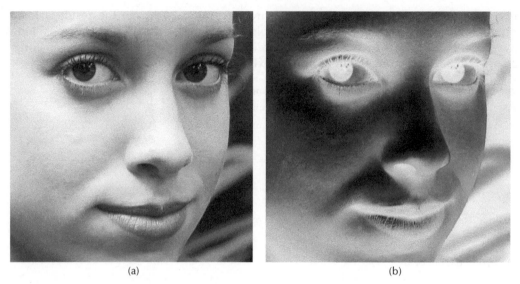

**FIGURE 0.2**
Reversing a monochrome image: (a) original [face.tif], (b) "inverse" or complement.

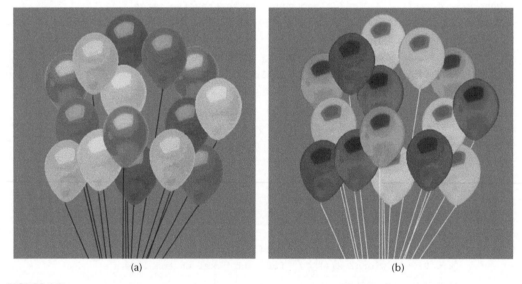

**FIGURE 0.3**
(See **Color insert following page 172.**) Reversing the values in a color image: (a) original [balloons.tif], (b) "inverse" or complement.

The code fragment shows the procedure for reading the image, one line of red-green-blue values at a time, followed by the actual processing of each pixel and, finally, rewriting each line to the host program. This line-by-line raster format corresponds to the way images are usually stored in memory, printed, displayed, and acquired. There are many reasons for this line-by-line format, some of them historical (the way televisions, cameras, and displays, and other hardware devices function) and some practical (the organization of computer memory and the efficiency that buffer memory in modern processors provides).

In the code fragment shown, a pointer is defined and memory allocated to hold one line of values (three floating point numbers for each pixel). Creating and disposing of pointers, and defining the other variables used in the code, is not shown explicitly in all of the code fragments in the text, but is assumed to be the responsibility of the student.

The interface convention used by Photoshop compatible plug-ins allows the original data to be read as many times as desired, but only to be rewritten once. (More specifically, you can write the values as many times as you want, but only the last time will matter, and reading will always obtain the original values, not reflecting any changes you have made.) Consequently, some procedures will need to keep a copy of the image data in memory to iterate upon it, and there are routines provided to create the necessary arrays in memory for that purpose. Those temporary image arrays may also contain three floating point values per pixel, but for some purposes it will be preferable to instead create arrays of a single brightness value per pixel, or a complex value with a real and imaginary part, according to the specific need.

## Input and Output

Many image processing programs incorporate elaborate user interfaces with interactive dialogs that can be manipulated via mouse and keyboard to control parameters and choices used in the routines. They may also write files of data intended for use in spreadsheet and statistical analysis programs. Although very useful, these capabilities are secondary to the central interests of this text, which are the actual image processing and analysis.

For some purposes, changing the numerical or logical values of variables used in the various routines can be accomplished satisfactorily by recompiling the routines. A greater degree of flexibility is afforded by the ability to read numeric values from a simple text file, such as can be created with most word processors (including the editor used for programming). The support routines provided with this book (and documented in the Appendix) include the ability to open a text file, and to read one or more floating point numbers from it. Several of the examples and problems use this capability to allow making adjustments to internal variables as the routines are run.

There is a corresponding routine for output that will create and open a text file, and one to write a floating point number to it (followed by a line feed and carriage return). This can be used to create a text file containing data that can be opened in a spreadsheet, or most other data analysis and plotting programs. It is suggested that students use this basic capability along with a spreadsheet program such as Excel® to prepare graphs such as some of those shown in this text.

## Compiling a Function

The following steps will allow you to create your own first image processing plug-in routine, by starting with an example project (the one that reverses or complements the image contrast, shown in the preceding code fragment) that is provided on the companion CD. The steps correspond to using Microsoft Visual Studio® on a Windows XP® or Windows 2K® computer, but can be easily adapted to other compilers or platforms. Additional details and information can be found in the Appendix. More information, late additions, instructions for compiling plug-ins for the Macintosh, and other resources, can be found at the support Web site <www.intro2imaging.com>.

1. Duplicate the **Example** folder containing a minimal example project, and give the folder an appropriate name.

2. Click on the **UserCode.sln** file to open the project.

3. Edit the **PiplData.h** file. Remember to **Save** the result after editing. This file contains several items that control the appearance of the plug-in:

   a. The main menu category (**R+R_Book** by default) defines the name that will appear in the "Photoshop Filters" menu. A submenu with the individual filter plug-ins opens when this menu entry is selected. This name is a Pascal string that is always 32 characters long, so pad out a shorter entry with spaces as shown in the example and place the actual length of the string in hex at the beginning.

   b. The submenu name for the plug-in. The format for this string is identical to that for the category.

   c. A description for the "about" box. This is a C-string and may contain any printable characters, including the programmer's name. Inserting a '\n' in the string will start a new line.

   d. A unique signature for the plug-in that is used by Photoshop for actions and history. The format is four printable characters, at least one of which must be uppercase and one lowercase. The signature must be entered in both forward and reverse order, as shown in the example.

4. Edit the **UserCode.c** file to create the plug-in. The **#include "PhotoshopShell.h"** statement is necessary to bring in the various interface and support subroutine calls that are described later and illustrated throughout this text.

5. Compile the plug-in. Be sure that **Release** is selected in Visual Studio and select **Build->Rebuild Solution**. Any errors and warnings will be reported. A Warning from the linker that the filename ends in .8BF is expected and can be ignored. Remember to select **File->Save All** and **File->Close Solution** after successful compilation.

6. Copy the compiled plug-in (**.8BF** file) to the folder in which plug-ins are stored for access by the host program, and rename the plug-in from **UserCode.8BF** (for example, **PlugInName.8BF**). Your host program must be told where this folder is located (in Photoshop, select **Edit->Preferences->Plug-Ins & Scratch Disks**). Photoshop scans this folder when it is launched, and recognizes the **.8BF** files there to build its menu. You can replace an existing plug-in file with another one having the same name while Photoshop is running, but to add a new one or to change the menu entry, you must quit and relaunch the host program.

## Problems

In each chapter, there are several sets of problems that relate to the procedures and examples shown. Within each set, there are some that can be implemented simply by putting together the example code fragments shown, and others that require a greater degree of effort on the part of the student. There is no single "correct" answer in terms of the exact code written, but the correctness of the result can be taken in many cases as evidence that the procedure has been properly implemented.

The text also describes and illustrates additional procedures that may be assigned as problems, depending on the skill level of the students, the time available and, of course, the whims and interests of the instructor.

Solutions to the problems marked with a (#) are provided both as source code and compiled plug-ins on the companion CD that is included with the book.

The images shown in the text are also provided on the CD. Each image is saved as a **.TIF** file, readable by all of the host programs listed above.

0.5.1#.  Implement a program that reads the image from the host, complements the values, and writes it back.

0.5.2#.  Implement a program that replaces the contents of an image with a horizontal (or vertical) linear ramp of grayscale values (red = green = blue). Optionally, modify this to generate a horizontal ramp in the red channel, a vertical ramp in the green channel, and a reversed vertical ramp in the blue channel, or other color combinations.

# The Authors

**John C. Russ** received his B.S. and M.S. degrees in engineering and solid-state physics from California Institute of Technology and his Ph.D. in engineering from California Coast University. At the Homer Research Labs of Bethlehem Steel Corp., in the 1960s, development of new steel alloys, such as those used in the Trans-Alaska Pipeline, was strongly linked to the microstructure as revealed by light and electron microscopy, microprobe, and x-ray analysis. In 1968, Dr. Russ became director of the Applications Laboratories at Japan Electron Optics Laboratories (JEOL), introducing the scanning electron microscope. From there, it was a natural step to join in the formation of EDAX, which became the leading supplier of microanalysis instrumentation for use on SEMs and TEMs. As senior vice president, he was deeply involved in the development of these devices, the creation of software for qualitative and quantitative interpretation of the spectra, and the imaging of elemental distributions. After the sale of EDAX to Philips, Dr. Russ joined the faculty of North Carolina State University in 1978. He also participated in research at the Danish Technological Institute. After retirement from formal teaching duties at NCSU in 1996, he accepted a position as research director of Rank Taylor Hobson, a British manufacturer of precision instrumentation. He continues to be active as an adjunct professor at NCSU, as well as a consultant and author.

As a professor in the Materials Science and Engineering Department, Dr. Russ and his students have used a broad array of microscope technologies to study materials microstructures and surfaces. These have included conventional and confocal light microscopes, electron and ion microprobes, scanning and transmission electron microscopes, x-ray and neutron tomography, and a variety of scanned probe microscopes. The need to process these images to obtain quantitative structural information led to the development of computer control for instruments and computer processing for the data. Dr. Russ has become widely known as a leader in the development and use of these tools for image analysis. At NCSU his collaborations have extended far beyond the materials science field, including food science, archaeology, biology, veterinary medicine, textiles, and others. Beyond the campus, he has worked with a worldwide range of companies in fields such as pharmaceutical and energy applications, and has been retained as an expert witness in forensic cases, both civil and criminal.

Through academic courses and workshops, Dr. Russ has presented image analysis methods to more than 4,000 students. He has taught acclaimed hands-on workshops worldwide, from Australia to Slovenia, Japan to South Africa. His more than 300 publications, including more than a dozen books, have reached thousands more. These books include *Computer Assisted Microscopy, Practical Stereology* (with Robert Dehoff), *Fractal Surfaces, The Image Processing Handbook* (now in its fifth edition), *Forensic Uses of Digital Imaging*, and *Image Analysis of Food Microstructure*. On November 16, 2006, Dr. Russ received the 2006 Ernst Abbe Memorial Award of the New York Microscopical Society for achievements made in the field of microscopy.

**J. Christian (Chris) Russ** has been writing image processing and image analysis software since 1979. His undergraduate degree is in computer science from the University of

Michigan, and he subsequently attended graduate school in biomedical engineering at the University of Texas at Austin. Presently, he owns Reindeer Graphics, Inc., a supplier of image processing and related software. He also works as a senior scientist on forensic imaging for Ocean Systems, Inc.

# 1

## Adjusting Pixel Values

Few images are acquired with perfect exposure and contrast settings, accurate color representation, uniform illumination, or an ideal point of view. Methods are available that can often correct for a variety of these shortcomings. It is important to understand and use the image histogram as a tool for examining the image contents, and to understand the differences between different representations of color and their uses. This chapter introduces those concepts, and illustrates how adjustments to pixel values can be performed to improve images.

## 1.1 Optimizing Contrast

Adjusting the image contrast to make best use of the display (and/or that of a hardcopy print) is one of the basic functions provided by virtually all image handling software. But, of course, there are several different ways to accomplish this task.

### 1.1.1 The Image Histogram

The image *histogram* is a vital tool for understanding and manipulating the pixel values in an image (Wall, Klinger, and Castleman 1974). It is simply a plot showing the number of pixels, or the fraction of the total, as a function of their value, which may be the mean intensity (the average of red, green, and blue values), the calculated luminance (weighted for the human perceptual response as described below), the individual red, green, and blue (R, G, and B) intensities, or the values of some of other color coordinates.

A common source of confusion arises from the misuse of the terms intensity, luminance, lightness, and brightness. Technically *intensity* is a linear measure of radiant power per unit area, but it is also commonly used to describe the numerical values stored in an image file. These values may or may not represent actual linear intensity data. *Luminance* is a weighted summation of intensities over the spectral range (blue to red) covered by human vision (for example, Equation 1.1). *Lightness* describes a perceptual response to light, and is defined by the Commission Internationale de L'Eclairage (CIE 1978) as being proportional to the cube root of luminance (Equation 1.5). Although, as shown in Figure 1.1, the functions are slightly different, the perception of lightness is often described as being roughly logarithmic. Human vision detects differences in lightness when regions differ by a few percent. This text will use the term *brightness*, which does not have a formal definition, for various quantities such as the average of the red, green, and blue stored intensity values.

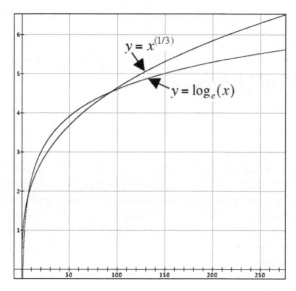

**FIGURE 1.1**
Comparison of cube root and logarithmic functions.

The formation of the histogram is shown in Code Fragment 1.1, which calculates the average value of the R, G, and B values. The number of bins into which the histogram is summed in this example is 256, corresponding to the integer range from 0 to 255 that the pixels in an 8-bit image can have. The resulting array of values is written to a disk file as a simple list that can be read by many programs, including Microsoft Excel®, for analysis or to plot it in graphical form, as shown in Figure 1.2.

```
// Code Fragment 1.1 - Calculate average brightness histogram
RGBPixel *Line;             // pointer to hold one line of pixel values
long      x, y, width, height, bright;
long      Histogram[256];   // uses a fixed number (256) of bins
long      FileID;           // file to write the histogram values

GetOriginalDimensions(&width, &height);        // read image dimensions
Line = CreateAPointer(width, sizeof(RGBPixel)); // allocate memory
for (bright = 0; bright < 256; bright ++)
   Histogram[bright] = 0;                       // clear the array
for (y = 0; y < height; y++)
{
   ReadOriginalLine(y, Line);
   for (x = 0; x < width; x++)
   { // average the R G B values
      bright = (long)((Line[x].red + Line[x].green + Line[x].blue)/3.0);
      Histogram[bright]++;                      // increment histogram bin
   } // for x
} // for y
DisposeAPointer(Line);
CreateTextFileToWrite("C:\Histogram.txt", &FileID);
for (bright = 0; bright < 256; bright++)
   WriteANumber(FileID, Histogram[bright]);
CloseTextFile(FileID);
```

(a)

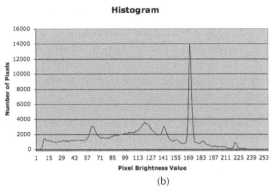

(b)

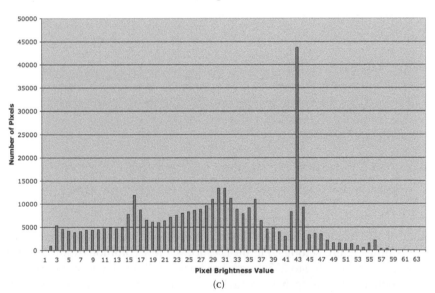

(c)

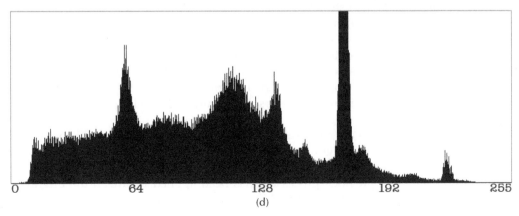

(d)

**FIGURE 1.2**
(a) The brightness (average of red, green, and blue) values of an image [harbor.tif], and the histogram plotted by Excel: (b) counted in 256 bins, (c) counted in 64 bins, (d) counted in 768 bins.

Because the images accessed by the plug-in architecture described in the Introduction may have other, greater bit depths, the floating point values that are available for calculation in the plug-ins may have greater precision, which would permit a greater number of bins. As the values still cover the range from 0 to 255, either the increment that corresponds to the width of each bin or the number of bins should be a variable in the code, as shown in Code Fragment 1.2. The advantage of a histogram with more than 256 bins is the ability to represent details for images that have more than 8 bits of tonal resolution (but the plot may become too wide to see in its entirety), whereas histograms with fewer bins will have more counts in each bin to minimize statistical fluctuations, which is particularly appropriate for smaller images.

```
// Code Fragment 1.2 — Calculate histogram with variable number of bins
{  // ... define N = number of bins
   long Histogram[N];     // array of N bins
   float bright;          // average of red, green, blue
   long index;            // address in histogram array
   // ... define variables, initialize histogram array,
   // ... create line pointer, as shown above
   for (y = 0; y < height; y++)
   {
      ReadOriginalLine(y, Line)
      for (x = 0; x < width; x++)
      {  // ... calculate bright = average of R, G, B as shown above
         index = (long)((float)N * bright / 255.0);
         Histogram[index]++;
      } // for x
   } // for y
   // ... dispose of pointer, write histogram file, as shown above
}
```

Examination of the histogram can reveal many problems with the image, such as poor exposure or contrast (gain) settings. Figure 1.3 shows a few common examples. If the histogram does not cover the full 0 to 255 range, the image will display low contrast and can be improved by either reacquiring it with higher gain or by image processing as discussed below. If the image histogram shows pile-ups at either end, it indicates that the image has many pixels that are black (0) and/or white (255). The original pixel brightness exceeded the full range of the sensor or electronics, and the values were clipped to these limits. Information has been lost from the image, and details in the white or black areas are gone and cannot be recovered. The only solution in these cases is to reacquire the images with better exposure or lighting. Comparison of the red, green, and blue histograms may reveal that the image has a color cast if one channel is substantially darker or brighter than the others.

Some other image problems are more subtle but can still be detected in the histogram. An image that is predominantly light or dark can be improved by the processing steps discussed below. If the image histogram shows gaps (sometimes called a "comb pattern"), it usually indicates that previous processing has been applied to stretch the brightness scale and spread some of the values out while compressing others, and because this may have caused some information to be lost, it may be desirable to also examine the original, unmodified image file. (A comb pattern could also be the result of an imaging device that recorded very few intensity levels, such as some Web cams.)

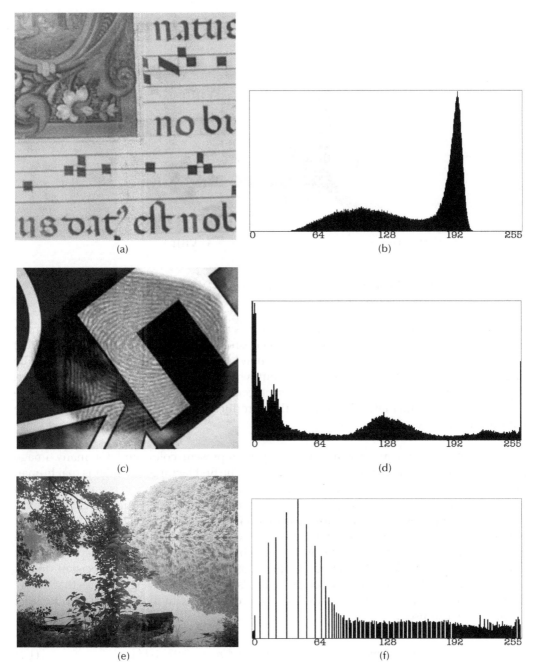

**FIGURE 1.3**
Several images with their histograms: (a,b) [manuscript.tif] showing low contrast, (c,d) [fingerprint.tif] illustrating clipping at ends, (e,f) [lake.tif] with missing values resulting from prior processing.

### 1.1.2  Other Color Coordinates

The RGB color values for each pixel correspond to the ways that many camera sensors work, the way most color images are stored in the computer, and the way that the images are displayed on color monitors. Most digital cameras use colored RGB filters in front of the individual sensors, so that each one records the intensity in a limited color band. A

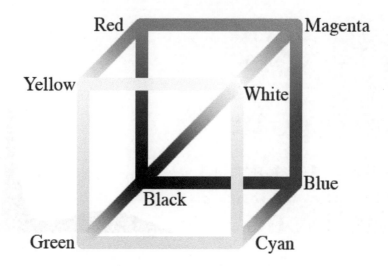

**FIGURE 1.4**
(See **Color insert following page 172**.) Diagram of RGB color space coordinates. The diagonal line through the cube is the grayscale axis from black to white.

few cameras use other filter combinations, but convert the results to RGB within the camera firmware. The computer display uses red, green, and blue phosphors (for a CRT monitor) or colored filters (for an LCD flat screen display) that are individually too small to be distinguished at normal viewing distances, but whose proportional brightnesses are visually blended to produce all of the various perceived colors. RGB color coordinates are very convenient mathematically, because they produce a color space that is a simple cube with orthogonal axes (Figure 1.4).

However, RGB values are not the only way to represent color, and for many image processing purposes are not the most appropriate. The first factor to consider is that human vision is not equally sensitive to red, green, and blue. Luminance is calculated as a weighted sum of these color components, as shown in Equation 1.1. (Figure 1.19 compares the result of the weighted luminance calculation to the average of the red, green, and blue.) There is some debate over what proper weighting values for the red, green, and blue should be used. Those shown in the equation were derived for the specific color phosphors used in color television sets, and in spite of their apparent precision do not take into account the variation between individuals. Other sets of weight values that follow the same general pattern, corresponding to the greatest sensitivity to green, and the least for blue, are also used.

$$Luminance = 0.2125 \cdot Red + 0.7154 \cdot Green + 0.0721 \cdot Blue \tag{1.1}$$

Color values may be represented in any of several ways. So-called *hue–saturation–intensity* (HSI) spaces use a polar coordinate representation for color. There are several specific versions such as HSV (hue-saturation-value), HLS (hue-lightness-saturation), and HSB (hue-saturation-brightness), which vary in details. The intensity (or brightness, lightness, or value) is calculated as some blending of the three original R, G, B values. The hue value is an angle, corresponding to the color wheel that many of us were introduced to in kindergarten. Starting at red (some versions of this space instead define blue as zero degrees), the hue changes progressively through orange, yellow, green, cyan, blue, magenta, and back to red (Figure 1.5).

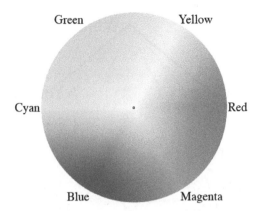

**FIGURE 1.5**
(See **Color insert following page 172**.) The color wheel, with uniformly spaced R, G, and B.

Points along the periphery of this wheel from red through green to blue have corresponding wavelength equivalents although they are not uniformly spaced, and the same perception of color can be produced by combinations of many discrete wavelengths. The colors in the magenta segment of the wheel from blue to red do not correspond to wavelengths of light but, rather, are perceived colors that represent a relative absence of green.

The saturation value is the radius, so that zero saturation removes color and leaves just a neutral gray, and maximum saturation corresponds to the pure colors. Thus, pink is a reduced saturation red, and sky blue is less saturated than ocean blue. Many people, particularly artists (who use different words such as shade and tone to represent these coordinates), find that hue–saturation–intensity spaces provide a useful and somewhat intuitive description of colors.

Equation 1.2 (Gonzalez and Woods 2002) shows a set of calculations that can be used to convert from RGB to a bi-conical HSI color space. Figure 1.6 shows an HSI color coordinate space graphically. All of the various color space coordinates are related to each other with some distortions. To see the relationship between the bi-conical hue–saturation–intensity space in Figure 1.6 and the RGB cube in Figure 1.4, tilt the cones so that their black-white central axis lies along the black-white body diagonal of the cube. Hue–saturation–intensity representation is sometimes shown as a single cone with the apex at black, or as a cylinder, which does not fit the cube as well.

$$Intensity = \frac{R+G+B}{3}$$

$$Saturation = 1 - \frac{\min\{R,G,B\}}{Intensity}$$

$$\theta = \cos^{-1}\left\{ \frac{R - \frac{(B+G)}{2}}{\sqrt{(R-G)(R-G)+(R-B)(G-B)}} \right\} \tag{1.2}$$

$$Hue = \left\{ \begin{array}{l} (B \le G):\theta \\ (B > G):360° - \theta \end{array} \right\}$$

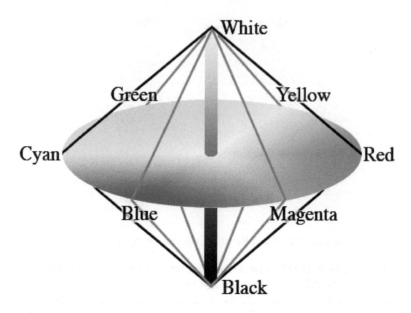

**FIGURE 1.6**
(See **Color insert following page 172**.) Diagram of bi-conical hue–saturation–intensity color space.

It is instructive to consider how the bi-conical hue–saturation–intensity space treats bright and dark color values. At the white point there is no possibility of increasing color saturation except by reducing the intensity. That matches the behavior of a computer monitor, which produces white by displaying the maximum intensity in all three colors. To increase the amount of any color (and hence the saturation), it is necessary to reduce the intensity of the complementary color values and, hence, the overall brightness. A corresponding effect occurs at the black point, where increasing saturation requires increasing the brightness. Yellow appears brighter than red or green when it is displayed on a computer monitor because it is produced by turning on both green and red. The same situation occurs for cyan (green and blue) and magenta (blue and red).

Mathematically, hue–saturation–intensity color spaces are awkward because of the polar coordinates, hue varying modulo 360°, and saturation that varies with intensity.

Converting the RGB values in the original image to another set of color coordinates, such as HSI, can be very useful for processing the image contents. However, if the values are written back to the host program, the display may be confusing. The host program and the hardware will still display the values as though they are RGB, often with bizarre results. By showing just a single color channel at a time, the data can be examined and interpreted. The display may still be labeled as red, green, or blue, but the intensities will then represent the calculated new color coordinate values. In the example shown in Figure 1.7 these are hue, saturation, and intensity (which in this example is just the average of the red, green, and blue and is visually a monochrome representation of the scene).

Another major class of color coordinate representations uses orthogonal axes rather than polar coordinates. A simplified ***spherical LAB*** space (Equation 1.3, Code Fragment 1.3) maps color values into a sphere as shown in Figure 1.8. The axis from the north to south poles is the intensity (*L*) axis. This *L* should not be confused with the CIE lightness value

(a)  (b)

(c)

**FIGURE 1.7**
(a) Hue, (b) saturation, and (c) intensity values calculated for the [harbor.tif] image displayed as grayscale intensities (note that the hue scale jumps abruptly from 255 to 0 at red).

below, which is denoted $L^*$. The $L$ shown for the spherical LAB space may be calculated either as a weighted sum of the RGB values or (as shown) their average. The $A$ and $B$ axes run from red to green ($A$) and yellow to blue ($B$) as shown.

Note that in the circle containing these axes, colors are distributed differently than in the color wheel shown in Figure 1.5. Instead of having red, green, and blue equally spaced around the wheel at 120° intervals, now green is opposite red and yellow is opposite blue. The spherical shape of the space limits the amount of color that can be present near the white (north pole) and black (south pole) ends of the $L$ axis. This is a relatively simple space mathematically, and is often convenient for image processing as will be seen below, but produces several perceptual problems. The most important problem is that on the

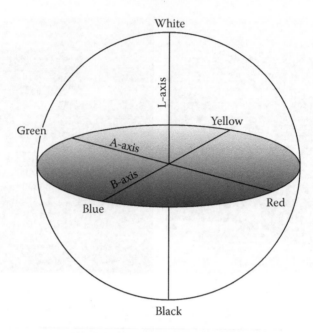

**FIGURE 1.8**
(See **Color insert following page 172**.) Diagram of a spherical LAB color space.

distorted color wheel, moving in a straight line does not produce a linear variation in the perceived color.

$$L = \frac{R+G+B}{3}$$

$$A = R - G$$

$$B = \frac{R+G}{2} - B$$

$$R = L + \frac{A}{2} + \frac{B}{3}$$

$$G = L - \frac{A}{2} + \frac{B}{3}$$

$$B = L - \frac{2 \cdot B}{3}$$

(1.3)

```
// Code Fragment 1.3 — Conversion between RGB and spherical LAB coordinates
    // convert RGB -> LAB
    L = (Red + Green + Blue) / 3;
    A = (Red - Green);
    B = (Red + Green) / 2 - Blue;
```

```
// convert LAB -> RGB
Red   = L + A / 2 + B / 3;
Green = L - A / 2 + B / 3;
Blue  = L - 2 * B / 3;
```

As with the HSI values shown in Figure 1.7, the LAB values can be examined as individual channels in grayscale displays. This is shown in Figure 1.9.

The **_YIQ_** space used in broadcast television is similar to LAB but uses weights as shown in Equation 1.4. Note that the $Y$ value (called Luma) calculated from the RGB values by these weights is different from the CIE lightness value ($L^*$) shown below (Equation 1.5) and from the luminance calculated in Equation 1.1 above. The weights in YIQ are adjusted for the fact that the RGB values in this application are nonlinear. They were defined for broadcast color television in the 1950s and do not properly describe the color phosphors used in current computer monitors, but are still used.

$$\begin{bmatrix} Y \\ I \\ Q \end{bmatrix} = \begin{bmatrix} 0.30 & 0.59 & 0.11 \\ 0.60 & -0.28 & -0.32 \\ 0.21 & -0.52 & 0.31 \end{bmatrix} \cdot \begin{bmatrix} R \\ G \\ B \end{bmatrix} \tag{1.4}$$

The color coordinates that are believed to best represent color perception are a modification of *LAB* space called **_CIELab_** or **_CIEL\*a\*b\*_**, in which the simple circular color wheel is replaced by a different shape (Figure 1.10) in which linear blending of colors is achieved; additional modifications are sometimes used to produce a space in which equal distances represent equal perceived color differences. In the CIELab space, if two colored phosphors

(a)  (b)

**FIGURE 1.9**
Display of the (a) red–green A channel and (b) yellow–blue B channel calculated for the [harbor.tif] image and displayed as grayscale intensities. The brightness (L) channel is identical to the result shown in Figure 1.7c.

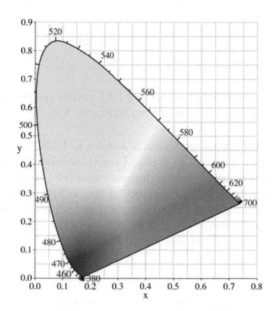

**FIGURE 1.10**
(See **Color insert following page 172**.) CIELab color. (The intensity axis is perpendicular to the plane shown.)

or filters are used, varying their individual intensities will produce colors that lie on the straight line between their coordinates. If three colors are used, controlling their proportional intensities can produce any color within the triangle they define. That triangle defines the *gamut* of the display; colors outside the triangular range cannot be generated. Colors along the curved periphery of the diagram are fully saturated and correspond to wavelengths of light. Colors along the straight edge from blue to red (called the "line of purples") are perceived colors resulting from a deficiency of green.

Unfortunately, converting measured RGB values to and from CIEL*a*b* space is not simple. It requires two steps (Equation 1.5), first going from RGB to an intermediate set of XYZ values, and then to CIEL*a*b*, and depends on the color coordinates of neutral white $(X_n\,Y_n\,Z_n)$ in the intermediate XYZ space. The $L^*$ value is the lightness, and the equation shown strictly applies only for $Y/Y_n > 0.08856$, but for practical purposes that is virtually black and below the threshold of visibility. The white point, in turn, depends on the scene illumination (and perhaps on the individual viewer). The scene illumination is often specified as a color temperature, which is approximately the temperature of an ideal black box radiator that would emit a particular spectrum of light. Reddish light (as from an incandescent bulb) has a lower color temperature than sunlight, whereas bluish light (as from a fluorescent tube) has a higher color temperature. Adjusting the colors of recorded images by defining the neutral color values may be performed through this transformation.

All of these color spaces require three values to specify the color, and they all have somewhat different shapes. Image histograms can be formed with any of these color coordinate systems. The luminance histogram (weighted values of R, G, and B) is interpreted in the same way as the brightness value shown before (average of R, G, and B). Histograms of hue and saturation may be examined to see whether images consist predominantly of a few colors, or are low or high in overall saturation.

$$\begin{bmatrix} X \\ Y \\ Z \end{bmatrix} = \begin{bmatrix} 0.412453 & 0.357580 & 0.180423 \\ 0.212671 & 0.715160 & 0.072169 \\ 0.019334 & 0.119193 & 0.950227 \end{bmatrix} \cdot \begin{bmatrix} R \\ G \\ B \end{bmatrix}$$

$$L^* = 116 \left( \frac{Y}{Y_n} \right)^{1/3} - 16 \qquad (1.5)$$

$$a^* = 500 \left[ \left( \frac{X}{X_n} \right)^{1/3} - \left( \frac{Y}{Y_n} \right)^{1/3} \right]$$

$$b^* = 200 \left[ \left( \frac{Y}{Y_n} \right)^{1/3} - \left( \frac{Z}{Z_n} \right)^{1/3} \right]$$

Each of these sets of color coordinates is optimized for some specific purpose. Some are more useful for image processing whereas others are used primarily for image display, television transmission, or printing. Printing images is a particularly complicated subject, in which at least four inks (cyan, magenta, yellow, and black, or CMYK) are used, and conversion from the stored RGB intensity values is required. Some printers employ additional inks (e.g., cyan, light cyan, yellow, magenta, light magenta, black, and gray) and have proprietary software to calculate the conversion from RGB that also takes into account the specific papers being used. There are many books and other references on color science that describe these applications in detail (e.g., Sharma 2003), but the preceding introduction will be sufficient for the purposes of this text.

### 1.1.3 Maximizing Contrast

If the image histograms show that the brightness or luminance values do not cover the full 0 to 255 range that can be displayed, stretching them to cover that range is a simple way of increasing the visual contrast in the image. It is important to do this only for the intensity, brightness, or luminance values, and not for the other color axes. For each of the color spaces introduced above, stretching the values along the intensity scale should be accomplished without altering the color values. This can be difficult when the color space has a shape in which the maximum saturation varies with intensity.

Code Fragment 1.4 shows an attempt to maximize image contrast by stretching the individual R, G, and B values for an image. It scans through the image to find the maximum and minimum values in each color channel, and then uses these values in a second scan through the image to scale the values in each color channel linearly between these limits. As shown in Figure 1.11, this produces color shifts in the resulting image because the relative proportions of R, G, and B are altered. Changing the logic to find only a single set of minimum and maximum values across all three color channels (Code Fragment 1.5) produces a better result (Figure 1.11c). Converting the image to a color space in which the weighted luminance values are stretched between their minimum and maximum while keeping the color values unchanged is shown in Figure 1.11d and Code Fragment 1.6.

```
// Code Fragment 1.4 — Stretch individual R, G, B values
{  // read image, find min and max R, G, B values
    float    Red, Green, Blue;
    float    minR, minG, minB;
    float    maxR, maxG, maxB;
    float    scaleR, scaleG, scaleB;
    // ... declare variables, create line pointer, get height & width
    minR = minG = minB = 255.0;
    maxR = maxG = maxB = 0.0;
    for (y = 0; y < height; y++)
    {
        ReadOriginalLine(y, Line);     // first find min and max
        for (x = 0; x < width; x++)
        {
            Red   = Line[x].red;
            Green = Line[x].green;
            Blue  = Line[x].blue;
            if (Red   > maxR) maxR = Red;    if (Red   < minR) minR = Red;
            if (Green > maxG) maxG = Green; if (Green < minG) minG = Green;
            if (Blue  > maxB) maxB = Blue;   if (Blue  < minB) minB = Blue;
        } // for x
    } // for y
    scaleR = 255.0 / (maxR - minR);
    scaleG = 255.0 / (maxG - minG);
    scaleB = 255.0 / (maxB - minB);
    for (y = 0; y < height; y++)      // now stretch each channel
    {
        ReadOriginalLine(y, Line)
        for (x = 0; x < width; x++)
        {
            Red   = Line[x].red;
            Green = Line[x].green;
            Blue  = Line[x].blue;
            Line[x].red   = scaleR * (Red   - minR);
            Line[x].green = scaleG * (Green - minG);
            Line[x].blue  = scaleB * (Blue  - minB);
        } // for x
        WriteResultLine(y, Line);
    } // for y
    // ... dispose pointer
}

// Code Fragment 1.5 — Stretch with single min and max for all channels
{  // ... read image, find min and max
    float Red, Green, Blue;
    float min, max, scale;        // single min, max and scale
```

```
   // ... declare variables, create line pointer, get height & width
   min = 255.0; max = 0.0;
   for (y = 0; y < height; y++)
   {
      ReadOriginalLine(y, Line); // first find min and max
      for (x = 0; x < width; x++)
      {
         Red   = Line[x].red;
         Green = Line[x].green;
         Blue  = Line[x].blue;
         if (Red   > max) max = Red;   if (Red   < min) min = Red;
         if (Green > max) max = Green; if (Green < min) min = Green;
         if (Blue  > max) max = Blue;  if (Blue  < min) min = Blue;
      } // for x
   } // for y
   scale = 255.0 / (max - min); // used for all channels
   for (y = 0; y < height; y++) // now stretch each channel
   {
      ReadOriginalLine(y, Line)
      for (x = 0; x < width; x++)
      {
         Red   = Line[x].red;
         Green = Line[x].green;
         Blue  = Line[x].blue;
         Line[x].red   = scale * (Red   - min);
         Line[x].green = scale * (Green - min);
         Line[x].blue  = scale * (Blue  - min);
      } // for x
      WriteResultLine(y, Line);
   } // for y
   // ... dispose pointer
}

// Code Fragment 1.6 — Stretch luminance, adjust RGB proportionately
{ // read image, find min and max
   float    Red, Green, Blue;
   float    min, max, scale, luminance, newluminance;
   // declare variables, create line pointer, get height & width
   min = 255; max = 0;
   for (y = 0; y < height; y++)
   {
      ReadOriginalLine(y, Line); // first find min and max
      for (x = 0; x < width; x++)
      {
         Red   = Line[x].red;
         Green = Line[x].green;
```

```
            Blue   = Line[x].blue;
            luminance = 0.25 * Red + 0.65 * Green + 0.10 * Blue;
                // or other similar weighting factors
            if (luminance > max) max = luminance;
            if (luminance < min) min = luminance;
        } // for x
    } // for y
    scale = 255.0 / (max - min);
    for (y = 0; y < height; y++)   // now stretch each channel
    {
        ReadOriginalLine(y, Line)
        for (x = 0; x < width; x++)
        {
            Red   = Line[x].red;
            Green = Line[x].green;
            Blue  = Line[x].blue;
            luminance = 0.25 * Red + 0.65 * Green + 0.10 * Blue;
            if (luminance < 0.5) luminance = 0.5;
            newluminance = scale * (luminance - min);
            Line[x].red   * = newluminance/luminance;
            Line[x].green * = newluminance/luminance;
            Line[x].blue  * = newluminance/luminance;
        } // for x
        WriteResultLine(y, Line);
    } // for y
    // ... dispose pointer
}
```

Using only the extreme values of brightness in the entire image as the limits for the contrast stretch makes the result very sensitive to the presence of small bright spots (reflections or glare) or dark shadows, which may occupy only a tiny fraction of the image and be relatively unimportant. Allowing a small percentage of the pixels in the image to be pushed out of the 0 to 255 range and clipped to those limits can produce greater visual contrast in the image. To do that requires constructing the image histogram instead of just scanning through the image for the extreme values. Code Fragment 1.7 shows an example in which a fixed percentage of the pixels is allowed to be clipped at each end of the spectrum, producing a result as shown in Figure 1.12. In many cases, the amount of clipping will vary from a fraction of a percent up to about one percent, and may be adjusted by the user. The histogram must have a large enough number of bins to provide sufficient precision.

(a)

(b)

(c)

(d)

**FIGURE 1.11**
(See **Color insert following page 172**.) Maximizing image contrast: (a) original image [rose.tif], (b) stretching RGB histograms individually (note the yellow color cast), (c) using a single maximum and minimum for all channels, (d) stretching the luminance.

```
// Code Fragment 1.7 — Stretch with clipping
{  // ... first construct histogram[N] of brightness or luminance values
   //  (N = number of bins, at least 1024)
   // ... define or read clip = fraction of pixels to be clipped at each end
   //  (e.g., 0.0025 = 0.25%)
   float min, max, sum = 0, total = height * width;
   long index, mindex = 0, maxdex = N;
```

```
for (index = 0; index < N; index++)
{
    sum += histogram[index];
    if ((mindex==0) && ((sum/total) > clip))      mindex = index;
    if ((maxdex==N) && ((sum/total) > (1-clip)))  maxdex = index-1;
} // for index
min = 255.0 * mindex / N;
max = 255.0 * maxdex / N;
// ... read image, adjust values using min & max as in fragment 1.5 or 1.6
// ... if new value < 0, set to 0; if new value > 255, set to 255
}
```

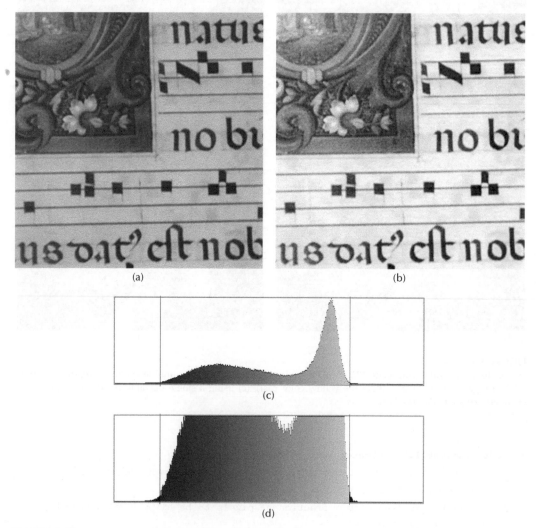

**FIGURE 1.12**
(See **Color insert following page 172.**) Stretching contrast with clipping: (a) original image [manuscript.tif], (b) result of stretching with clipping, (c,d) histogram with limits marked to show 0.1% of the pixels that are clipped to black and white, respectively.

### 1.1.4 Nonlinear Stretching

The procedures described above perform a linear stretch of the pixel values between the selected white and dark points. There is no fundamental reason for this limitation. Applying a nonlinear function can be accomplished in many ways, but the most common methods mimic the behavior of photographic darkroom techniques, and are described by a single constant called **_gamma_**. Photographic film responds logarithmically to light intensity (similar to the response of human vision). A plot of film density versus light intensity is a straight line on a logarithmic scale over the central, useful range. This is shown in Figure 1.13. The slope of the line is the gamma value. Film with a high gamma value produces high-contrast images, whereas low gamma values accommodate a wider range of brightness.

Computer processing of a digitally recorded image can alter brightness values by adjusting the gamma value. Varying the value of gamma using Equation 1.6 can brighten dark areas of the image, or darken bright areas, as shown in Figure 1.14.

$$Intensity = 255 \cdot \left( \frac{Original - Minimum}{Maximum - Minimum} \right)^{(1/\gamma)} \qquad (1.6)$$

Photoshop, and most of the other host programs, provide several built-in tools that can adjust the gamma value to modify the pixel brightnesses in an image. Many of these function in RGB space, and may therefore alter colors by changing the relative proportions of colors as shown above. By first converting the image from RGB to a different color space, and adjusting only the luminance or brightness channel, these shifts can be avoided. The purpose here is not to make the adjustments using Photoshop's tools, but to understand the underlying principles by programming the adjustments directly. Code Fragment 1.8 shows the procedure for altering the gamma value for an image. It assumes that the appropriate maximum and minimum (white and black points) have been determined

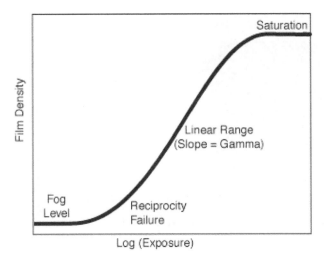

**FIGURE 1.13**
Diagram of film response to light intensity. High-contrast film has a steep slope in the central linear section, as compared to low-contrast film, which can cover a greater exposure range.

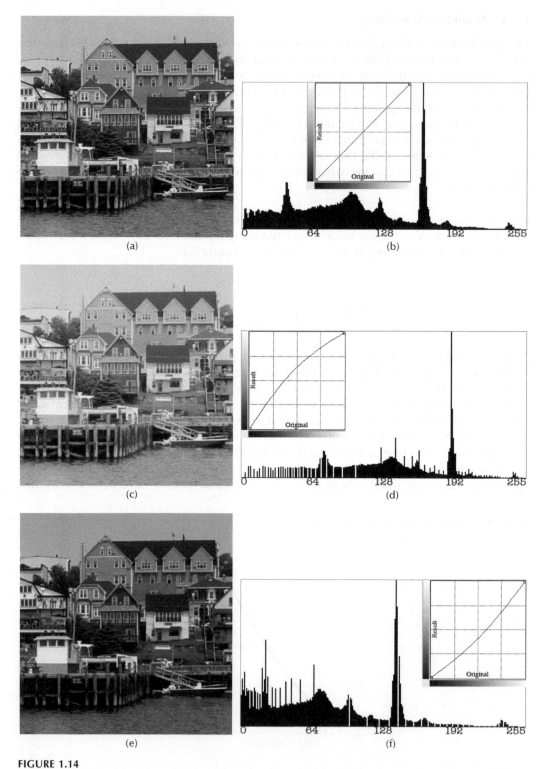

**FIGURE 1.14**
Varying gamma: (a,b) original image [harbor.tif] and its histogram, (c,d) adjusting gamma greater than 1 with transfer function and resulting histogram, (e,f) adjusting gamma less than 1 with transfer function and resulting histogram.

beforehand, for example, using the method in Code Fragment 1.7, and that the pixel RGB values are converted to a set of color values that separate the intensity and color values, using any of the methods described above.

```
// Code Fragment 1.8 — Applying a gamma adjustment
{  // ... determine min and max as above
   // ... read image, for each pixel get value = brightness or luminance
   value = (value - min) / (max - min);// convert to range 0..1;
   if (value < 0) value = 0; if (value > 1) value = 1; // clip extremes
   value = pow(value, 1.0/gamma);        // apply gamma function
   value = min + (max - min) * value;   // convert back to 0..255 range
   // ... proceed to adjust R G B as shown above
}
```

The examples shown in Figure 1.14 include a plot of the relationship between the original pixel brightness values and the new, adjusted values. This plot is called the ***transfer function*** or curve, and in principle can have any shape, not just the gamma curves shown in the figure. A method that has particular utility creates a transfer function whose shape is specific to each image (Stark and Fitzgerald 1996; Stark 2000). The technique is called ***histogram equalization***, and is best understood by again looking at the image histogram. As shown in Figure 1.15, the histogram of an image can be replotted as a cumulative sum, representing the fraction of the image area (the total number of pixels) that has values less than or equal to each brightness level. Such a cumulative plot can be constructed using a spreadsheet, or with a calculation such as that shown in Code Fragment 1.9. The conventional histogram

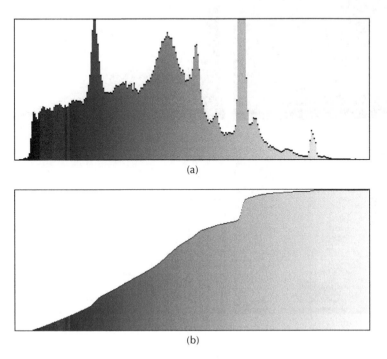

(a)

(b)

**FIGURE 1.15**
Example of (a) the conventional histogram for the [harbor.tif] image and (b) the corresponding cumulative or integral histogram.

usually plots the number of pixels as a function of brightness, whereas the cumulative histogram has a vertical axis that ranges from 0 to 100% of the image area.

```
// Code Fragment 1.9 — Summing the cumulative histogram
// ... assumes the Histogram[N] array has already been created
// ... and allocate Cumulative[N] to hold result
total = height * width;
sum = 0;
for (i = 0; i < N; i++)
{
    sum += Histogram[i];
    Cumulative[i] = sum / total; // values range from 0..1
} // for i
```

The cumulative histogram can be used as the transfer function, as shown in Figure 1.16, to produce an image in which the pixel brightness values are uniformly distributed across the available 0 to 255 range. In the resulting image the histogram is flat (on the average), and the cumulative histogram is a straight line, as shown. Note that because of the stretching of contrast in some portions of the brightness range, and compression in others, the histogram has some gaps in it, and the "straight line" description is only true as an average.

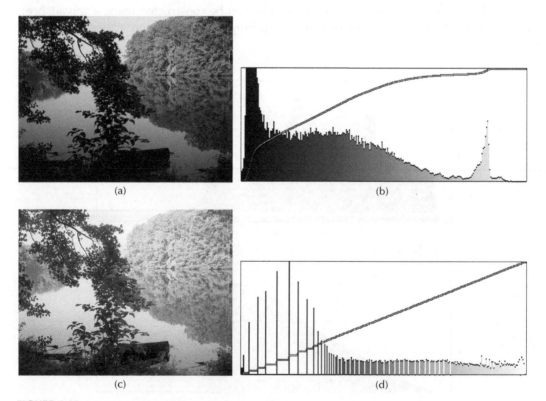

(a)                                    (b)

(c)                                    (d)

**FIGURE 1.16**
Histogram equalization: (a) original image [lake.tif], (b) superimposed histogram and cumulative histogram of (a), (c) resulting image after processing, (d) superimposed histogram and cumulative histogram of (c).

Histogram equalization is a useful shortcut method to selectively expand the contrast in predominant areas of the image. If there are many pixels that are bright, or dark, or have any other particular range of values, the contrast is expanded in those areas. Code Fragment 1.10 shows how the cumulative histogram can be used as a transfer function to alter the pixel brightness values.

```
// Code Fragment 1.10 — Applying a transfer function
{ // ... assume transfer[256] contains the brightness values to be assigned
   //     (for histogram equalization this is just the cumulative histogram)
   // ... read image as shown above; for each pixel calculate the brightness
   index = (long)brightness;
   fraction = brightness - (float)index;
   // interpolation is needed if the original precision exceeds 8 bits
   value = (1 - fraction) * transfer[index];
   // perform interpolation
   if (index < 255) value += (fraction * transfer[index + 1]);
   // ... scale red, green, blue values times (value/brightness)
   // ... write modified image data back
}
```

### 1.1.5  Problems

1.1.5.1#. Implement a program to construct and save the histogram of brightness or of the individual red, green, or blue channels. Modify this to save the histogram of the luminance using several different weighting factors for red, green, and blue, and compare the results.

1.1.5.2. Implement a program to convert an R, G, B image to each of the other color spaces, and a second one to convert it back. Note that when the channels have been converted to another color space, the display of the image on the computer screen will not be correct (the program and hardware are still interpreting the values as though they are RGB). Examine the channels one at a time to better see the information.

1.1.5.3. Produce a negative image by replacing each R, G, B value with a (255 – value) as in Problem 0.5.1. Compare this to a negative image produced by replacing the brightness or luminance value with (255 – value) while preserving the color values, using any of the color models shown above.

1.1.5.4#. Implement a program to stretch contrast linearly between brightness values that are determined by clipping a fixed percentage (e.g., 0.25%) of the pixels to black and white, respectively. Optionally, read the clipping percentage from a text file.

1.1.5.5. Implement a program to apply a gamma function to the brightness values determined by converting the original RGB values to another color space, and then converting back. Read the gamma value from a text file.

1.1.5.6#. Implement a program to apply histogram equalization to the brightness values in an image. Compare your result (both the image and its histogram) to that produced by the Photoshop equalization routine.

**FIGURE 1.17**
Superimposing the histogram onto an image [rose.tif].

1.1.5.7.   Combine the histogram equalization procedure in which the transfer function is derived from the cumulative histogram with a gamma correction. Modify the transfer function with the gamma values before applying it to the pixel values.

1.1.5.8.   In addition to saving data to a text file on disk, it is sometimes useful to create a graph for immediate visual interpretation. One simple but effective way to do this is to use the image itself as a place to do the drawing. In the example shown in Figure 1.17, the luminance histogram of an image has been superimposed directly onto the original image. In the example, the tallest bar rises to half the height of the image, the number of bins in the histogram is set equal to the width of the image, and each bar is drawn with a black and white dot at the top, with grayscale values below to show the variation from dark to light. These variations are not necessary, but illustrate some of the graphic options that are possible. Implement a program that draws the image histogram into the image window. Note that this overwrites image data, and therefore it is important to perform this on a duplicate of the original image.

## 1.2   Color Correction

### 1.2.1   Neutral Gray Methods

Human vision is quite tolerant of images in which the colors are "off." This occurs, for example, when a scene is illuminated by different light sources. Viewing or photographing the same scene under sunlight, incandescent lighting, fluorescent lighting, etc., produces different spectral values for the resulting image, but the overall interpretation of the scene and of the relative colors of the objects in the scene remains remarkably consistent. This is particularly true when the scene includes significant areas of neutral grays, which are used to visually compensate for the illuminant color. Museums display art well separated on neutral colored walls for this reason.

However, for many purposes ranging from catalog advertising to scientific imaging, it is desirable to adjust the colors in the recorded image to accurately represent the actual colors in the scene. This can also be important when showing an image to someone who is not familiar with the scene and could misinterpret the colors. For the best results, it is necessary to record an image of known reference colors under the same illumination, a procedure that has long been used in conventional photography.

As described in the preceding section, the use of CIEL*a*b* color coordinates requires knowing the color of the illumination. That will ensure that neutral colors (white and grays) are represented without any residual color cast, although the brightness scale may still be incorrect, and other colors may still be in error. If the color in an image of a neutral gray target can be measured, it can be used to define the white point in CIEL*a*b* space, and the color values reconverted to an adjusted set as shown above.

A procedure that uses information from the image itself can also be used to adjust the neutral tones. In its simplest form, it relies on the user to locate regions in the image that should be nearly white (brightest intensity and without color) and nearly black (lowest intensity and without color). These regions are adjusted to have equal values of R, G, and B. The RGB values of these regions are used as upper and lower limits, respectively, to perform linear stretching of the individual RGB values. Code Fragment 1.11 illustrates this procedure by reading from a text file two sets of values, representing the RGB values for regions selected to become neutral white and black. If in addition there are regions with intermediate brightness values that are known to be neutral grays, the RGB values at those locations could also be used to calculate individual gamma corrections for each color channel, producing neutral gray values in the image by adjusting the RGB values to be exactly equal.

```
// Code Fragment 1.11 — Linear stretching RGB values between neutral limits
{  // read maxred, maxgreen, maxblue, minred, mingreen, minblue values
   ErrType   err;
   File_ID   myfile = 0;
   err = OpenTextFileToRead("c:\\RGBLimits.txt", &myfile);
   if (err == noErr) err = ReadANumber(myfile, &maxred);
   if (err == noErr) err = ReadANumber(myfile, &maxgreen);
   if (err == noErr) err = ReadANumber(myfile, &maxblue);
   if (err == noErr) err = ReadANumber(myfile, &minred);
   if (err == noErr) err = ReadANumber(myfile, &mingreen);
   if (err == noErr) err = ReadANumber(myfile, &minblue);
   if (myfile != 0)  CloseTextFile(myfile);
   // ... use these in the same procedure as code fragment 1.4
}
```

This neutral-tone approach to color correction is widely used, and sometimes made automatic by finding the brightest and darkest values in the image and making the assumption that they correspond to neutral white and black and, consequently, should have equal R,

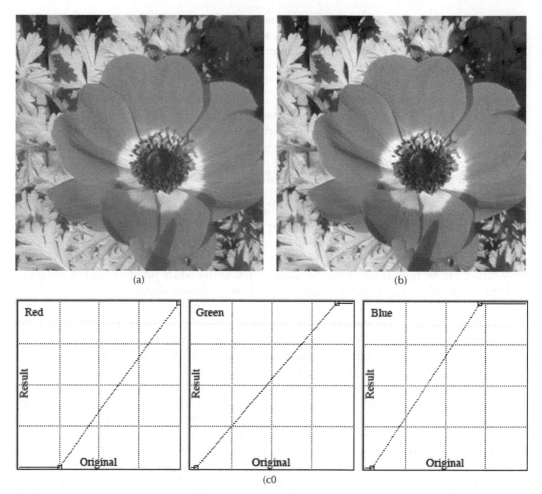

**FIGURE 1.18**
(See **Color insert following page 172**.) Setting the limits for each channel on the RGB values for the darkest and lightest pixels in the image: (a) original [anemone.tif], (b) result, (c) transfer functions for each color channel.

G, and B components. Figure 1.18 shows an example. As with the contrast stretching methods described above, this may be done by finding the extreme pixel values or by constructing a histogram and averaging some fraction of pixels at either end. Note that once the brightest and darkest pixels are located, this is the same procedure as shown in Code Fragment 1.4, and it can result in significant color shifts if the assumption that the brightest and darkest values are indeed neutral is not met.

## 1.2.2 Color Filters

Converting a color image to a grayscale (monochrome) image can be accomplished in several different ways, with different results. The brightness or luminance values shown in the preceding section can be used to produce a grayscale image by assigning the same calculated value to the red, green, and blue channels as shown in Code Fragment 1.12 and illustrated in Figure 1.19. Depending on the color space coordinates used, the results will differ somewhat. Much more dramatic (and sometimes useful) results can be obtained by using arbitrary multipliers for the red, green, and blue values. It is even possible to set some of these weighting factors to negative values, as shown in Figure 1.19c.

(a)

(b)                                                    (c)

**FIGURE 1.19**

Reducing a color image to grayscale (the original image [rose.tif] is shown in Figure 1.11a): (a) average of RGB, (b) luminance (0.25 * red + 0.65 * green + 0.10 * blue), (c) arbitrary weights (0.75 * red + 1.20 * green – 0.65 * blue).

```
// Code Fragment 1.12 — Producing a grayscale image
// ... declare variables, read dimensions, allocate line pointer
for (y = 0; y < height; y++)
{
   ReadOriginalLine(y, Line);
   for (x = 0; x < width; x++)
   {
      Red   = Line[x].red;
      Green = Line[x].green;
      Blue  = Line[x].blue;
```

```
    Gray   = (Red + Green + Blue) / 3; // assign equal weights to R, G, B
    // or use other weights as desired, may be read from a file
    // it is helpful to normalize the sum of the weights to 1.0
    Line[x].red = Line[x].green = Line[x].blue = Gray;
    // assign equal results to red, green, and blue to display grays
  } // for x
  WriteResultLine(y, Line);
} // for y
```

Another approach to reducing an image to a grayscale version that produces the maximum contrast uses least-squares fitting. This is also known as a principal components approach. The basic idea is shown in Figure 1.20, using the red-green-blue color coordinate system (similar methods can be applied in other color spaces, but with somewhat more complicated arithmetic). Each pixel in the image has RGB color values and is represented by a point in this space. It is possible to fit a straight line through the space by regression that minimizes the distance of the points from the line. The cosines of the angles between the line and each of the axes then become the weighting factors used to combine the RGB values to produce the derived grayscale image. This is equivalent to projecting each pixel's RGB values onto the line and using the position along the line as a grayscale value, scaled to the 0 to 255 range.

Code Fragment 1.13 shows the calculation of the regression line, and Figure 1.21 illustrates the results. The procedure produces grayscale images that in some cases provide excellent contrast for some of the structures present, but as with all such automatic techniques, cannot predict which structures or detail are of interest to the viewer.

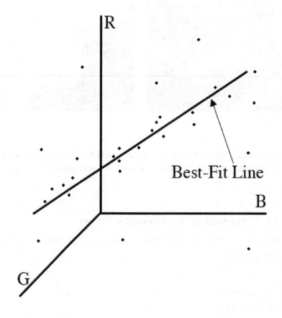

**FIGURE 1.20**
Diagram of best-fit regression line in RGB space. Each point represents the color coordinates of one pixel in the image (for a typical real image there are, of course, many thousands, or even millions, of points).

**FIGURE 1.21**
Example of maximum contrast (same original as Figure 1.19) produced by regression (0.46 * red + 0.19 * green + 0.35 * blue).

```
// Code Fragment 1.13 — Regression fit of line
// ... declare variables, etc., initialize the sums to 0
// building the covariance sums
   for (y = 0; y < height; y++)
   {
      ReadOriginalLine(y, Line);
      for (x = 0; x < width; x++)
      {
         r = Line[x].red;
         g = Line[x].green;
         b = Line[x].blue;
         sumn++;
         sumx += r;
         sumy += g;
         sumz += b;
         sumx2 += r * r;
         sumy2 += g * g;
         sumz2 += b * b;
         sumxy += r * g;
         sumxz += r * b;
      } // for x
   } // for y
   // compute best-fit line
   mx  = sumx2 - (sumx * sumx) / (float)sumn;
```

```
my  = sumy2 - (sumy * sumy) / (float)sumn;

mz  = sumz2 - (sumz * sumz) / (float)sumn;

mxy = sumxy - (sumx * sumy) / (float)sumn;

mxz = sumxz - (sumx * sumz) / (float)sumn;

txy = my - mx;

txz = mz - mx;

rxy = sqrt(txy * txy + 4 * mxy * mxy);

rxz = sqrt(txz * txz + 4 * mxz * mxz);

sxy = (txy + rxy) / (2 * mxy);

sxz = (txz + rxz) / (2 * mxz);

yp  = 1 / sxy;

zp  = 1 / sxz;

len = sqrt(1 + yp * yp + zp * zp);

cosx = 1 / len;

cosy = yp / len;

cosz = zp / len;

// second pass, calculate the gray value from the cosines
{  // ... read each line, access each pixel as above
   r = Line[x].red;
   g = Line[x].green;
   b = Line[x].blue;
   val = r * cosx + g * cosy + b * cosz; // weighted combination
   // ... assign value to pixel, write line back
}
```

### 1.2.3   Tristimulus Correction

The red, green, and blue filters used in digital cameras and scanners cover broad, and somewhat overlapping, wavelength ranges (just as the sensitivity of the cones in the human eye cover broad, overlapping ranges). The result is that an image of a color target with pure red, green, and blue areas will contain finite intensities of red in the green area, green in the blue area, and so on. Measuring these intensities allows for a different type of color correction, which can restore the image of the color target to pure colors and also correct other images taken with the same camera and illumination to produce correct colors. The method is called a ***tristimulus correction***.

The procedure followed is shown in Table 1.1 and Figure 1.22. First, the average intensities of red, green, and blue in each of the color target areas are measured. Using a spreadsheet such a Excel, these are normalized by dividing by 255 to produce the initial matrix. The inverse of this matrix is then calculated, and becomes the tristimulus matrix. It is applied as shown in Equation 1.7, to calculate new values of red, green, and blue as linear combinations of the original R, G, and B values for each pixel.

$$\begin{bmatrix} R^{\mathrm{T}} \\ G^{\mathrm{T}} \\ B^{\mathrm{T}} \end{bmatrix} = \begin{bmatrix} c_{RR} & c_{RG} & c_{RB} \\ c_{GR} & c_{GG} & c_{GB} \\ c_{BR} & c_{BG} & c_{BB} \end{bmatrix} \cdot \begin{bmatrix} R \\ G \\ B \end{bmatrix} \qquad (1.7)$$

**TABLE 1.1**

Tristimulus Correction: (a) Measured Intensities
in the Red, Green, and Blue Standards in Figure 1.22a;
(b) Normalized C Matrix; (c) Tristimulus Matrix
Calculated by Excel MINVERSE Function

|  |  | Red | Green | Blue |
|---|---|---|---|---|
| (a) Measured intensities: | | | | |
| Area: | Red | 124.0 | 34.7 | 45.2 |
|  | Green | 32.9 | 175.8 | 51.7 |
|  | Blue | 27.3 | 53.7 | 153.3 |
| (b) Normalized intensity matrix (above values divided by 255) | | | | |
|  |  | 0.4863 | 0.1361 | 0.1773 |
|  |  | 0.1290 | 0.6894 | 0.2027 |
|  |  | 0.1071 | 0.2106 | 0.6012 |
| (c) Inverse matrix calculated by Excel MINVERSE function | | | | |
|  | Red | 2.254 | −0.270 | −0.574 |
|  | Green | −0.339 | 1.658 | −0.459 |
|  | Blue | −0.283 | −0.533 | 1.926 |

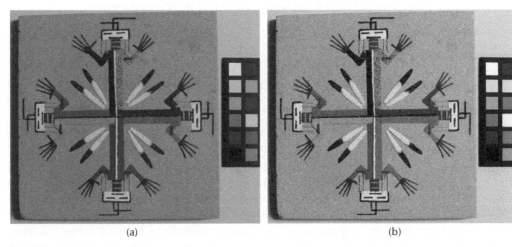

(a)  (b)

**FIGURE 1.22**
(See **Color insert following page 172**.) Application of tristimulus correction: (a) original image with color standards [sandpaint.tif], (b) corrected result.

In many practical cases, such as studio photography, an image of a color standard is acquired at the start of an imaging session under the same lighting conditions as will be used, and the derived tristimulus correction is then applied to each of the subsequent images. Commercially available devices for calibrating image scanners, cameras, displays and printers rely on the same approach, using additional known color swatches so that nonlinear terms can be introduced into the correction equations.

## 1.2.4  Problems

1.2.4.1#. Implement a program to find the average red, green, and blue values for the brightest and darkest pixels in an image; assume that they should represent neutral white and black (equal proportions of red, green, and blue), and construct and apply linear color correction curves.

1.2.4.2#.   Implement a program to read three weight values from a file and apply them to red, green, and blue channels to produce a grayscale image. Investigate the effects of different combinations of values on the contrast that can be obtained in various images. Note that some of the weights can be negative. It is convenient to include in the program a normalization of the weights as they are read from the file so that they total exactly 1.0.

1.2.4.3.    Implement a program that calculates the optimum-contrast grayscale image by fitting a regression line through the points in RGB space. Compare this to the result that is obtained by fitting the regression line in another color space, such as spherical LAB. Note that attempting this in a space such as HSI is quite difficult because of the polar coordinates and the variation of saturation with brightness.

1.2.4.4.    Implement a program to apply a tristimulus correction. Have it read in the nine values in the tristimulus matrix from a text file, and apply them to the measured RGB values.

1.2.4.5.    Implement a program to calculate and apply a tristimulus correction. Have it read in the nine measured intensity values from a color standard (i.e., the red, green, and blue intensities measured in each of the red, green, and blue target areas from a text file). The calculation requires inverting a matrix.

---

## 1.3   Correcting Nonuniform Illumination

### 1.3.1   Calculating a Correction

All cameras, whether film, video, or digital, share a problem with *vignetting*, which is a gradual darkening of the corners of the image, particularly with wide-angle (short focal length) lenses. The amount of light that passes through the optics to the center of the field of view is greater than that which reaches the edges and corners, and those parts of the image are somewhat darker. For most visual observation, this is not objectionable, because viewers tend to concentrate their attention on the center of the image rather than the corners. Also, gradual variations in brightness are not visually noticeable. In some cases, however, it can be important to correct this problem and level the image brightness. This is needed, for example, when thresholding is to be used to select features according to brightness, or if a series of images is to be combined as a mosaic or panorama. It also improves display and printing of images as it allows expanding the overall contrast of the picture.

Similar problems arise from a different cause when images are obtained with a copy stand or microscope, because the illumination may not be uniform across the field of view. Arranging lighting to produce perfectly uniform brightness values across the image area can be very difficult, and in these cases it is often quite important to produce results that are corrected for nonuniformities in illumination. Image processing offers several possible solutions. One that requires comparisons of pixels with their neighbors is shown in the next chapter (Chapter 2, Section 2.2.4). Another method, shown below in Section 1.3.2, requires capturing a second image of a uniform background with no features present.

Except for controlled situations such as the use of a copy stand, it is rarely possible to record an image of a scene, such as a cloudless sky, that is uniformly illuminated and contains no features and thus allows direct measurement of the vignetting produced by

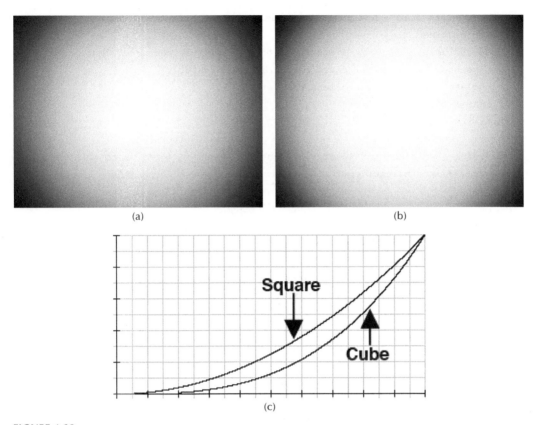

**FIGURE 1.23**
Vignetting correction models. The progressive darkening of pixels is calculated proportional to the radius from the center raised to a power: (a) square, (b) cube, (c) the correction models for brightness that must be added as a function of radius, showing the square and cube functions.

the imaging optics. Furthermore, changing the lens focal length or aperture settings changes the amount of the vignetting and also its functional form. However, a useful model can be constructed by using a power-law relationship of brightness as a function of radius, as shown in Figure 1.23 and Equation 1.8. The two examples show brightness as the square and the cube, respectively, of the radius. Typical real lenses fall between those cases.

$$Background = K \cdot \left( \frac{Radius}{Max\ Radius} \right)^{Power} \tag{1.8}$$

In actual use, the exponent (*Power* in the equation) may be determined by measurement of a few locations on an image, or (more commonly) by a user-adjustable input that allows the viewer to interactively make the adjustment. Likewise, the magnitude of the correction (*K* in the equation) can be determined by either a measurement of the average brightness of similar regions in the center and a corner of the image, or by an interactive user adjustment. The correction is applied to the image by adding the brightness generated by the model to the intensity of pixels in the scene. Code Fragment 1.14 illustrates the case in which the power-law model is used to add an increment of brightness to the image. Notice that this is done by scaling up the red, green, and blue values proportionately. An alternative method would be to convert the pixel values to a LAB or HSI space, increase the brightness values, and convert back.

```
// Code Fragment 1.14 — Add power law model to pixel brightness values.
{  //main routine
   float radius;           // distance of each pixel from center of image
   long xcenter, ycenter;  // coordinates of image center
   // declare variables, get dimensions, etc.
   // read K and power values
   xcenter = width / 2;
   ycenter = height / 2;
   maxradius = sqrt(xcenter*xcenter + ycenter*ycenter); // radius to corner
   for (y = 0; y < height; y++)
   {
       float r, g, b;
       ReadOriginalLine(y, Line);
       for (x = 0; x < width; x++)
       {   // apply equation 1.8
           radius = sqrt((x-xcenter)*(x-xcenter) + (y-ycenter)*(y-ycenter))
           value = K * pow((radius / maxradius), power);
           r = Line[x].red;
           g = Line[x].green;
           b = Line[x].blue;
           luminance = 0.25*r + 0.65*g + 0.10*b; // or other weights
           value = value + luminance; // boost corner brightness
           r = r * (value / luminance); if (r > 255) r = 255;
           g = g * (value / luminance); if (g > 255) g = 255;
           b = b * (value / luminance); if (b > 255) b = 255;
           Line[x].red   = r;
           Line[x].green = g;
           Line[x].blue  = b;
       } // for x
       WriteResultLine(y, Line);
   } // for y
}
```

Figure 1.24 shows a scene in which vignetting is present. Using the power-law method, the darkening of the corners of the scene can be corrected. As shown in the code fragment, the correction is made only to the brightness data, because the colors are not affected by the vignetting.

## 1.3.2   Measuring the Background

The case of imaging on a copy stand or similar situations can be handled in a different way. Because the copy stand has fixed lights, and the specimens or objects being photographed can be removed, it is possible to capture an image of the base of the stand, which is usually painted a uniform neutral color. This image can be used to level the brightness of the original image as shown in Figure 1.25, instead of modeling the nonuniformity with an equation. Notice in this example that the lighting is off center (brighter at the top) and that there is a color nonuniformity as well because of a light reflection from a surface next to the copy stand that appears at the lower right. In this case the background image is an

**FIGURE 1.24**
(See **Color insert following page 172.**) Correcting vignetting: (a) original image [ocean.tif], (b) corrected result.

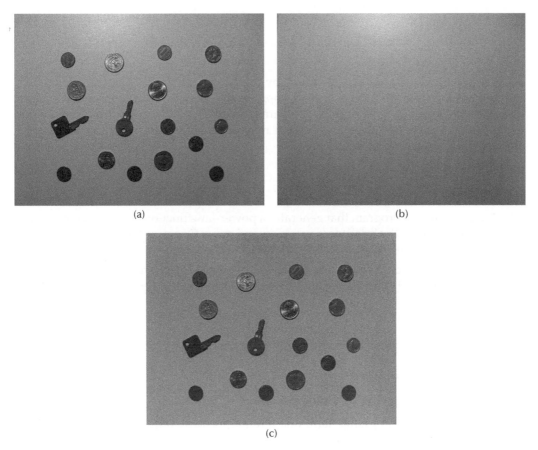

**FIGURE 1.25**
(See **Color insert following page 172.**) Correcting nonuniform lighting: (a) original [coins.tif], (b) background [background.tif], (c) leveled by subtraction.

RGB color image, which is subtracted pixel by pixel from the RGB color original as shown later in this chapter. Notice that in the result (Figure 1.25c) the correct neutral gray color is recovered for the background.

The background removal may be accomplished (as in this example) by subtraction, but in some instances it is more appropriate to use division instead. Section 1.5 shows the procedure for performing general arithmetic combinations of images, including both subtraction and division. The question of interest here is which operation to choose, and why. The answer lies in the nature of the image sensor and its associated processing.

The fundamental relationship that underlies the background removal operation is the ratio of the intensity of features to the intensity of background. If the image has recorded the intensities linearly, division is the appropriate function. Solid-state detectors are inherently linear, so it might be assumed that division is the natural choice. However, film (and human vision) responds logarithmically to light intensity, and many video and digital still cameras have built-in circuitry or processing that converts the original linear signals to logarithmic ones, in order to produce images that are similar to the familiar results obtained with film cameras. In that case, as shown in Equation 1.9, subtraction produces the equivalent result.

$$Result = \frac{Intensity}{Background}$$

$$\log(Result) = \log(Intensity) - \log(Background)$$

(1.9)

In many practical cases, raw images stored by digital still cameras retain the linear values, and if these are used division is the appropriate selection. The final processed images, which may be stored as TIFF or JPEG files, are more likely to have logarithmic values and subtraction is appropriate. In a few cases, it may be necessary to try both and see which procedure produces correct results.

### 1.3.3   Problems

1.3.3.1#.  Implement a program that generates a power-law function to model vignetting and add it to the brightness values in an image. Compare the effect of powers of 2.0, 2.5, and 3.0.

*Note:* Problem 1.5.4.1 performs subtraction of a reference image from the current image, and can be used to subtract a measured background to level image illumination.

## 1.4   Geometric Transformations

### 1.4.1   Changing Image Size and Interpolation

Digital images can be very large, with many more pixels than the computer display can show. In order to see the full image, it is necessary to reduce the image size by displaying only some of the image pixels. Most computer programs do this by selecting every second pixel vertically and horizontally (50% size), every third pixel (33%), every fourth (25%) and so on. This is very fast, often handled directly by the display hardware, and serves the intended purpose but, of course, it can hide small details and cause visual artifacts such as the breakup of lines and the stair-stepping of edges.

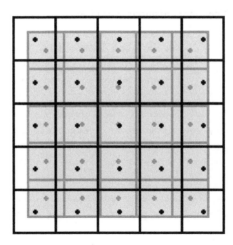

**FIGURE 1.26**
Schematic diagram of image enlargement. New pixel centers (black) lie between the centers of the original pixels (gray).

Conversely, the individual display pixels are too small for convenient viewing, and enlarging an image to permit the details to be easily seen is also important. This is most often done by repeating each image pixel vertically and horizontally. Showing each image pixel as a $2 \times 2$ block on the screen produces a 200% display, a $3 \times 3$ block produces a 300% display, and so on. Again, this is fast, often handled by the display hardware, and serves the intended purpose.

However, the general case of enlarging or reducing the size of an image, for example, to fit it into an allotted space in a presentation document, usually requires an enlargement or reduction factor that is not an integer. The general case of changing image size, as well as the tasks of rotation and translation, requires a more flexible procedure. Likewise, any geometric distortion of the image, for instance, to correct for a tilted point of view, or correction for lens distortions, must make use of these techniques.

When an image is enlarged, rotated, translated, or otherwise altered geometrically, the new pixel addresses rarely align exactly with the old ones. As shown in Figure 1.26 for the simple case of image enlargement, the new pixels lie "between" the old pixel addresses. Several different procedures for determining the values to assign to the new pixel can be used. The least complicated is just to take the values from the pixel that is nearest to the new location, called the nearest-neighbor method. This has the advantage that pixel values are preserved and so edge contrast is not blurred, but in most cases it produces aliasing or stair-stepping artifacts along edges that are visually distracting.

Code Fragment 1.15 shows how the nearest-neighbor procedure can be applied to enlarge or reduce an image by an arbitrary amount. The program keeps the original center point and enlarges or reduces the image. In the case of enlargement, some of the image is lost beyond the image boundaries, and in the case of reduction, the borders of the image are filled with black.

```
// Code Fragment 1.15 — Nearest-neighbor scaling
long   xcenter, ycenter;
float scale;                    // may be read from a text file
float OrigY, OrigX;             // exact address of original pixel
long  FetchY, FetchX;           // integer addresses used for interpolation
long  x, y, c;
RGBPixel *LineA, *Result;       // lines to read, write
// ... allocate line memory, get dimensions, read scale value
xcenter = width / 2; ycenter = height / 2;
for (y = 0; y < height; y++)
{
   FetchY = (long)(ycenter + scale * (y - ycenter) + 0.5); // round off
   if ((FetchY >= 0) && (FetchY < height-1))        // within boundaries
   {
      ReadOriginalLine(FetchY, LineA);
      for (x = 0; x < width; x++)
      {
         FetchX = (long)(xcenter + scale * (x - xcenter) + 0.5);
         if ((FetchX >= 0) && (FetchX < width-1))
         {
            Result[x].red   = LineA[FetchX].red;
            Result[x].green = LineA[FetchX].green;
            Result[x].blue  = LineA[FetchX].blue;
         }
         else
            Result[x].red = Result[x].green = Result[x].blue = 0;
      } // for x
      WriteResultLine(y, Result);
   } // if Fetchy in boundaries
   else                                             // outside boundaries
   {
      for (x = 0; x < width; x++)
         Result[x].red = Result[x].green = Result[x].blue = 0;
      WriteResultLine(y, Result);
   } // else not in boundaries
}// for y
// ... dispose of pointers
```

*Interpolation* is usually a preferable method for determining the new pixel value. The simplest procedure, bilinear interpolation, uses the four original pixels that surround the new pixel address to linearly interpolate new values at a fractional position $\delta_x, \delta_y$ between the original pixels, as shown in Equation 1.10. Intermediate values are calculated by linear interpolation along each line, and then these values are used to calculate a final result by interpolation between the lines. This technique is called bilinear interpolation and is illustrated in Figure 1.27c. The use of additional surrounding neighbors allows fitting higher order functions. Bicubic interpolation (Figure 1.27d) is one of the most widely used methods, producing less blurring of edges and detail than bilinear. Equation 1.11 shows the procedure; interpolation along each of four lines (the subscript $i$ takes on values between −1 and +2) is followed by interpolation between the lines to obtain the final result.

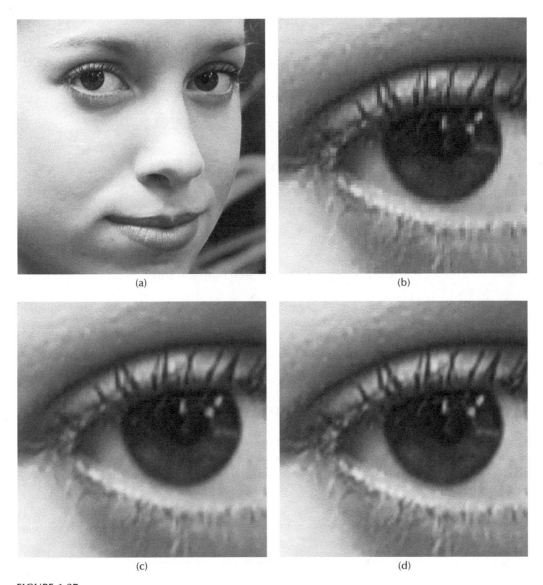

(a)

(b)

(c)

(d)

**FIGURE 1.27**
Enlargement of an image (a) original image [face.tif], (b) detail enlarged 266% by nearest-neighbor method, (c) bilinear interpolation, (d) bicubic interpolation.

The coefficients used in the bicubic equation can be modified to produce somewhat sharper or smoother results. Any of these procedures can be used for either enlarging or reducing an image. Applying an interpolation method requires access to more than a single row of pixels from the original image, as shown in Code Fragment 1.16. For bicubic interpolation, four lines of pixels must be read from the original image and four values taken from each line. As discussed in detail in Chapter 2, it is important to protect against attempting to read values outside the boundaries of the image. Usually, any regions in the final image that correspond to locations outside the original image boundaries are set to a default value, typically either black or white.

$$Value_{j+\delta x,k} = \left(1-\delta x\right)\cdot Pixel_{j,k} + \delta x\cdot Pixel_{j+1,k}$$

$$Value_{j+\delta x,k+1} = \left(1-\delta x\right)\cdot Pixel_{j,k+1} + \delta x\cdot Pixel_{j+1,k+1} \qquad (1.10)$$

$$Result_{j+\delta x,k+\delta y} = \left(1-\delta y\right)\cdot Value_{j+\delta x,k} + \delta y\cdot Value_{j+\delta x,k+1}$$

$$Value_{j+\delta x,k+i} =$$

$$\tfrac{1}{6}\left(R_1\cdot Pixel_{j-1,k+i} + R_2\cdot Pixel_{j,k+i} + R_3\cdot Pixel_{j+1,k+i} + R_4\cdot Pixel_{j+2,k+i}\right)$$

$$Result_{j+\delta x,k+\delta y} = \qquad\qquad (1.11)$$

$$\tfrac{1}{6}\left(R_1\cdot Value_{j+\delta x,k-1} + R_2\cdot Value_{j+\delta x,k} + R_3\cdot Value_{j+\delta x,k+1} + R_4\cdot Value_{j+\delta x,k+2}\right)$$

$$R_1 = \left(3+\delta x\right)^3 - 4\cdot\left(2+\delta x\right)^3 + 6\cdot\left(1+\delta x\right)^3 - 4\cdot\delta x^3$$

$$R_2 = \left(2+\delta x\right)^3 - 4\cdot\left(1+\delta x\right)^3 + 6\cdot\delta x^3$$

$$R_3 = \left(1+\delta x\right)^3 - 4\cdot\delta x^3$$

$$R_4 = \delta x^3$$

```
// Code Fragment 1.16 — Bilinear interpolation for enlargement/reduction
long xcenter, ycenter;
float scale;            // may be read from text file
float OrigY, OrigX;     // exact address of original pixel
long FetchY, FetchX;    // integer addresses used for interpolation
float dx, dy;           // fractional part of addresses
long x, y, c;
RGBPixel *LineA, *LineB, *Result; //two lines to read, one to write
// ... allocate line memory, get dimensions, read scale value
xcenter = width/2; ycenter = height/2;
for (y = 0; y < height; y++)
{
   OrigY = ycenter + (y - ycenter) / scale;
   FetchY = (long)OrigY;
   dy = OrigY - FetchY;
   if ((FetchY >= 0) && (FetchY < height-1))
   {
     ReadOriginalLine(FetchY,   LineA);
     ReadOriginalLine(FetchY+1, LineB);
     for (x = 0; x < width; x++)
     {
        OrigX = xcenter + (x - xcenter) / scale;
        FetchX = (long)OrigX;
        dx = OrigX - FetchX;
        if ((FetchX >= 0) && (FetchX < width-1))
        { // process r, g, b individually
           ValueA = (1-dx)*LineA[FetchX].red + dx*LineA[FetchX+1].red;
```

```
        ValueB = (1-dx)*LineB[FetchX].red + dx*LineB[FetchX+1].red;
        // horizontal interpolation along each line
        Value = (1 - dy) * ValueA + dy * ValueB;
        // vertical interpolation between lines
        Result[x].red = Value;
        // ... repeat for .green and .blue
      } // if FetchX in boundaries
      // ... else use black value outside boundaries
    } // for x
    WriteResultLine(y, Result);
  } // if FetchY in boundaries
  // ... else write black outside boundaries
} // for y
// ... dispose of pointers
```

The preceding example keeps the center of the image in its original location. In the more general case there may also be an offset applied to shift the image. The *x*- and *y*-offset values may not be an integral number of pixels, so even if no enlargement or reduction is applied, interpolation of the pixel values would be required. When pixel values are interpolated between neighbors, the sharpness of edges and detail is inevitably reduced, but some methods are better than others.

### 1.4.2   Rotation

Rotation of images, for instance, to correct for camera tilt or to align several images, also requires interpolation as indicated in Figure 1.28. The address of each new pixel is calculated as shown in Equation 1.12, and Code Fragment 1.17 shows how the procedure can be applied. Because the code fragment must read an entire line of pixels in order to access just the two pixels in that line that are neighbors of the calculated position, this method can be slow for a large image. Reading the entire image into a temporary array

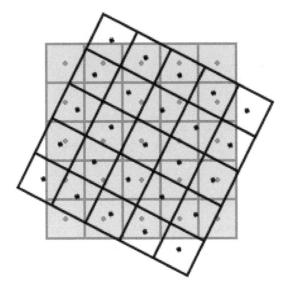

**FIGURE 1.28**
Schematic diagram of image rotation. New pixel centers (black) lie between the centers of the original pixels (gray).

in memory may be useful as a way to speed access times. As shown in Figure 1.29, the appearance of lines, edges, and detail varies somewhat with angle, depending on the interpolation method used.

$$x' = x \cdot \cos\theta + y \cdot \sin\theta$$

$$y' = y \cdot \cos\theta - x \cdot \sin\theta$$

(1.12)

```
// Code Fragment 1.17 — Rotation (bilinear interpolation)
long  xshift, yshift;    // centered coordinates
float rotation;          // read from text file
float xorig, yorig;      // exact address of original pixel
float xprime, yprime;    // rotated coordinates
long  x2, y2;            // integer addresses used for interpolation
float dx, dy;            // fractional part of addresses
long  x, y, c;
RGBPixel *LineA, *LineB, *Result; //two lines to read, one to write
// ... allocate line memory, get dimensions, read rotation angle
for (y = 0; y < height; y++)
{
    for (x = 0; x < width; x++)
    {
        xshift = (x - width/2);  // convert to centered coords
        yshift = (y - height/2); // so rotation is about center of image
        xprime = xshift * cos(rotation) + yshift * sin(rotation);
        yprime = yshift * cos(rotation) - xshift * sin(rotation);
        xorig = xprime + (width/2);
        yorig = yprime + (height/2);
        y2 = (long)yorig;
        x2 = (long)xorig;
        dx = xorig - (float)x2;//fractional address;
        dy = yorig - (float)y2;
        if ((y2>0) && (y2<height-1) && (x2>0) && (x2<width-1))
        {
            ReadOriginalLine(y2  , LineA);
            ReadOriginalLine(y2+1, LineB);
            // process red, green and blue
            partial1 = (1.0 - dx)*LineA[x2].red + dx*LineA[x2+1].red;
            partial2 = (1.0 - dx)*LineB[x2].red + dx*LineB[x2+1].red;
            Result[x].red = ((1.0-dy)*partial1 + dy*partial2);
            // ... repeat for .green and .blue
        }
        else
        { // outside image area
            Result[x].red = Result[x].green = Result[x].blue = 0;
        }
    } // for x
    WriteResultLine(y, Result);
} // for y
// ... dispose of pointers
```

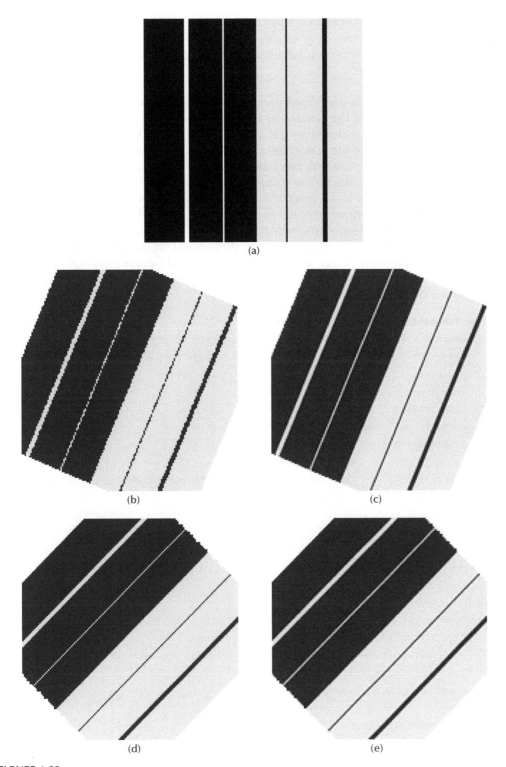

**FIGURE 1.29**
Effect of varying the angle and interpolation method on the appearance of lines and edges: (a) original image, (b) rotated 23° using nearest-neighbor interpolation, (c) rotated 23° using bicubic interpolation, (d) rotated 45° using nearest-neighbor interpolation, (e) rotated 45° using bicubic interpolation.

In many texts, translation, rotation, and scaling operations are combined into a single matrix notation that provides very efficient execution as well as a certain elegance of notation. However, because the goal here is clarity, the separation of the various operations into discrete steps is shown.

Rotation of an image can be performed very efficiently by successive shearing distortions, each one operating in just the horizontal or vertical direction. This offers a significant speed advantage because it simplifies the addressing of pixels, loading rows or columns of pixels to operate on all of the pixels along the row or column at a time. There are several variants of the technique; the one described by Paeth (1986) and Tanaka et al. (1986) requires three shear operations but no scaling, which (as indicated in Code Fragment 1.18) allows a single offset value and interpolation constant to be used for each row or column, and interpolation is performed only in one direction at a time. Figure 1.30 illustrates the effect of each of the three shear steps. Note that the image in the example is clipped because part of it extends beyond the original image dimensions in the intermediate steps. To avoid this, the temporary image array (which is required because multiple passes are needed) should be created with a width equal to the sum of original image width plus (height times the tangent of half the rotation angle).

```
// Code Fragment 1.18 — Rotate by shearing
// ... declare variables, read dimensions, allocate pointers, temp memory
//     note that dimension of Line and Result arrays should be
//     whichever is larger of temp image width and height.
// ... copy image to temporary memory array
// ... read angle A
xc   = width/2; yc = height/2; // center coordinates
A    = 3.14159265/180;    // convert to radians
TanA = tan(A / 2);        // for horizontal shears
B    = atan(sin(A));      // B is the angle whose tangent equals the sine of A
TanB = tan(B);            // for vertical shear
// pass 1 - horizontal shear
for (y = 0; y < height; y++)
{
   xoffset = (y - yc) * tanA;
   // calculate dx if interpolating
   ReadTempImageLine(y, LineA);
   for (x = 0; x < width; x++)
      if (((x - xoffset) > 0) && ((x-xoffset < width))
         Result[x] = LineA[xoffset]; // or perform interpolation
      else Result[x] = 0;
   WriteTempImageLine(y, LineA)
} // for y
// pass 2 - vertical shear
for (x = 0; x < width; x++)
{
   yoffset = (y - yc) * tanB;
```

```
   // calculate dy if interpolating
   ReadTempImageColumn(x, LineA);
   for (y = 0; y < height; y++)
      if (((y - yoffset) > 0) && ((y-yoffset < height))
         Result[y] = LineA[xoffset]; // or perform interpolation
      else Result[y] = 0;
   WriteTempImageColumn(x, LineA);
} // for x
// pass 3 - horizontal shear - identical to pass 1
for (y = 0; y < height; y++)
{
   xoffset = (y - yc) * tanA;
   // calculate dx if interpolating
   ReadTempImageLine(y, LineA);
   for (x = 0; x < width; x++)
      if (((x - xoffset) > 0) && ((x-xoffset < width))
         Result[x] = LineA[xoffset]; // or perform interpolation
      else Result[x] = 0;
   WriteTempImageLine(y, LineA)
} // for y
// ... rewrite image to host, dispose pointers, etc.
```

### 1.4.3   Alignment

A combination of rotation, scaling (enlargement or reduction), and translation is usually required to align one image to another. This is needed in many situations, including combining multiple images into a mosaic (by aligning the portions of the images that overlap), and as a precursor to applying the image arithmetic described in the next section to add, subtract, etc., images. It is also used when images are captured with different signals (e.g., different colored filters in a microscope, different wavelength bands in a satellite, or x-ray tomography and magnetic resonance images in medical imaging) so that they can be compared.

Note that it is important to consider the order of applying the scaling, rotation, and translation, and that the enlargement factors in the vertical and horizontal directions may not be the same. The minimum information needed to perform the alignment is the location of three corresponding points in each image. The $x_i$, $y_i$ locations of these points in the current image and in the reference image with which it is to be aligned are used to solve three simultaneous equations that give the coordinates of each new pixel in the original image. These coordinates are then used (as shown above) to interpolate the values for each new pixel. Equation 1.13 shows the equations that calculate the new coordinates. The coordinates of three points in the image to be adjusted and in the reference image to which it is to be aligned are $(x_i, y_i)$ and $(xref_i, yref_i)$, respectively. These are used to calculate the $a_i$ and $b_i$ coefficients that relate the new coordinates of each point to the old coordinates. Figure 1.31 shows the results of performing this general alignment procedure.

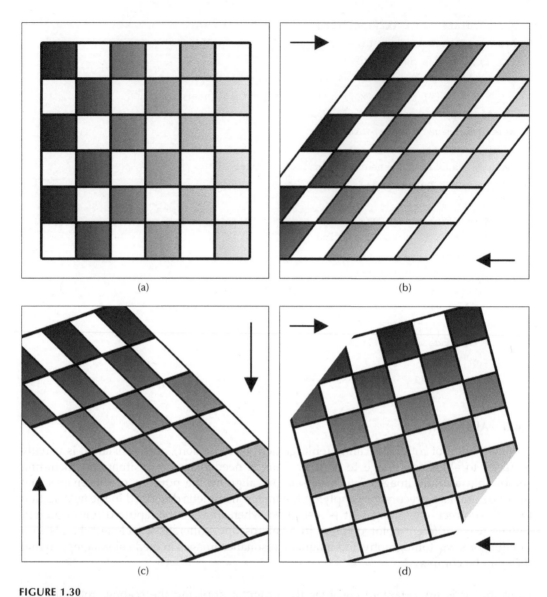

(a)

(b)

(c)

(d)

**FIGURE 1.30**
Steps in performing an image rotation by shearing (in the example, the final rotation angle is 75°): (a) original image [mesh.tif], (b) horizontal shear, angle = 37.5° (half of the total rotation angle), (c) vertical shear, angle = 44.0° (the angle whose tangent equals the sine of the rotation angle), (d) horizontal shear, angle = 37.5° (half of the rotation angle).

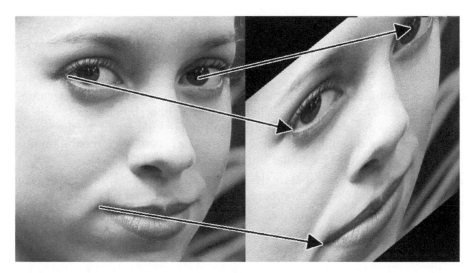

**FIGURE 1.31**
Translating, rotating, and scaling an image based on three reference points (indicated by the arrows).

$$x' = a_0 + a_1 x + a_2 y$$

$$y' = b_0 + b_1 y + b_2 x$$

$$V_1 = \frac{x_1 - x_2}{yref_1 - yref_2} - \frac{x_1 - x_3}{yref_1 - yref_3}$$

$$V_2 = \frac{xref_1 - xref_2}{yref_1 - yref_2} - \frac{xref_1 - xref_3}{yref_1 - yref_3}$$

$$V_3 = \frac{x_1 - x_2}{xref_1 - xref_2} - \frac{y_1 - y_3}{xref_1 - xref_3}$$

$$V_4 = \frac{yref_1 - yref_2}{xref_1 - xref_2} - \frac{yref_1 - yref_3}{xref_1 - xref_3}$$

$$a_1 = \frac{V_1}{V_2}$$

$$a_2 = x_1 - x_3 - a_1 \cdot \left( \frac{xref_1 - xref_3}{yref_1 - yref_3} \right)$$

$$a_0 = x_3 - a_1 \cdot xref_1 - a_2 \cdot yref_1$$

$$b_1 = \frac{V_3}{V_4}$$

$$b_2 = y_1 - y_3 - b_1 \cdot \left( \frac{yref_1 - yref_3}{xref_1 - xref_3} \right)$$

$$b_0 = y_3 - b_1 \cdot yef_1 - b_2 \cdot xref_1$$

(1.13)

General image warping or projective geometric translation is handled by the same interpolation procedures, but with other, often more complex, equations describing the relationship between pixel locations in the new image and the corresponding positions in the old one.

### 1.4.4 Problems

1.4.4.1#. Implement a program that enlarges or reduces an image by an scale factor read from a text file. Use and compare nearest-neighbor and bilinear interpolation.

1.4.4.2. Implement a program that translates an image by $x$- and $y$-offset values (in pixels) read from a text file, using nearest-neighbor, bilinear, or bicubic interpolation as defined by a value also read from the file. Compare the sharpness of edges and detail in the result when the shift is an exact integral number of pixels compared to fractional shifts of 0.1, 0.25, and 0.5 pixels.

1.4.4.3#. Implement a program that rotates an image around its center by an angle read from a text file. Use and compare nearest-neighbor and bilinear interpolation.

1.4.4.4. Implement a program that rotates an image around its center using the shearing method. Note that this requires copying the image into a temporary storage array to perform the sequential operations.

1.4.4.5. Implement a program that scales, rotates, and translates an image using values for the vertical and horizontal expansion factors (which may not be the same), rotation angle, and $x,y$ offset that are read from a text file. *Note that the order in which the scaling, rotation, and translation are applied makes a difference in the result.* Use and compare nearest-neighbor, bilinear, and bicubic interpolation.

1.4.4.6. Implement a program that reads in a text file containing a list of 12 numbers, representing the $x,y$ locations of three reference points in the original image and the $x,y$ locations to which those points are to be aligned in the final image. Then solve the equations and apply the transformation to the image, using bilinear or bicubic interpolation.

---

## 1.5 Image Arithmetic

In order to carry out the arithmetic operations that combine the current image with a second one, it is necessary to first specify the reference image. This is done by selecting a separate routine beforehand, shown in Code Fragment 1.19. It copies the current image into a disk file, where it remains as the reference image until it is specifically replaced with another image. Note that the reference image file will persist even if the original image is not currently displayed or loaded into the host program, and even if the computer is shut down and restarted.

```
// Code Fragment 1.19 — Create reference image file
// ... declare variables, read dimensions, allocate pointer
Err = InitializeRefImageFile(width, height);
Line = CreateAPointer(width, sizeof(RGBPixel);
for (y = 0; y < height; y++)
{
   ReadOriginalLine(y, Line);
   WriteRefImageLine(y, Line);
} // for y
DisposeAPointer(Line);
```

The format for this file is a six-byte header followed by a sequence in raster order of floating point RGB values in the 0 to 255 range for each pixel. The file is named **Reference Image.xzy** and is stored in the user's private space, which by default is **C:\\Documents and Settings\UserName\Local Settings\Temp\** on Windows computers.

### 1.5.1 Adding and Subtracting

The reference image may be added to or subtracted from the current image, for example, as in the subtraction illustrated in Section 1.3.2 to correct for nonuniform illumination. There are several factors to consider in performing this apparently simple operation. The first is whether the two images are of the same size. If they are not, there are several possible options:

1. Warn the user and abort the operation.
2. Perform the operation on a rectangle having the smaller vertical and horizontal dimensions, either centered in each of the two images or located at the upper-left corner.
3. Apply the scaling and interpolation procedures shown in the preceding section to create a copy of the reference image with dimensions that match the current image, and then perform the arithmetic operation.

The code examples that follow use the first option, but one of the problems introduces the third choice.

A second factor is the need to deal with the range of values that can be produced by the arithmetic operation. Adding two pixel values together can produce a range of values from 0 to 510. This exceeds the 0 to 255 range that can be displayed. The easiest method for dealing with this range is to divide the result of the addition by 2. Likewise for subtraction, the possible range of values is from –255 to +255, and the easiest way to convert them to the legal 0 to 255 range is to add 255 and then divide by 2. Code Fragment 1.20 carries out the addition or subtraction procedures using this logic.

```
// Code Fragment 1.20 - Add or subtract the reference image with fixed limits
long height, width, h, w;
GetOriginalDimensions(&width, &height); // check dimensions
GetRefDimensions(&w, &h);
if ((height != h) || (width != w)
{
   ErrorMessage("Image dimensions don't match stored reference");
   return;
}
// ... declare variables, create line pointers RefLine, OrigLine
for (y = 0; y < height; y++)
{
   ReadRefImageLine(y, RefLine)
   ReadOriginalLine(y, OrigLine)
   for (x = 0; x < width; x++)// operate on R G B values per pixel
   {
```

```
      if (adding) //Boolean value defined previously or read from a file
      {
         OrigLine[x].red   = (OrigLine[x].red   + RefLine[x].red)/2;
         OrigLine[x].green = (OrigLine[x].green + RefLine[x].green)/2;
         OrigLine[x].blue  = (OrigLine[x].blue  + RefLine[x].blue)/2;
      }
      else
      {
         OrigLine[x].red   = (OrigLine[x].red   - RefLine[x].red + 255)/2;
         OrigLine[x].green = (OrigLine[x].green - RefLine[x].green + 255)/2;
         OrigLine[x].blue  = (OrigLine[x].blue  - RefLine[x].blue + 255)/2;
      }
   } // for x
   WriteResultLine(y, OrigLine)
} // for y
```

For some purposes, it is preferable to perform an automatic scaling of the results of the arithmetic operation so that they fill the available 0 to 255 range. There are two ways of accomplishing this. One is to make two passes through the images, the first one to find the maximum and minimum values that will result from the operation, and the second to perform the arithmetic and apply the automatic scaling, as shown in Code Fragment 1.21.

```
// Code Fragment 1.21 - Subtract reference image with two pass autoscaling
// ... check dimensions as above
// ... declare variables and allocate line pointers
// ... initialize max = -255, min = 255
for (y = 0; y < height; y++) // first pass - get min and max
{
   ReadRefImageLine(y,RefLine)
   ReadOriginalLine(y,OrigLine)
   for (x = 0; x < width; x++)
   {
      value = (OrigLine[x].red - RefLine[x].red);
      if (value > max) max = value; if (value < min) min = value;
      value = (OrigLine[x].green - RefLine[x].green);
      if (value > max) max = value; if (value < min) min = value;
      value = (OrigLine[x].blue - RefLine[x].blue);
      if (value > max) max = value; if (value < min) min = value;
   } // for x
} // for y
scale = 255.0 / (max - min); // note single max and min for R, G, and B
for (y = 0; y < height; y++) // second pass, autoscale and write back
{
   ReadRefImageLine(y,RefLine)
   ReadOriginalLine(y,OrigLine)
   for (x = 0; x < width; x++)
   {
```

```
      value = (OrigLine[x].red - RefLine[x].red);
      OrigLine[x].red = scale * (value - min);
      value = (OrigLine[x].green - RefLine[x].green);
      OrigLine[x].green = scale * (value - min);
      value = (OrigLine[x].blue - RefLine[x].blue);
      OrigLine[x].blue = scale * (value - min);
   } // for x
   WriteResultLine(y, OrigLine)
} // for y
```

The drawback of this method is that single-pixel outliers may be responsible for establishing the limits, so that the full possible contrast is not achieved for the final image. The second method requires creating a temporary image to hold the results. The values in this image can then be used to construct a histogram, and find limits that clip a small percentage (e.g., 0.1%) of the brightest and darkest pixels to black and white, as shown in Section 1.1.3. Code Fragment 1.22 outlines that procedure.

```
// Code Fragment 1.22 - Subtract reference image with autoscaling
// create a temporary image to hold the intermediate results
// read the original and reference images line by line
// subtract the reference from the original, pixel by pixel
// write the difference values to the temporary image array
// construct a histogram of the values in the temporary array
// find the values that correspond to the 0.1% and 99.9% values
// read the temporary image array line by line
// scale the values from 0 to 255 using the limit values (and clip outliers)
// write the values back to the host program
```

### 1.5.2  Multiplication and Division

Multiplication and division by the values in the reference image introduce some additional factors for consideration. First, the range of values that can be produced must be taken into account. Multiplication of values in the 0 to 255 range can produce results that range from 0 to 65,025. An obvious way to rescale these results to fit into the 0 to 255 range is to calculate the result as (Result = Original * Reference/255). Alternately, one of the autoscaling procedures shown above could be invoked. A third possibility, rarely used but visually very useful, is to handle the very large possible range of the resulting values but display instead a value derived from the actual product of the values. Two possibilities for this procedure are the square root of the result or the logarithm of the result. Figure 1.32 illustrates the result of this and other arithmetic combinations.

Division likewise has a range problem, and there is the additional possibility of division by zero. This latter issue is generally dealt with by performing the division using (1 + value) rather than the actual pixel values. The possible results of dividing with values that can vary from 1 to 256 is a range of results from 256 (=256/1) to 0.0039 (=1/256). Most of the possible values are very small, and as shown in Figure 1.32f when this range is linearly rescaled to the 0 to 255 legal range (Equation 1.14) these values are displayed as nearly black. Autoscaling with a small percentage of clipping may provide a useful result. Another approach uses log or square root functions to alter the results, or (as shown in Figure 1.32h) to use the logarithms of the original values and calculate the difference.

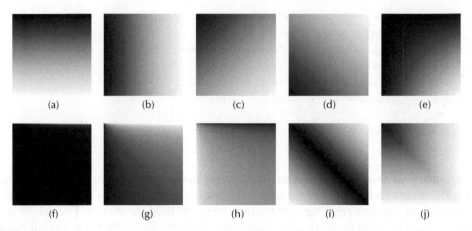

**FIGURE 1.32**
Arithmetic image combinations: (a) reference image [vramp.tif] (a vertical ramp with values that range from 0 to 255), (b) original image [hramp.tif] (a horizontal ramp with values that range from 0 to 255), (c) sum (addition), (d) difference (subtraction), (e) product (multiplication), (f) ratio (division), (g) sum of the logs (log a + log b), (h) difference of the logarithms (log b − log a), (i) absolute difference, (j) brighter value.

The rather thin justification for this approach is that human vision responds approximately logarithmically to intensity, and that it produces viewable results in many instances.

$$Result = \frac{\left( \dfrac{(Original + 1)}{(Reference + 1)} \right) - Minimum}{(Maximum - Minimum)} \tag{1.14}$$

### 1.5.3 Other Possibilities

The typical pocket calculator offers only addition, subtraction, multiplication, and division as arithmetic functions, but there are several other functions that can be useful for combining images. These include the absolute difference between values, keeping the brighter or darker value at each location, and Boolean operations such as AND, OR, and Exclusive-OR, which are typically applied to binary images after thresholding and will consequently be discussed in Chapter 4.

Taking the absolute difference involves subtracting the reference image value from the original image, and then if the result is less than zero, the sign is changed to make it positive. This is particularly useful for detecting any changes that may have occurred in a scene, as shown in Figure 1.33. In this example, two pictures of the same scene were taken slightly over a minute apart. The absolute difference shows a background of slight differences in random noise in the two pictures, as well as the motion of the minute hand, and even the slight motion of the hour hand of the clock. The latter is not visually evident in a side-by-side comparison of the pictures but becomes noticeable in the absolute difference result.

All of the foregoing procedures have combined the pixel values in an image with those in a reference image. They are typically performed individually on the RGB values, although in a few instances such as the application of texture (a reference image containing a pattern) to an image, it may be preferable to modify only the brightness values and

**FIGURE 1.33**
(See **Color insert following page 172**.) Absolute difference: (a) reference image [clock1.tif], (b) original image [clock2.tif], (c) absolute difference.

preserve the color information, by first converting the RGB values to a different set of color coordinates, and then converting back afterward as shown in preceding sections.

There are some other arithmetic functions that do not involve the stored reference image which are sometimes useful, and which also typically operate only on the brightness values. It may be convenient to think of these as histogram modification tools, because several of them operate on the pixel values in the image in ways that are similar to changing the gamma value (Section 1.14). However, they are included here for completeness, and because they use common arithmetic functions such as inverse, square root, square, logarithm, and exponential. Figure 1.34 illustrates several functions applied to a linear brightness ramp.

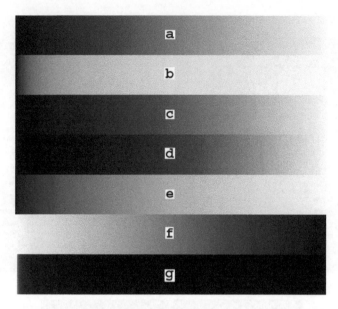

**FIGURE 1.34**
Application of several arithmetic functions to a linear ramp of brightness values: (a) original, (b) logarithm, (c) exponential, (d) square, (e) square root, (f) reverse or complement (255 – value), (g) inverse (1/value).

In order to apply these functions, it is often best to convert the original 0 to 255 range of values to a normalized 0.0 to 1.0 range, and then to restore the values afterward as shown in Equation 1.15. (Adding 1 to the Original value avoids an error that would arise in attempting to take the logarithm of zero.) Figure 1.35 shows an example that compares the effect of applying a reversal to the brightness values only and retaining the original color to the conventional procedure of reversing each of the RGB channels by calculating (255 – value) as shown in the Introduction. The figure also shows application of an arithmetic function to the brightness values.

$$Result = 255 \cdot \log\left(\frac{Original + 1}{256}\right) \tag{1.15}$$

**FIGURE 1.35**
(See **Color insert following page 172.**) Applying functions to pixel values: (a) original image [rose.tif], (b) conventional reverse applied to RGB channels, (c) reversing the intensity only, (d) square root function applied to intensity values.

### 1.5.4 Problems

1.5.4.1#. Implement programs to save and recall a reference image. When recalling a reference image that does not have the same dimensions as the current image into which it is written, center the portion that will fit.

1.5.4.2. Implement a program that carries out addition and subtraction of a reference image, selecting the operation to apply by reading a value from a text file.

1.5.4.3#. Modify the program in Problem 1.5.4.2 to apply automatic scaling of the results by performing two passes through the data.

1.5.4.4. Modify the program in Problem 1.5.4.2 to create a temporary image with the results, and perform autoscaling by forming a histogram and clipping a fixed percentage (e.g., 0.2%) of the values to black and white.

1.5.4.5#. Implement a program to apply inverse, logarithmic, exponential, square, and square root functions to the brightness values in an image.

# 2

## Neighborhood Operations

In Chapter 1, individual pixel values were considered independently, without regard to their location or the values of neighboring pixels. For example, the histogram presents a summary of pixel values that ignores their locations, and manipulation of the pixel values (RGB, HSI, or any other color space representation) is also independent of the color or brightness information in other nearby pixels. These operations are generally called pixel *point processes*, to distinguish them from *neighborhood operations* that treat each pixel as part of a local region.

## 2.1 Convolution

The neighborhood operations dealt with in this chapter do take into consideration the values of pixels that are adjacent to or near the pixel being modified. The first of these operations that will be taken up, convolution, is selected not because it is necessarily the most powerful or simplest, but it is perhaps the most widely used, and it introduces and illustrates the basic techniques of pixel and neighborhood access.

### 2.1.1 Neighborhoods and Kernels

*Convolution* can be performed in several ways that are fundamentally identical but initially appear to be very different. When Fourier domain image processing is introduced in Chapter 3, convolution will be revisited. In this chapter, convolution is described and implemented as an arithmetic process that involves each pixel and a small set of its close neighbors. Initially, only the eight immediately adjacent neighbors will be used.

The simplest of all these operations is averaging of a pixel with its neighbors, but for illustrative purposes it is better instead to introduce from the outset the concept of scalar weights that are multiplied by the pixel values. It is common to represent a convolution operation as a *kernel* or small array of numbers, often integers, such as the example in Table 2.1.

**TABLE 2.1**

Example of a 3 × 3 Convolution Kernel

| +1 | +2 | +1 |
|----|----|----|
| +2 | +4 | +2 |
| +1 | +2 | +1 |

The kernel is understood to be centered on the original pixel. The table shown is a shorthand representation of the following procedure:

For each pixel in the image:

1. Multiply the value of the pixel by 4.
2. For each of the edge-sharing pixels (the ones to the left and right, and above and below), multiple their values by 2.
3. For each of the corner-sharing pixels (the ones in diagonal directions), multiply their values by 1.
4. Add together all of the values from steps 1, 2, and 3, and divide the result by the sum of the multipliers, in this example, 16.
5. Write the resulting value back into the original pixel location but in a new image. This uses the original pixel values, not the newly calculated ones, to calculate results for each successive pixel.

The output of such a kernel is shown in Figure 2.1. The original image has quite visible random variations in the pixel values, which are not part of the original scene but originate in the camera. Photographs in dim light, for example, show this kind of noise because of the higher gain that must be used to amplify the signal. This amplification is usually expressed by digital camera manufacturers as a high effective ISO rating, an analogy to the speed ratings of high-speed film used for available light photography, which also have visible grain that produces similar random variations in images.

The amount of random variation in pixel values produced by the camera sensor and electronics can be seen and measured by the width of the peak in the histogram of an image of a uniformly illuminated gray test card. Figure 2.2 shows an example. The camera was mounted on a tripod and pointed at a test card and purposely adjusted to be out of focus. A series of images was captured with different effective ISO settings, and the histogram of a $1024 \times 1024$ region in the center of each image (positioned to avoid any corner vignetting) obtained. The increase in the width of the peak is shown, along with a plot of the standard deviation as a function of the camera gain.

For color images, the greatest amount of this type of noise is usually found in the blue channel because the silicon-based sensors used are least sensitive to blue light, and so that channel requires the highest amplification. Regardless of the source, one of the ways (not always the best, as will be seen later on) to reduce this type of random, or "speckle," noise is by a convolution similar to the one shown that averages the pixel together with its neighbors.

The convolution process can be written as shown in Equation 2.1.

$$Result_{x,y} = \sum_{i,j} \left( Kernel_{i,j} \cdot Pixel_{x+i,y+j} \right) \tag{2.1}$$

It is important to be aware that in the process of carrying out this procedure, the pixel values from the original image are always used in the calculations, not ones that have been computed and stored in the result image. In principle, the image pixels can be traversed in any order. The usual procedure for processing an image is to proceed through

(a)

(b)

(c)

**FIGURE 2.1**
(See **Color insert following page** 172.) Example of smoothing random noise by applying the kernel in Table 2.1 to an image: (a) original color image [bird.tif], (b) brightness values, (c) result of the convolution. Notice that the detailed enlargement showing individual pixels includes a background region that is not sharply focused in the original image, and that the visible speckle noise results from the camera and electronics.

it in raster order, starting at the upper-left corner, processing each line from left to right, and working down from top to bottom.

For the pixel located at address *x,y* the neighbors to the left and above will already have been processed by the convolution procedure. Consequently, it is necessary to keep the results separate from the original values. The Photoshop plug-in interface handles part of this chore for us, because the interface for plug-ins operates in a special way as described in the Introduction. Reading values from the image always provides the original values. Because an entire line of pixels will be read from the image at a time (and a full line written back), a separate line is required for the results of the calculations.

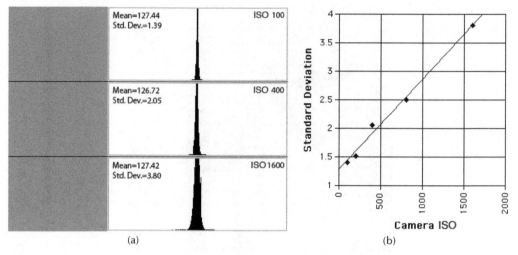

(a)                                                      (b)

**FIGURE 2.2**
Measured random noise in a digital camera as a function of gain (expressed as equivalent ISO rating): (a) image histograms and statistics, (b) plot of standard deviation versus ISO rating.

The process can be handled with code similar to the example in Code Fragment 2.1.

```
// Code Fragment 2.1 - Apply smoothing kernel
// ... declare variables, get dimensions, create line pointers
for (y = 0; y < height; y++)
{  // Read 3 lines for neighborhood
   ReadOriginalLine(y-1, LineA);
   ReadOriginalLine(y  , LineB);
   ReadOriginalLine(y+1, LineC);
   for (x = 0; x < width; x++)
   {
      Sum  = 1 * LineA[x-1].red + 2 * LineA[x].red + 1 * LineA[x+1].red;
      Sum += 2 * LineB[x-1].red + 4 * LineB[x].red + 2 * LineB[x+1].red;
      Sum += 1 * LineC[x-1].red + 2 * LineC[x].red + 1 * LineC[x+1].red;
      Result[x].red = Sum / 16;
      // ... repeat for .green and .blue
   } // for x
   WriteResultLine(y, Result);
} // for y
// ... dispose pointers
```

In the example, **LineA**, **LineB**, **LineC**, and **Result** are pointers, each allocated to hold a line of values from the image, as shown in Chapter 1. It is possible to rewrite this procedure so that each line can be read only once, and not three times as shown here, but the emphasis in these code fragments is always on clarity rather than efficiency or speed.

### 2.1.2   Colors

The routine in Code Fragment 2.1 processes the red, green, and blue values for each pixel individually. As discussed in the preceding chapter, these are stored as three separate values for each pixel along each line through the image, and the **x** index is used to access them. For the smoothing operation shown, the results of treating the data this way are not too bad. But, for many of the other sets of kernel values used in convolution (examples of which are shown above), working with the individual red, green, and blue channels produces very undesirable results. Altering the proportions of the red, green, and blue values independently results in the appearance of new, unexpected, and unwanted colors in the image (see the example in Figure 2.6b). The preferred procedure is to modify only the intensity for each pixel, while leaving the color information unchanged.

One way to accomplish this is to

1. convert the RGB values for each pixel to a different color space representation,
2. process just the intensity or luminance value, and then
3. transform the results back to RGB before rewriting the results.

Any of several color space representations can be used, because the color values are not affected and are restored after the modified intensity value is calculated, as indicated in Figure 2.3. In the example of Code Fragment 2.2 a conversion from RGB to the spherical LAB values introduced in Chapter 1 is used. The RGB values along each line of pixels are transformed to LAB values, processed, and converted back. Similar routines to use other color representations from the preceding chapter may also be used.

```
// Code Fragment 2.2 - Applying kernel to color image
typedef struct
{
    float red, green, blue;
} RGBPixel;    // this is defined in the glue code

typedef struct
{
    float L, A, B;
} LABPixel;    // _not_ defined in the glue code

// declare variables, read dimensions, etc...
// allocate memory to access image and process values
RGBLine = CreateAPointer(width, sizeof(RGBPixel));
LineA =   CreateAPointer(width, sizeof(LABPixel));
LineB =   CreateAPointer(width, sizeof(LABPixel));
LineC =   CreateAPointer(width, sizeof(LABPixel));
Result =  CreateAPointer(width, sizeof(LABPixel));

for (y = 0; y < height; y++)
{   // Read 3 line neighborhood and convert to LAB
    ReadOriginalLine(y-1 , RGBLine);
    ConvertRGBtoLAB(width, RGBLine, LineA);
    ReadOriginalLine(y   , RGBLine);
    ConvertRGBtoLAB(width, RGBLine, LineB);
    ReadOriginalLine(y+1 , RGBLine);
    ConvertRGBtoLAB(width, RGBLine, LineC);
    //Apply kernel to line
    for (x = 0; x < width; x++)
    {
```

```
   Sum  = 1 * LineA[x-1].L + 2 * LineA[x].L + 1 * LineA[x+1].L;
   Sum += 2 * LineB[x-1].L + 4 * LineB[x].L + 2 * LineB[x+1].L;
   Sum += 1 * LineC[x-1].L + 2 * LineC[x].L + 1 * LineC[x+1].L;
   Result[x].L = Sum / 16;
   // process the L values
   Result[x].A = LineB[x].A; // copy over the A value
   Result[x].B = LineB[x].B; // and the B value
   } // for x
   ConvertLABtoRGB(width, Result, RGBLine); // convert back to RGB
   WriteResultLine(y, RGBLine);
} // for y

ConvertLineRGBtoLAB(width, RGBLine, LABLine)
{
   long  x;
   float L, A, B;
   for (x = 0; x < width; x++)
   { // Convert to LAB values
     L = (RGBLine[x].red + RGBLine[x].green + RGBLine[x].blue)/3;
     A = (RGBLine[x].red - RGBLine[x].green);
     B = (RGBLine[x].red + RGBLine[x].green)/2 - RGBLine[x].blue;
     LABLine[x].L = L;
     LABLine[x].A = A;
     LABLine[x].B = B;
} // ConvertLineRGBtoLAB

ConvertLineLABtoRGB(width, LABLine, RGBLine)
{
   long  x;
   float L, A, B;
   for (x = 0; x < width; x++)
   { // Access LAB values
     L = LABLine[x].red;
     A = LABLine[x].green;
     B = LABLine[3].blue;
     // Convert to RGB values
     RGBLine[x].red   = L + A/2 + B/3;
     RGBLine[x].green = L - A/2 + B/3;
     RGBLine[x].blue  = L - 2 * B/3;
} // ConvertLineLABtoRGB
```

### 2.1.3   Boundary Effects and Value Limits

> *Warning: The vast majority of bugs and errors in code arise from improperly handled boundary conditions!*

The observant reader will note several flaws in the preceding examples of code. The most deadly is that as the **x** and **y** variables cover the full dimensions of the image, the pixels

```
for (y = 0; y < height; y++)...
   for (x = 0; x < width; x++)...
```

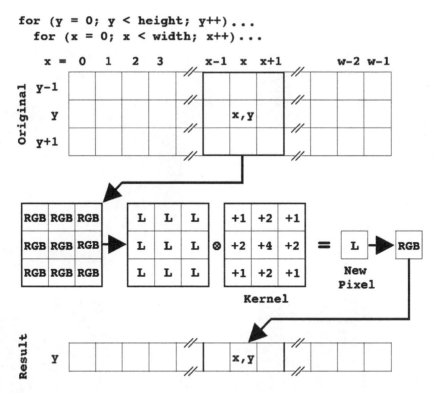

**FIGURE 2.3**
Schematic diagram of the convolution procedure described in the text.

along the boundaries have neighbors that lie outside the image. When **y-1** or **y+1** exceeds the range from **0** to **height-1**, or **x-1** or **x+1** exceeds the range from **0** to **width-1**, this code will try to access locations beyond the ends of the line, and the routine will probably crash. That raises the subject of dealing with the boundaries of images. The pixel point processes in Chapter 1 do not have this problem, because each pixel is dealt with in the same way regardless of its location or neighbors, and there is no attempt to address locations outside the image area.

There are several different approaches that can be used to deal with boundary pixels; they are illustrated in Figure 2.4. Some of them are mathematically or programmatically simpler than others, whereas some are less likely to cause problems for the image contents. For example, one technique assigns a default value, typically white, black, or an intermediate gray, to any location beyond the image boundaries. This is usually called "padding" an image.

A quite different method is called "wrap-around" or modulo addressing. Any address value that is beyond the right side of a line, or the top of the image, for example, is treated as being located the same distance from the opposite side. In other words, the top and bottom boundaries of the image are considered to be contiguous, as are the left and right sides (see Figure 2.5). Modulo arithmetic is easy for the computer to handle, and will be important in the Fourier-space procedures in Chapter 3, but there aren't many images in which the contents actually wrap around in this way and it can lead to significant errors along the image boundaries.

**FIGURE 2.4**
Possible methods of providing values beyond the boundaries of an image [girl.tif]: (a) padding, (b) wrap around, (c) mirroring, (d) duplicating.

A very limited (and limiting) approach to deal with image boundaries is to restrict the loops in the code example shown above so that instead of being

```
for (y=0; y<height; y++)
    for (x=0; x<width; x++)
```

the range of values is

```
for (y=1; y<height-1; y++)
    for (x=1; x<width-1; x++)
```

This is another way of saying that the boundary pixels may be used in calculations, but are never themselves actually modified. It amounts to discarding the boundary pixels in

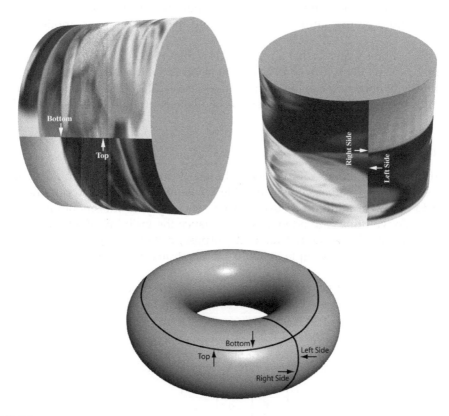

**FIGURE 2.5**
Wrap-around or modulo addressing is equivalent to considering the image to be cylindrical so that the top and bottom sides are contiguous, and also so that the left and right sides are contiguous. This is mathematically equivalent to wrapping the entire image around a torus. This formulation is used for the Fourier-based techniques in Chapter 3.

the operation, and slightly shrinking the image. The loss of one line of pixels around the periphery of a large image may not seem to be very important, but as will be seen below, neighborhoods can be significantly larger than just the 3 × 3 kernel shown above and access neighbor pixels farther from the center. Also, multiple operations may be used in sequence, so the actual loss might be much greater. So, that method is not a preferred solution.

Another approach that is sometimes adopted is to use only neighbors that lie within the image area, which requires special versions of the kernels to use along the edges and in the corners. The special-case logic this involves can dominate the code and obscure the underlying principles, and is therefore not considered here.

That leaves methods that try to process the boundary pixels but make some assumptions about the image to decide what values to use for their neighbors. The most common solutions are to treat each edge as a mirror (Figure 2.4c), or to duplicate the boundary pixels (Figure 2.4d), or to somehow extrapolate the pixel values beyond the boundaries of the image. There are several different ways the extrapolation can be performed, but generally some form of polynomial is fit to the values within the image and used to estimate the pixels beyond the boundary. Whether the results are reasonable or not depends very much on the nature of the image and the function being applied.

Mirroring the boundary means that an address less than zero is used as an absolute value, and one that exceeds the width or height dimension is subtracted from that dimension. Duplicating the boundary means that any address exceeding the image bounds uses the value from the boundary pixel. The latter method is used in the following examples.

### 2.1.4   Other Kernels

Another difficulty with the example code shown above is that it is too specific. The particular set of kernel values shown in Table 2.1 is by no means the only one that we may want to apply. In addition to using other sets of values for smoothing to reduce speckle noise, convolutions are often used to enhance the visibility of edges and boundaries. A typical convolution kernel for that purpose, called a **_Laplacian_** or **_sharpening_** kernel, might look like the example in Table 2.2 (illustrated in Figure 2.6).

Some of the other types of kernels of interest include directional **_derivatives_** (Table 2.3), which can produce an embossed effect such as shown in Figure 2.7.

**TABLE 2.2**

A Laplacian or Sharpening Kernel

| −1 | −1 | −1 |
|----|----|----|
| −1 | +9 | −1 |
| −1 | −1 | −1 |

(a)                                        (b)

**FIGURE 2.6**

(See **Color insert following page 172.**) Example of sharpening detail in the same image as Figure 2.1 using the Laplacian kernel shown in Table 2.2: (a) application to the L channel only, (b) application to the individual R, G, and B channels (note the introduction of extraneous colors for the pixels).

**TABLE 2.3**

Example of a Directional Derivative
or Embossing Kernel

| +2 | +1 | 0 |
|----|----|----|
| +1 | 0 | −1 |
| 0 | −1 | +2 |

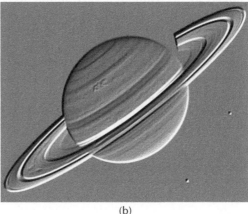

(a)                                              (b)

**FIGURE 2.7**
Example of embossing using the kernel shown in Table 2.3: (a) original image [saturn.tif], (b) result.

In order to produce a function that can easily handle any of the possible kernels that might be encountered, some other changes are needed. One is to rewrite the summations as loops, use an array of pointers for the lines of pixels, and store the kernel itself as a matrix of values. This will make the resulting code much more general.

Another necessary change is to protect the resulting value for the pixel from exceeding the 0 to 255 legal range. Historically, image values have been stored as unsigned bytes with values ranging from 0 (black) to 255 (white). With the progression to greater dynamic range from modern cameras and scanners, and more precise values from other devices that produce images, images may be stored with more than one byte (e.g., two bytes, or even floating point). In this text, as described in the Introduction, we use the conventional 0 to 255 numerical range, but use floating point values so that the extra precision is represented as a fractional part of the number. Consequently, all of the internal calculations are handled in floating point arithmetic, and conversion to and from the format of the original image is handled automatically.

In the smoothing kernel shown in Table 2.1, the presence of only positive values and the division by the sum of the values guarantee that the result will not lie outside of the 0 to 255 range. However, kernels with negative values can produce negative results. Furthermore, the divisor for kernels in Tables 2.2 and 2.3 is not the sum of the values (which may be zero) but is often set equal to one, which can produce results greater than 255. Sometimes, a divisor equal to the magnitude of the largest value in the kernel is used to reduce this overflow (for instance, 9 in the case of the kernel in Table 2.2).

Kernels similar to Table 2.3 that have both positive and negative weight values produce results in which many of the summation values will be negative and so would be clipped to zero. It is usual to add an offset value such as 127.5 (medium gray) to the results, so that a zero result is shown as gray and either positive or negative results from the calculation can be seen as brighter or darker values. The offset value, similar to the divisor, is usually specified along with the contents of the kernel. The offset value is added after the divisor is applied.

The most common answer to the problem of exceeding the legal bounds is to limit the resulting values, so that negative results are clipped to zero and those that exceed 255 are clipped to 255. Although very simple to accomplish, this is neither the only nor always the best solution as it can lead to loss of information. Automatic scaling of the results can also be performed, but requires two passes through the image. Either the original results are saved for the entire image in a temporary array, and then rescaled to the 0 to 255 range afterward before rewriting the values, or on the first pass through the original data, only the minimum and maximum results are saved. They are then used during the second pass through the image to rescale the resulting values using an expression similar to that in Equation 2.2. (It is important to check that the maximum is greater than the minimum, to avoid division by zero.)

$$FinalResult = 255 \cdot \frac{\left( RawResult - Minimum \right)}{\left( Maximum - Minimum \right)} \qquad (2.2)$$

Of course, it is most efficient to calculate the scaling factor once, outside the loop. As usual, considerations of clarity are given precedence over efficiency.

Although *autoscaling* offers the advantage of always fitting the data into the 0 to 255 range for display, without any need to test and clip the data, it creates a few problems of its own. First, it is sensitive to the presence of outliers, a very few values that are much larger or smaller than the majority of the results but are used to determine the overall scaling factor. A second consequence of this sensitivity is that different images of the same, or comparable, scenes will usually have different minimum and maximum values, resulting in different scaling factors that make direct comparisons difficult or impossible. These problems can be avoided by either having the autoscaling routine suggest appropriate scaling factors but leaving the final decision up to the user, or to examine the distribution of results and make more robust decisions about the scaling factors based on the statistics of the distribution (for example, the mean and standard deviation, or the 5 and 95% points, etc.) In these cases, protection against outliers is still needed.

Taking these various factors together, the basic convolution code shown above becomes the more flexible version of Code Fragment 2.3.

```
// Code Fragment 2.3 — Apply convolution kernel
// ... declare variables, read dimensions
// ... assign values to Kernel[3][3] array, Divisor and Offset
// ... create line pointer array Line[3] and pixel array Pixel[3][3]
for (y = 0; y < height; y++)
{  // Read neighborhood lines into Line[] pointers with boundary protection
   if (y-1 < 0)     ReadOriginalLine(0, RGBLine);         // line above
   else             ReadOriginalLine(y-1, RGBLine);
```

```
  ConvertLineRGBtoLAB(width, RGBLine, Line[0]);
  ReadOriginalLine(y, RGBLine);                          // current line
  ConvertLineRGBtoLAB(width, RGBLine, Line[1]);
  If (y >= height) ReadOriginalLine(height-1, RGBLine);  // line below
  else            ReadOriginalLine(y+1, RGBLine);
  ConvertLineRGBtoLAB(width, RGBLine, Line[2]);
  // traverse each line horizontally
  for (x = 0; x < width; x++)
  { // fill Pixel[][] array with neighborhood L values
    Sum = 0;
    if (x-1 < 0)                                 // test left boundary
       for (j = 0; j < 3; j ++)
          Pixel[j][0] = Line[j][0].L;
    else
       for (j = 0; j < 3; j ++)
          Pixel[j][0] = Line[j][x-1].L;
    for (j = 0; j < 3; j ++)
       Pixel[j][1] = Line[j][x].L;
    if (x+1 >= width)                            // test right boundary
       for (j = 0; j < 3; j ++)
          Pixel[j][2] = Line[j][width-1].L;
    else
       for (j = 0; j < 3; j ++)
          Pixel[j][2] = Line[j][x+1].L;
    // apply convolution Kernel[][] to Pixel[][]
    for (j = 0; j < 3; j++)
       for (i = 0; i < 3; i++)
          Sum += Kernel[j][i] * Pixel[j][i];
    Sum = (Sum / Divisor) + Offset;              // scale result
    // limit output range to 0..255
    if (Sum < 0) Sum=0;
    if (Sum > 255) Sum=255;
    Result[x].L = Sum;                           // the new L value
    Result[x].A = Line[1][x].A;                  // copy the A, B color values
    Result[x].B = Line[1][x].B;
  } // for x
  ConvertLineLABtoRGB(width, Result, RGBLine);
  WriteResultLine(y, RGBLine); // write results to host program
} // for y
// ... dispose pointers
```

This code fragment presumes that the **Kernel[3][3]** array and the **Divisor** and **Offset** values have been established beforehand, as well as the various pointers dimensioned (and disposed afterwards).

A more versatile kernel convolution routine would allow larger arrays of weight values to be used, so that neighbors a greater distance away from each pixel may be included in the calculation. The built-in Photoshop procedure handles kernels up to $5 \times 5$ in dimension, but restricts the values to integers. This is particularly limiting for the **Divisor** value. In order to deal with images of various bit depths, real number values and arithmetic are desirable, and acceptably fast with modern processors. The problems suggest several ways that the basic kernel convolution code can be extended.

To expand the size of the neighborhood to N × N (where N is almost always an odd number so that the kernel is symmetrical about the pixel being processed), several alterations in the example in Code Fragment 2.3 are useful. The first is to use a **for** loop to read in the lines from the image using an array of pointers as shown in Code Fragment 2.4.

```
// Code Fragment 2.4 - Read N lines with boundary protection
for (j = 0; j < N; j++)
{
   lineaddr = y - N/2 + j;
   if (lineaddr < 0)          ReadOriginalLine(0, Line[j]);
   else if (lineaddr > height-1) ReadOriginalLine(height-1, Line[j]);
   else                       ReadOriginalLine(lineaddr, Line[j]);
} // for j
```

A loop can also be used to extract the pixel values as shown in Code Fragment 2.5.

```
// Code Fragment 2.5 - Access N pixels with boundary protection
for (i = 0; i < N; i++)
{ // access L values
   pixaddr = x - N/2 + i;
   if (pixaddr < 0)          Pixel[j][i] = Line[j][0].L;
   else if (pixaddr > width-1)  Pixel[j][i] = Line[j][(width-1].L;
   else                      Pixel[j][i] = Line[j][pixaddr].L;
} // for i
```

Equally important to achieving a routine of useful flexibility is allowing arrays of weights for the kernel to be changed without having to recompile the code. Instead of hard-coding values into the routine, they can be read from a text file, as shown in Code Fragment 2.6. The code projects on the companion CD provide basic routines to open text files and to read in real number values. Of course, this requires that the N × N dimension of the neighborhood and the number of values in the file are in agreement. It is even possible to read N from the file, but the code must dimension the arrays large enough to handle the actual values and provide protection and appropriate error messages.

```
// Code Fragment 2.6 - Read values from text file
long  FileID;
ErrType  theerr;
if (OpenTextFileToRead("C:\\Kernel.txt", &FileID)) == noErr)
{
   for (i = 0; i < N; i++)
     for (j = 0; j < N; j++)
        theerr = ReadANumber(FileID, &Kernel[i][j]);
   theerr = ReadANumber(FileID, &Divisor);
   theerr = ReadANumber(FileID, &Offset);
   CloseTextFile(FileID);
} // if noErr
```

Because the pixel values are scaled to the 0 to 255 range and saved as R, G, and B floating point numbers, the kernel values are read as floating point (even if they are entered in the text file as integers) and all of the arithmetic is floating point, this routine will work in exactly the same manner regardless of the original image format.

Kernels can be displayed graphically for visualization by convolving them with an image of a single white pixel (an impulse function). Figure 2.8 shows an example. The result of the convolution is an image of the kernel, rotated by 180°. Applying the kernel (rotated by 180°) to an image of itself can be used to locate features shaped similar to the image of the kernel, often called a template in this context. This operation is called ___cross-correlation___, and is more commonly carried out in the Fourier domain as discussed in the next chapter, but is shown here to emphasize the fact that cross-correlation and convolution are the same except for the 180° rotation of the kernel. If the kernel shown in Table 2.4 is cross-correlated with the image in Figure 2.9, which contains several differently shaped features, the result is an image containing a set of peaks whose heights measure the similarity of each feature to the template and whose locations correspond to the feature

**FIGURE 2.8**
The result of convolving the kernel in Table 2.4 with an image containing a single white pixel.

**TABLE 2.4**

A 9 × 9 Kernel Shaped Like the Letter "R"

| 0 | 0 | 0 | 0 | 0 | 0 | 0 | 0 | 0 |
|---|---|---|---|---|---|---|---|---|
| 0 | 0 | 1 | 1 | 1 | 1 | 0 | 0 | 0 |
| 0 | 0 | 1 | 0 | 0 | 0 | 1 | 0 | 0 |
| 0 | 0 | 1 | 0 | 0 | 0 | 1 | 0 | 0 |
| 0 | 0 | 1 | 1 | 1 | 1 | 0 | 0 | 0 |
| 0 | 0 | 1 | 0 | 1 | 0 | 0 | 0 | 0 |
| 0 | 0 | 1 | 0 | 0 | 1 | 0 | 0 | 0 |
| 0 | 0 | 1 | 0 | 0 | 0 | 1 | 0 | 0 |
| 0 | 0 | 0 | 0 | 0 | 0 | 0 | 0 | 0 |

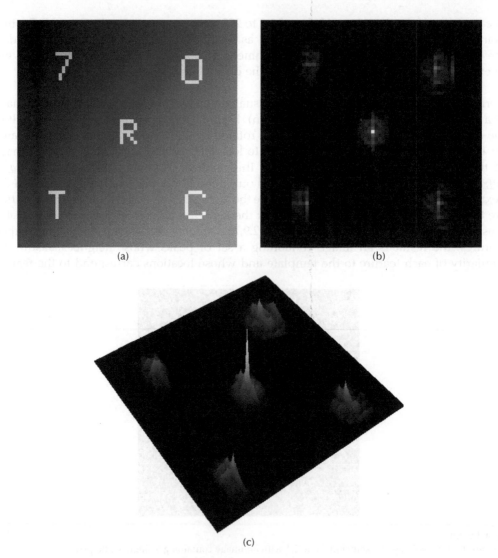

**FIGURE 2.9**
Cross-correlation used to locate a specific feature: (a) image containing several letters on a varying background, (b) result of cross-correlation with the kernel in Table 2.4, (c) isometric display of the result in (b) showing the maximum response at the location of the target feature.

positions. Note that the varying brightness of the background in the original image is ignored by the cross-correlation procedure.

### 2.1.5 Uses of Gaussian Convolutions

Many different sets of kernel values can be devised, and it is instructive to consider some of the ones that are typically used. Smoothing an image to reduce noise or introduce intentional blur is best done with a **_Gaussian_** kernel. A Gaussian kernel is one that has numeric values that correspond to the shape of a Bell curve or Gaussian, as shown in Figure 2.10. The small array of integer values used in Table 2.1 produces a very crude approximation to this shape. It is straightforward to calculate real number values for such a kernel using the expression in Figure 2.10, for any desired value of the standard

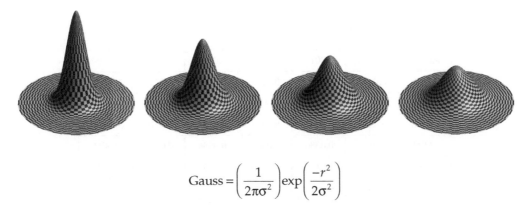

$$\text{Gauss} = \left(\frac{1}{2\pi\sigma^2}\right)\exp\left(\frac{-r^2}{2\sigma^2}\right)$$

**FIGURE 2.10**
The shape of Gaussian kernels with increasing values of $\sigma$, the standard deviation; $r$ is the radial distance from the center. Note that the sum of all values in the kernel (the volume under the graphed surface) remains fixed, and so the maximum weight applied to the central pixel (the height of the central peak) decreases as the standard deviation increases.

deviation $\sigma$. Larger values of $\sigma$ produce greater smoothing but require a larger kernel and a larger neighborhood dimension. Of course, this increases the number of pixel accesses, multiplications, and additions, which increases the time needed to perform the operation. There are practical alternative methods of implementation (shown below) that deal with this problem, but considering the full kernel of weights provides the best basic understanding of the method. Historical methods of designing kernels with small integers to approximate Gaussian kernels of various sizes were difficult, and given the floating point hardware in modern computers, are no longer necessary.

Representative sets of weights are shown in Table 2.5, along with the results of their application (Figure 2.11). Note that the two kernels in the table are shown in different formats. In one case (Table 2.5a) the full kernel is shown and the values are scaled so that they total 1.0, so that no divisor is required. In the second case (Table 2.5b), only one quadrant of the larger symmetrical kernel is shown, and the values are scaled so that the central weight is 1.0. Both formats and both scaling conventions are commonly used.

Because pixels are not points but have a finite area, and in most cases contain values that average the light intensity over that area, there is a subtle effect on the values of the Gaussian kernels. Instead of calculating the value of the Gaussian function at addresses that correspond to the centers of each pixel location, the values should actually be integrated over the pixel areas. The differences are small, and are usually ignored.

Although Gaussian smoothing produces the most reduction of additive random noise possible for a given amount of blurring of edges and detail (or conversely, the least amount of blurring for a given amount of noise reduction), it is not the best or most widely used method for dealing with random noise. Several superior methods will be discussed and illustrated later on. Why, then, is so much attention given to this method here?

One answer is that convolution with arrays of other weights can be used for other purposes, such as the Laplacian or sharpening example and the derivative or embossing example shown above. These can also benefit from larger arrays of weights, and a few examples will be shown below. Another answer is that the idea of convolution is revisited

**TABLE 2.5**

Representative Gaussian Kernels

| 0.0000 | 0.0002 | 0.0011 | 0.0018 | 0.0011 | 0.0002 | 0.0000 |
|--------|--------|--------|--------|--------|--------|--------|
| 0.0002 | 0.0029 | 0.0131 | 0.0215 | 0.0131 | 0.0029 | 0.0002 |
| 0.0011 | 0.0131 | 0.0586 | 0.0965 | 0.0586 | 0.0131 | 0.0011 |
| 0.0018 | 0.0215 | 0.0965 | *0.1592* | 0.0965 | 0.0215 | 0.0018 |
| 0.0011 | 0.0131 | 0.0586 | 0.0965 | 0.0586 | 0.0131 | 0.0011 |
| 0.0002 | 0.0029 | 0.0131 | 0.0215 | 0.0131 | 0.0029 | 0.0002 |
| 0.0000 | 0.0002 | 0.0011 | 0.0018 | 0.0011 | 0.0002 | 0.0000 |

*a\**

| *1.0* | 0.8825 | 0.6065 | 0.3247 | 0.1353 | 0.0439 |
|-------|--------|--------|--------|--------|--------|
| 0.8825 | 0.7788 | 0.5353 | 0.2865 | 0.1194 | 0.0388 |
| 0.6065 | 0.5353 | 0.3679 | 0.1969 | 0.0821 | 0.0266 |
| 0.3247 | 0.2865 | 0.1969 | 0.1054 | 0.0439 | 0.0143 |
| 0.1353 | 0.1194 | 0.0821 | 0.0439 | 0.0183 | 0.0059 |
| 0.0439 | 0.0388 | 0.0266 | 0.0143 | 0.0059 | 0.0019 |

*b\*\**

\* Standard deviation = 1 pixel (entire $7 \times 7$ array, normalized to sum to 1.0) with the central value shown in bold italics.

\*\* Standard deviation = 2 pixels (lower right quadrant of an $11 \times 11$ array, normalized to make the central pixel weight 1.0, shown in bold italics).

Note that a kernel dimension that includes the weights with values large enough to affect the calculation is from four to six standard deviations wide. The values in the corners of the arrays shown are very small, so those distant neighboring pixels have little influence on the result.

again in Chapter 3 on Fourier transform methods, where identical results to a Gaussian smoothing can be achieved with what appears to be a different approach (one that is actually just a different execution of the same mathematics, and which is more efficient for large neighborhoods).

But in practical terms, perhaps the best answer is that the Gaussian blur, while it is not used much by itself, is a key part of one of the most widely used (and misused) of all image processing methods, the so-called ***unsharp mask***. Photoshop and practically all other image processing programs use this tool, because it improves the visual appearance (the visual "sharpness") of many images and also may make them easier to print by reducing the dynamic range of the brightness values (Kim and Allebach 2005).

Photoshop places the "unsharp mask" routine under a submenu called "sharpen," which seems contradictory. The name actually comes from a darkroom technique used for at least a century in conventional photography. Photographic negatives have a much greater dynamic range than photographic prints, and there was interest in finding ways to print pictures so as to preserve or enhance the visibility of fine detail and local contrast while reducing the overall contrast or dynamic range so that bright and dark regions (and detail

(a)                                              (b)

**FIGURE 2.11**
Smoothing the *L* channel of the image from Figure 2.1 with Gaussian kernels: (a) standard deviation = 1 pixel, (b) standard deviation = 2 pixels.

in those regions) could be seen on the print. This was particularly true for astronomical images, but also for night scenes and interior photos.

The unsharp mask was implemented in the darkroom by making a print from the original negative onto another piece of film (or originally on glass plates), at 1:1 scale, but slightly out of focus (hence, "unsharp"). When developed, this film was a positive image, dark where the original negative was light and vice versa. Aligning the two pieces of film and printing (onto paper) through both of them produced a result that allowed less light through either the bright or dark areas of the original negative, except at edges and around fine detail, where the out-of-focus film (the "mask") was blurred. The result was a final image with reduced overall dynamic range, but which still preserved local contrast for details.

The computer implementation of the traditional unsharp mask is accomplished in the same way, by subtracting the Gaussian blurred mask from the original image. Selecting the degree of blurring (the standard deviation of the Gaussian kernel values) is done using the following logic: If blurring removes from the image whatever details are visually important, then the difference between the original and the blurred copy must be just those details. The general strategy to isolate features or structures of interest by removing them from a copy of the image by various processing methods, and then subtracting from the original to leave just those features, is often used in image processing.

In most computer programs, the details obtained by the subtraction are added back to the original image to increase the contrast of fine detail. Often, there is an adjustment for the amount of the detail image to add back, which can range from a few percent to several hundred percent. Figure 2.12 shows a step-by-step example.

Figure 2.13 shows this process applied to an image containing light and dark lines and a brightness step. The intensity profiles show the step-by-step process in which the original image is smoothed, and then subtracted from the original to leave just the local changes.

(a)

(b)

(c)

(d)

**FIGURE 2.12**
Example of an unsharp mask procedure: (a) original image [saturn.tif], (b) Gaussian blurred intermediate, (c) difference between (a) and (b), (d) result of adding (c) to (a).

Adding these back to the original image produces increased local contrast for the details while retaining some of the original contrast. The behavior of the unsharp mask at steps is particularly interesting, as the gradual change in brightness approaching the step on either side can enhance the visibility of the step even though the actual magnitude of the brightness change at the step is reduced.

The kernel shown in Table 2.2 and illustrated in Figure 2.6 is a very limited form of an unsharp mask. It can be understood as subtracting a small $3 \times 3$ smoothing kernel that averages the nine pixels in a $3 \times 3$ neighborhood from nine times the original central pixel, which would produce a zero result if the neighborhood was uniform, followed by adding the result back to the original pixel value. Combining the three individual kernels shown in Table 2.6 produces the sharpening kernel shown in Table 2.2. It is, of course, possible to construct a larger kernel that uses a Gaussian set of weights for the smoothing operation and performs the subtraction and addition.

The logic behind the unsharp mask, as noted above, is that whatever important details in the image can be removed by the smoothing operation are the difference between the original and the smoothed copy, and subtraction recovers these differences so that they become more visible. The advantage of the unsharp mask over the simple sharpening

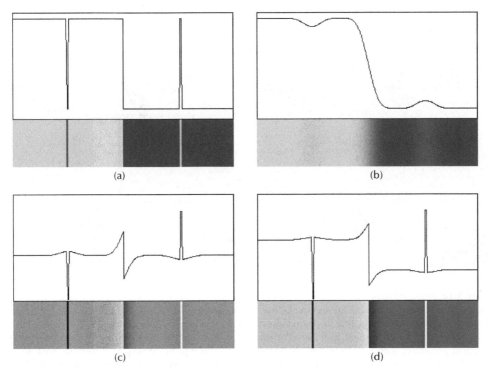

**FIGURE 2.13**
Example of an unsharp mask procedure: (a) original image (b) Gaussian blurred intermediate, (c) difference between (a) and (b), (d) result of adding (c) to (a).

**TABLE 2.6**

Combining Three Kernels

| 0 | 0 | 0 | | +1 | +1 | +1 | | 0 | 0 | 0 |
|---|---|---|---|----|----|----|---|---|----|---|
| 0 | +9 | 0 | − | +1 | +1 | +1 | + | 0 | +1 | 0 |
| 0 | 0 | 0 | | +1 | +1 | +1 | | 0 | 0 | 0 |

kernel is that the degree of smoothing can be adjusted so that details larger than a single pixel in width can be enhanced, and the degree of enhancement can be controlled by adjusting the proportions of the addition.

The unsharp mask is very widely used, but is not the most general form of this procedure. Random speckle noise in an image produces pixel-to-pixel differences that are also made more visible by subtracting the smoothed image from the original. Better results can be obtained by using two copies of the original, one smoothed by a small amount to reduce the random noise and the second smoothed by a larger amount to remove the fine details as well. The difference between these, called the ***Difference of Gaussians*** or ***DoG***, provides a superior result with more control, as shown in Figure 2.14.

The DoG procedure is believed to be a close analog to the processing performed on images in the human retina (Marr and Hildreth 1980). The first few layers of neurons that are connected to the light-sensing cells perform averaging over different size neighborhoods,

(a)                                                                  (b)

**FIGURE 2.14**

Comparison of the unsharp mask (a) and the Difference of Gaussians (b) applied to the L channel of the image in Figure 2.1. Both use an image that has been Gaussian-blurred with a standard deviation of 3 pixels. The DoG subtracts this from an image that was Gaussian-blurred with a standard deviation of 0.5 pixels so that the sharpening procedure is applied to features between about 2 and 9 pixels in size. The unsharp mask subtracts the blurred image from the original and thus sharpens all details smaller than about 9 pixels in size (3 times the standard deviation of the Gaussian smoothing convolution).

and detect the differences between these averages, in order to isolate the edge detail in images that becomes the important information transmitted to the visual cortex. This is usually described as a "center–surround" model in vision research, because it responds to differences between a center region and its surroundings. In the same way, image processing in the computer enhances the fine details and edges using these operations.

The Gaussian weighting used in computer processing is better behaved mathematically than the averaging that takes place in the retina. It is also isotropic, meaning that the Gaussian smoothing is radially symmetric, and so the procedure enhances details equally, regardless of their orientation.

By combining two Gaussian kernels with different standard deviations, a single kernel can be produced as shown in Figure 2.15 that would produce the DoG result in a single operation. This is useful as an illustration of the procedure, but it is generally more flexible to carry out the two smoothing operations individually to provide more control over the amount of blurring and the amount of the difference to add back to the original image. Of course, changing the values for the two standard deviations can produce quite a wide range of shapes for the kernel. In most cases the standard deviation for the larger (blurring) Gaussian is 3 to 6 times larger than the smaller (noise reducing) Gaussian from which it is subtracted.

Simplified forms of the DoG procedure that use a large kernel containing positive weights applied to a central region and negative weights for a surrounding set of pixels, but in which the values are not necessarily Gaussian, are sometimes referred to as "Mexican Hat" kernels.

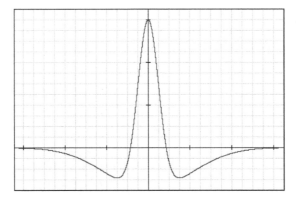

**FIGURE 2.15**
Cross section through a DoG kernel, showing the central positive peak and the broad negative annulus around it.

### 2.1.6   More about Gaussians

As indicated by the various examples above, Gaussian kernels are useful in a variety of applications, but because the kernels become quite large, they are slow to apply because of the large number of pixels that must be accessed and the large number of arithmetic operations that must be performed. Fortunately, there are several ways to mitigate these limitations.

One technique used primarily in older systems based on slower processors with only integer arithmetic was to apply a basic 3 × 3 averaging kernel (a kernel containing nine +1 values), but to apply it many times. With this approach, only additions (no multiplications) are required. Repeating this operation multiple times is similar in effect to applying a larger Gaussian kernel, but extremely inefficient. With most modern computers, the time required to access neighborhood pixels is of greater importance than the arithmetic processing. Multiple passes with a small kernel does not help with this limitation and, in fact, makes it somewhat worse. Fortunately, there is another, better solution.

The expression in Figure 2.10 can be separated into two terms, one having only $x$ terms and one only $y$ terms, as shown in Equation 2.3.

$$\exp\left(\frac{-r^2}{2\sigma^2}\right) = \exp\left(\frac{-x^2}{2\sigma^2}\right) \cdot \exp\left(\frac{-y^2}{2\sigma^2}\right) \tag{2.3}$$

Consequently, convolution with a two-dimensional Gaussian kernel can be separated into two orthogonal convolutions with a simpler array, a one-dimensional array of weights that is applied to just the neighbors in a line and in a column. As indicated in Figure 2.16, the procedure makes two passes through the image, one performing a convolution with Gaussian weights applied to just the vertical column of neighbors and the other to just the horizontal row of neighbors (in either order). This reduces the number of math operations from the order of $N^2$ to the order of $N$ for an $N \times N$ kernel. Of course, it requires storing the image in a temporary array so that both passes can be applied.

Most convolution kernels are not separable, so this is not a general approach to convolution operations, but because the Gaussian is so widely used, the special code to perform this function may be useful. We will also see, in Chapter 3 on Fourier transform processing,

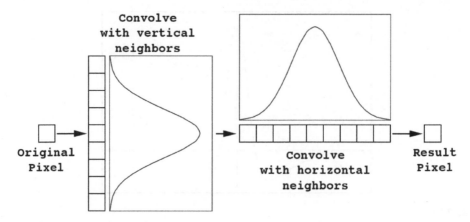

**FIGURE 2.16**
The separable Gaussian requires two passes through the image with storage of the intermediate results. In one pass the pixel is convolved with its neighbors in a vertical column, and in the second it is convolved with its neighbors in a horizontal row. The final result is identical to a single pass using convolution with a full two-dimensional kernel.

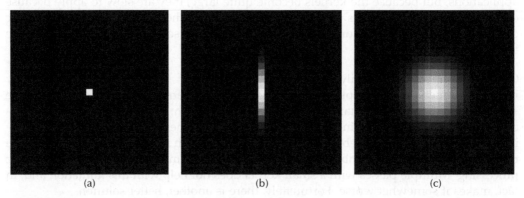

**FIGURE 2.17**
The separable Gaussian: (a) original image consisting of a single white pixel, (b) after the first (vertical) convolution, (c) after the second (horizontal) convolution.

that another approach to the efficient application of large convolution kernels, and particularly the Gaussian, is available.

Figure 2.17 illustrates the process of performing the Gaussian smooth in two separate passes. The first pass applies the vertical convolution, resulting in a column of pixels with a Gaussian intensity profile. The second pass then applies the horizontal convolution, which spreads the values out horizontally to form the final result. In the example, the pixel brightness values have been rescaled so that the central pixel remains at the maximum value, for visibility. In actual practice, the resulting pixel values would be reduced so that their sum corresponded to the magnitude of the original central pixel.

### 2.1.7 Derivatives

The embossing filter shown in Table 2.3 and illustrated in Figure 2.7 is an example of a directional derivative. By calculating the difference between intensity values to the left and right, or above and below each pixel, a discrete approximation to the derivative of

**TABLE 2.7**

Examples of Kernels That Produce a Horizontal Derivative

| −1 | 0 | +1 |
|----|---|----|
| −2 | 0 | +2 |
| −1 | 0 | +1 |

| −1 | 0 | 0 | 0 | +1 |
|----|----|---|----|----|
| −1 | −2 | 0 | +2 | +1 |
| −1 | 0 | 0 | 0 | +1 |

brightness with respect to direction can be obtained. By adjusting the values of the weights proportionately, the derivative in any direction can be calculated. Averaging a few pixels reduces the noise inherent in this local difference. The averaging may be done in several different ways, a few of which are shown in Table 2.7. Note that in both cases the value of the central pixel is not used. If the central pixel is also assigned a positive weight, the derivative is added back to the original image. In all of these cases, it is necessary to incorporate either autoscaling of the result or appropriate divisor and offset values to minimize the number of pixels that exceed the 0 to 255 range, requiring clipping.

Figure 2.18 shows the application of two directional derivatives to an image, one in the horizontal direction, and one in the vertical. Each of these enhances the visibility of lines and edges in the image that are oriented in a direction perpendicular to the derivative (e.g., horizontal features are enhanced by a vertical derivative). However, the enhancement produces light and dark edges around the features (which are visible because an offset of 127.5 has been used so that both positive and negative values are retained rather than being clipped).

Details that are oriented in a direction parallel to that of the derivative disappear in the result image. That ability to suppress details having a selected orientation can be useful in some cases, such as the illustration in Figure 2.19. Removing the horizontal line with a horizontal derivative makes the details in the handwriting more visually evident. This technique is sometimes used in the forensic examination of handwriting.

If the absolute value of the derivative result is used instead of the signed value, a somewhat different result is obtained, as shown in Figures 2.18d and 2.18e. The edges, lines, and steps that are discontinuities in the original image are now outlined, but still just in directions perpendicular to the derivative orientation.

It is reasonable to consider the two directional derivatives as projections in the horizontal and vertical directions of the vector that is the spatial derivative of the brightness, or the local brightness gradient, in the image. Combining the two magnitudes as shown in Figure 2.20 calculates the vector magnitude of that gradient. Displaying that value (Figure 2.18f) produces a complete set of outlines for the image. The magnitude of the gradient (shown as the brightness of the lines) is independent of the orientation of the lines and edges.

This method for outlining edges and detail in an image is widely used, and is called the *Sobel* gradient method after its originator (Sobel 1970). There are other techniques for obtaining similar edge delineation, which will be described below, but the Sobel technique is the most commonly used because it is relatively fast, using just a small (usually 3 × 3) neighborhood (but applying two kernels—one for the vertical and one for the horizontal derivative—to every pixel location). The Sobel method also offers an additional piece of information giving the orientation of lines and edges as shown in Figure 2.20 and illustrated in Figure 2.21.

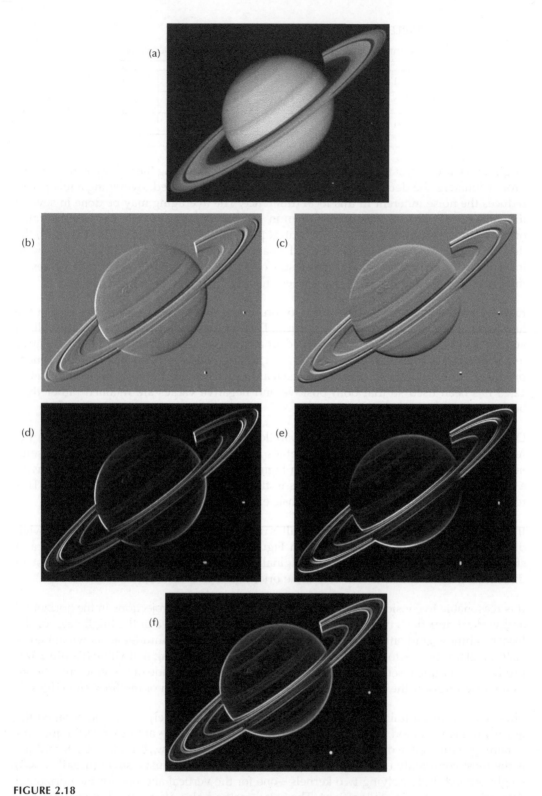

**FIGURE 2.18**
Directional derivatives used to outline edge features: (a) original image [saturn.tif], (b) horizontal derivative,
(c) vertical derivative, (d) absolute value of the horizontal derivative, (e) absolute value of the vertical derivative,
(f) vector magnitude of the brightness gradient.

(a)            (b)

**FIGURE 2.19**
Application of a horizontal derivative to suppress horizontal lines: (a) original image [writing.tif], (b) result.

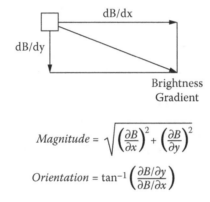

$$Magnitude = \sqrt{\left(\frac{\partial B}{\partial x}\right)^2 + \left(\frac{\partial B}{\partial y}\right)^2}$$

$$Orientation = \tan^{-1}\left(\frac{\partial B/\partial y}{\partial B/\partial x}\right)$$

**FIGURE 2.20**
Combining the horizontal and vertical derivatives to calculate the vector magnitude and angle of the local brightness gradient.

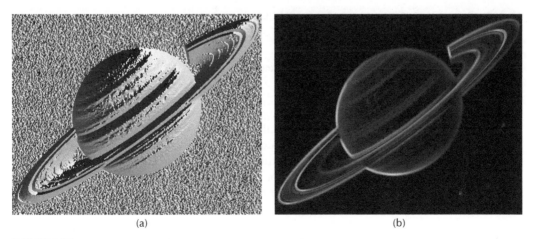

(a)            (b)

**FIGURE 2.21**
(See **Color insert following page 172**.) Displaying the vector orientation for the Sobel brightness gradient: (a) grayscale values, (b) hue values (with the brightness representing the vector magnitude).

Implementing the Sobel requires only a single pass through the image, using code much like that shown above. The results from the two kernels are combined to produce a pixel for the new image as shown in Code Fragment 2.7. It is worth noting that some antique implementations of the Sobel method did not use the square root function, but made crude approximations such as the sum or greater of the two directional derivative values.

```
// Code Fragment 2.7 — Sobel magnitude filter
for (x = 0; x < width; x++)
  { // ...
    HDeriv = VDeriv = 0;
    for (j = 0; j < 3; j++)
      for (i = 0; i < 3; i++)
        { // calculate results for two directional derivatives
          HDeriv += HKernel[j][i] * Pixel[j][i];
          VDeriv += VKernel[j][i] * Pixel[j][i];
        } // for i
    Magnitude = sqrt(HDeriv*HDeriv + VDeriv*VDeriv);
    // ...
```

Since the local brightness gradient is a vector, it also has an orientation as well as a magnitude. This is perpendicular to the local orientation of the line or edge detail in the image, and can be determined from the two derivative values as shown in Figure 2.20. The gradient magnitude and gradient orientation are both useful for image processing and measurement. Normally, an angle is measured in degrees, from 0 to 359, but the result of the Sobel orientation operation is most conveniently displayed in an image by scaling the numeric range of the values from 0 to 255 (meaning that each integer change in brightness corresponds to about 1.4 degrees).

There is an additional complication in determining the angle from the two derivative values. The atan function (arctangent) returns values from $-\pi$ to $+\pi$. It is necessary to examine the sign values of the derivatives to decide in which quadrant the vector lies, as shown in Code Fragment 2.8. This test may be omitted to map the orientation of lines into the range from 0 to 180 degrees.

```
// Code Fragment 2.8 — Sobel orientation filter
  // ...
  Angle = atan(VDeriv / HDeriv);                    // -pi to + pi
  Angle = 63.75 + (127.5/3.14159265) * Angle;  // 0 to 180 degrees
  if (HDeriv < 0) Angle += 127.5;                   // quadrants 3 and 4
  // ...
```

Notice in Figure 2.21a that the grayscale values around the limb or rings of the planet vary smoothly with orientation. These values can be used in a variety of situations to measure feature orientation in imaged structures, as will be seen in Section 2.3.2. However, the image also contains orientation values for other pixels that are part of the background. These are locations where the magnitude of the gradient vector is very small, but where the vector still has an orientation.

Another way to display the orientation values is to make use of the fact that hue values also are conveniently related to an angle. Assigning the vector orientation of the brightness

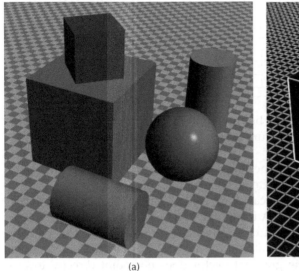 

(a)           (b)

**FIGURE 2.22**
Application of the Sobel operator to highlight the edges of objects in an image [boxworld.tif]: (a) original (L channel), (b) vector magnitude.

gradient to the hue, and the vector magnitude to the brightness, produces a result (Figure 2.21b) that is visually interpretable as well as aesthetically appealing. In this case, instead of the spherical LAB color space conversion used in the method shown in Code Fragment 2.2, making use of the hue–saturation–intensity coordinates is appropriate. The vector magnitude is assigned to the intensity and the vector angle to the hue, with the saturation value set to maximum, and then the values are converted to RGB for display. Of course, this technique processes the intensity values of the pixels and ignores (and replaces) any color information in the original image.

One of the important uses of edge detection is in robotics vision (Davies 1990). As shown in Figure 2.22, the edges of features provide a set of outlines much like a sketch or cartoon. The lengths and orientation of these lines, and the angles at which they meet, may be interpreted to build a model of the identity and positions of the objects that are present. This process is a higher-level computational task beyond the scope of this text, but it typically begins with simple image processing steps such as those shown here.

### 2.1.8 Other Edge-Detecting Convolutions

The Sobel operator, although widely used, is not the only convolution used to locate edges in images (Berzins 1984). In addition, other techniques that do not rely on convolution will be shown below. Before computers easily handled the floating point arithmetic used in the Sobel operator's square root function, the **_Kirsch_** operation was used. A $3 \times 3$ derivative kernel was applied in eight orientations in 45° increments, and the largest result was kept as the gradient magnitude. This required only integer arithmetic and comparisons. A typical kernel (in just two of the eight possible orientations) is shown in Table 2.8.

Because the weights in these kernels are large, the result is more sensitive to small discontinuities in the image (including noise). Also, because the numeric results may exceed

**TABLE 2.8**

Kernels That Produce Directional Derivatives

| +5 | −3 | −3 |
|----|----|----|
| +5 | 0  | −3 |
| +5 | −3 | −3 |

| +5 | +5 | −3 |
|----|----|----|
| +5 | 0  | −3 |
| −3 | −3 | −3 |

···

the 0 to 255 range for more pixels, these will be clipped to the maximum value, producing broader lines along the principal edges. Figure 2.24a shows an example.

For both the Sobel and Kirsch functions, the resulting lines are typically several pixels wide (typically as wide as the sum of the extent of the actual line or edge plus the width of the kernel). Sometimes, it is desirable to locate the edge to the nearest pixel, and produce delineations that are one-pixel-wide lines. A technique that can accomplish this is called *__non-maximum suppression__*. It is carried out as indicated in Figure 2.23. The gradient magnitude at each pixel is compared to the value at the next pixel in the gradient direction, and kept only if it is larger. When the gradient vector does not point directly toward another pixel, the value may be interpolated between the neighboring pixels.

This method for producing thinned lines delineating edge positions is used in an edge-finding procedure called the *__Canny__* operation (Canny 1986). This is a powerful but multistep function that utilizes several techniques that have been described and illustrated above. It begins by performing Gaussian smoothing on the image, then calculates derivatives in orthogonal directions using quite large kernels, and combines them to obtain a vector magnitude and direction. The resulting lines along edges and discontinuities are then thinned by the non-maximum suppression method and, finally, only lines that somewhere have a gradient value exceeding a preset threshold are retained.

The Canny method has several arbitrary user-defined constants that limit its generality, and the full programming is not undertaken here. But the step of non-maximum suppression

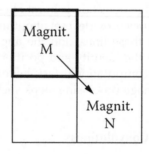

$$\left\{\begin{array}{l} \text{Let } \vec{M} \text{ be the gradient vector at } x, y \\ \text{Let } \vec{N} \text{ be the gradient vector at the adjacent} \\ \qquad \text{neighbor to which } \vec{M} \text{ points} \\ \text{IF } |\vec{M}| < |\vec{N}| \text{ set } \vec{M} = 0 \end{array}\right\}$$

**FIGURE 2.23**

Non-maximum suppression: The gradient vector at the black-bordered pixel has a magnitude $M$ and a direction as shown by the arrow. If the neighboring pixel in that direction has a gradient magnitude $N$ that is smaller than $M$, the result for the original (black-bordered) pixel is kept, otherwise it is erased.

(a)   (b)

**FIGURE 2.24**
(a) Edge delineation produced by the Kirsch function (Code Fragment 2.9a), (b) the result of applying non-maximum suppression (Code Fragment 2.9b).

operation can be usefully studied as an example of an adaptive or conditional filter, one in which intermediate results are used to determine the final outcome. In Code Fragment 2.9, non-maximum suppression is paired with the Kirsch operator to illustrate the procedure; Figure 2.24 illustrates the result.

The important point to note in the code is that two passes through the image are required. The first determines the vector magnitude and direction at each pixel, both of which must be saved, and the second uses those values to perform comparisons between pixels. Some adaptive filters do not require multiple passes and storage of intermediate results, but many do, and this requires creating a data structure in which to place those temporary values. Tools for creating such temporary arrays in memory are included with the support routines and are described in the Appendix.

```
// Code Fragment 2.9a - Kirsch filter
// ... create eight Kernel arrays (eight orientations)
// ... create two temporary images to hold magnitude and direction values
for (y = 0; y < height; y++) // first pass - build vectors
  for (x = 0; x < width; x++)
  { // ...
    for (j = 0; j < 3; j++)
      for (i = 0; i < 3; i++)
      {
        Deriv[0] += Kernel0[j][i] * Pixel[j][i];
        Deriv[1] += Kernel1[j][i] * Pixel[j][i];
        // ...
      } // for j
    magnitude = Deriv[0];
    direction = 0;
    for (k = 1; k < 8; k++)
    {
      if (magnitude < Deriv[k])
      {
        magnitude = Deriv[k];
```

```
            direction = k;
        }
    } // for k
    // ... save both magnitude and direction for every pixel
} // for x
```

**// Code Fragment 2.9b — Nonmaximum suppression**
```
for (y = 0; y < height; y++)
    for (x = 0; x < width; x++)
    { // ... load direction[][] and magnitude[][]
        switch (direction[x][y])
        {
            case 0:if magnitude[x][y] < magnitude[x+1][y]
                        magnitude[x][y] = 0;
                  break;
            case 1:if magnitude[x][y] < magnitude[x+1][y+1]
                        magnitude[x][y] = 0;
                  break;
            // ... and so on for other directions
        } // switch
    } // for x
    // ... save modified magnitude result as pixel brightness
```

### 2.1.9   Conditional or Adaptive Filters

The preceding example is an extreme case of a ***conditional*** or ***adaptive filter***. Another procedure, which is often used to improve the visual appearance of pictures, uses a conditional approach to increase the contrast at the edges by using a Laplacian sharpening convolution, while at the same time reducing random noise elsewhere using a smoothing convolution. The blending of the two results, each calculated for every pixel, is controlled by the Sobel edge-detecting gradient, also calculated for every pixel. All of these calculations have been shown above, and can be performed during a single pass through the image without the need for any temporary image array. An outline of the necessary procedure is shown in Code Fragment 2.10, and the separate components and result are illustrated in Figure 2.25.

**// Code Fragment 2.10 — Conditional smoothing and sharpening**
```
for (y = 0; y < height; y++)
{ // ...
    for (x = 0; x < width; x++)
    { // ...
        // for each pixel
        Sobel     = magnitude of gradient vector from Figure 2.20;
        Laplacian = convolution with Table 2.2;
        Smooth    = convolution with Table 2.1;
        BlendFraction = Sobel / 255;
        if (BlendFraction > 1.0) BlendFraction = 1.0;
        Result[x] = Laplacian * BlendFraction + Smooth * (1-BlendFraction);
        // ...
    } // for x
} // for y
```

**FIGURE 2.25**
Conditional or adaptive filtering to sharpen edges while smoothing random noise: (a) L channel of original image [rose.tif], (b) smoothed result obtained by applying kernel Table 2.1, (c) sharpened result obtained by applying kernel Table 2.2, (d) Sobel gradient magnitude obtained by applying the procedure in Figure 2.20, (e) blended final result obtained by using the Sobel gradient magnitude for each pixel to combine the smoothed and sharpened results.

## 2.1.10   Problems

2.1.10.1#.   Implement a routine that applies a $3 \times 3$ convolution using a kernel that is read from a text file. Your file should also include the divisor and offset values.

2.1.10.2.   Instead of using the boundary duplication technique shown in Figure 2.4d and Code Fragment 2.3, implement the edge-mirroring technique.

2.1.10.3.   Instead of the simplified spherical LAB color space conversion shown in Figure 2.3 and Code Fragment 2.2, implement code that uses a different color space, such as HSI space, to separate grayscale brightness from color when performing a convolution. Compare the results.

2.1.10.4#.   Instead of fixed scaling or clipping the results of a convolution calculation to fit the 0 to 255 range of values, implement a two-pass method that autoscales the output. Compare the results.

2.1.10.5.   Autoscaling as shown in Equation 2.2 will cause a shift in the middle gray level of an image. By adding a comparison such as

```
if ((max - 128) > (128 - min)) min = (255 - max);
else                           max = (255 - min);
```

it is possible to keep the middle gray value unchanged. Implement this method and compare the results.

2.1.10.6#.   Implement a program that calculates Gaussian kernels with various standard deviations. Apply a convolution kernel that is large enough to compare the results from several different kernels. Also compare the results to the built-in Photoshop function.

2.1.10.7.   Implement a separable Gaussian convolution, and compare the results to those from Problem 2.1.10.6, and to the built-in Photoshop function. It may be helpful to optionally display the result after the first linear convolution.

2.1.10.8.   Use a spreadsheet to combine two Gaussian kernels with different standard deviations, to produce a DoG kernel. Apply these and compare the result to that obtained by performing the two Gaussian smoothing operations individually on two copies of the image and subtracting one smoothed image from the other. (If applied to an 8-bit image, you may see differences resulting from limited numeric precision.)

2.1.10.9.   Implement a convolution that processes the individual RGB color channels using different kernels. A typical method used in some cameras applies modest sharpening to the green channel and smoothing to the (noisier) blue channel. Examine the results to detect color shifts. The [manuscript.tif] image shown in color plate 8 provides a suitable test for this procedure.

2.1.10.10.   Implement a convolution that applies sharpening to the intensity data while blurring the color channels. The method used will depend on which color space is used. For a hue–saturation–intensity space, the hue must be treated modulo 360°. In LAB space this is not required. Compare the results to those from Problem 2.1.10.9.

2.1.10.11#.   Implement a conditional convolution that applies an unsharp mask to the intensity data only when the difference between the new value and the original pixel value exceeds a preset threshold (e.g., 25 brightness levels). This is the same procedure that Photoshop's unsharp mask procedure uses when the "threshold" slider is set to a value greater than zero.

2.1.10.12.  Implement a conditional convolution that applies an unsharp mask to the intensity data only when the new pixel is darker (or lighter, depending on a logical variable read from a disk file) than the original value. Compare the visual enhancement of contrast at edges and details in bright and dark regions of an image.

2.1.10.13#.  Implement a Sobel gradient filter. Use a logical variable to select whether to save the magnitude or the orientation angle.

2.1.10.14.  Implement a conditional convolution that uses a gradient vector operator to determine the orientation of edges and then applies a smoothing operation along the edge (perpendicular to the gradient) and a sharpening operation in the gradient direction. This procedure is called an adaptive anisotropic filter, and is sometimes applied to images such as fingerprints.

2.1.10.15#.  Implement the conditional filter outlined in Code Fragment 2.10.

2.1.10.16.  Using a black image with a single white pixel, determine the kernels used by the following blur filters in Photoshop: (a) Filter > Blur > Blur, (b) Filter > Blur > Blur More. You may have to adjust the brightness levels (Image > Adjustments > Levels) to see some of the pixel values. Using a dark gray image with a light gray dot, determine the kernels used by the following sharpen filters in Photoshop: (c) Filter > Sharpen > Sharpen, (d) Filter > Sharpen > Sharpen More. Why is it important to use a 16-bit grayscale image to examine the sharpening kernels?

## 2.2   Other Neighborhood Operations

Convolutions are described as linear operations. They are mathematically equivalent to procedures that can be carried out in the Fourier domain (discussed in Chapter 3), and they use the values of all pixels in the neighborhood. Because of the variety of kernels that can be designed for various purposes, employing the same underlying code to perform the operation, they are widely used. However, they are not the only important class of neighborhood functions that are applied to images.

There is another very important category of neighborhood operations that are nonlinear, do not have Fourier domain equivalents, and discard some of the pixel values in the neighborhood. These operations are based on ranking the pixels in the neighborhood, typically (but not always), based on the intensity value (usually the average of the red, green, and blue, or sometimes, the weighted luminance). The most commonly encountered function of this type is the ***median filter***. It is usually superior to the smoothing convolution shown above for reducing random speckle noise, and much better for rejecting shot or salt-and-pepper noise.

### 2.2.1   Median Filter

Random noise ("speckle" or "texture") arises in many stages of image acquisition. This includes statistical variations in the charge produced in the detector, cumulative errors in chip readout, thermal noise in the amplification of the signal, and random electronic fluctuations in every step of the signal processing and digitization process. The result is a variation in the brightness and color values of pixels in regions that should be uniform. A measure of the amount of noise can be obtained by examining the width of the peak in a histogram of a uniform region as illustrated above in Figure 2.2. Smoothing random noise reduces its magnitude (and narrows the histogram peaks), but as shown in the preceding examples, blurs fine detail and edges.

The median filter preserves the amplitude and location of edges and steps better than smoothing, while reducing the noise. It also produces images whose histograms have narrower peaks and reduced background levels, because extreme values are eliminated. Figure 2.26 compares the effect of Gaussian smoothing and median filtering on the brightness histogram.

Shot noise is another type of noise defect in images. It typically results from discrete defects in the detector. A small number of transistors that are dead (inactive) or locked (always outputting the maximum value) may be present without making the device unusable. These result in pixels in the stored image that are black or white, or (depending on which color the defective transistor detects) pixels that are different in color from their surroundings. Applying a smoothing filter to shot noise averages the incorrect value with the surroundings, producing a larger, blurred defect. The median filter can replace the defective value with one from a nearby pixel. (Most cameras incorporate logic similar to that shown below to reduce the effects of such defects.)

The operation of the median filter is diagrammed in Figure 2.27.

1. The values of the pixels in the neighborhood are placed into an array.
2. The values in the array are then sorted into rank order from brightest to darkest.
3. The value in the center of the sorted list (the median value) is placed into the original central pixel location in a new image.
4. Then, the neighborhood is shifted over to the next pixel, and the process is repeated.

As with other neighborhood operations, the procedure always uses the pixel values from the original image and writes the results to produce a new image.

The same image boundary conditions as used for the convolution operations in Section 2.1 apply to the median filter as well. There is no need to protect against values exceeding the legal 0 to 255 range, because no new values are calculated; they are just copied from one location to another. The median filter does not blur edges because it must select a value that is already present on one side of the discontinuity or the other, unlike convolution with a Gaussian kernel, which can average the values. Consequently, the magnitude of the discontinuity at the edge is preserved.

However, the median filter does discard information from the image. Pixel values that are extreme are not selected as the median and so are replaced with other values from nearby pixel locations. In addition to rejecting single-pixel shot noise, the median will remove any feature that is smaller than half the area of the neighborhood, because the pixels in the feature cannot fill up the ranked list to reach the midpoint. That means the size of the neighborhood is an important parameter for controlling the median. Figure 2.28 illustrates this in one dimension. Applying the median filter with a neighborhood having a radius

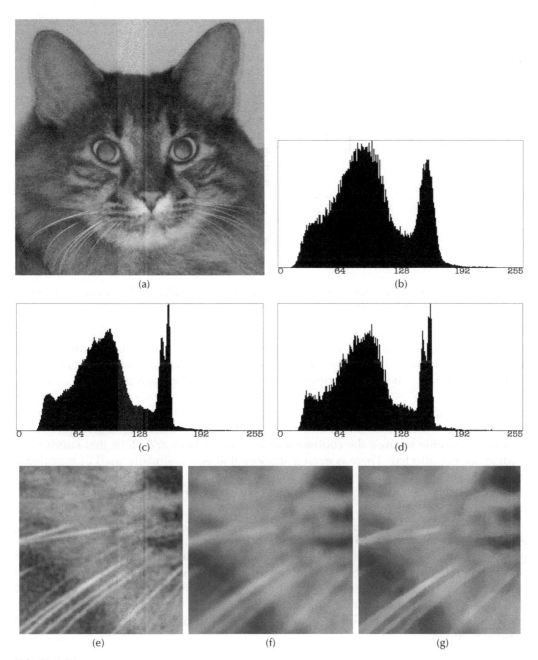

**FIGURE 2.26**
Effect of noise reduction on image histogram: (a) original image [cat.tif]; (b) histogram of (a) with broad peaks due to random speckle noise; (c) result of Gaussian smoothing with standard deviation = 2 pixels; (d) result of median filtering with radius of 2 pixels — both methods produce narrower, better-separated peaks; (e) detail of original image; (f) detail of Gaussian smoothed image; (g) detail of median filtered image.

wider than the line removes the line cleanly, replacing it with values from the nearby background. The step is retained in all cases.

Figure 2.29 illustrates the effect of increasing the size of the neighborhood used in the median. A radius of 1 pixel corresponds to just the original central pixel and its four

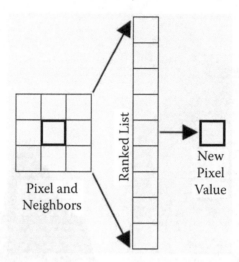

**FIGURE 2.27**
Schematic diagram of the median filter. The values of the pixels in the neighborhood are ranked in order and the median value (at the center of the ordered list) is selected for the pixel in the new image.

edgesharing neighbors that lie within a radial distance of one pixel. A radius of 2 or 3 pixels, respectively, includes pixels at a greater distance and progressively removes lines of increasing width. These neighborhoods are circular, as discussed below.

Because the larger neighborhoods remove more noise but also remove fine detail, another strategy is to use a small neighborhood but repeat the operation as shown in Figure 2.29e. This is permissible because the contrast and location of lines and steps that survive the median are not affected. There is even a strategy that is occasionally used of repeatedly applying a median filter until no further changes occur in the image, called an iterated median. This typically produces a result that consists of large regions of uniform brightness.

As mentioned above, the median is particularly effective for removing shot or "salt-and-pepper" noise. The original image in Figure 2.30a was corrupted by setting 200 of the pixels (randomly selected) to black and another 200 to white, to simulate an extreme case of locked or dead pixels in a camera. Applying the median filter with a $3 \times 3$ neighborhood (radius = 1.5 pixels) replaces all of the noise pixels with a value from an adjacent pixel. The result in Figure 2.30b is visually indistinguishable from the image before the shot noise was added.

The procedure shown in Code Fragment 2.11 performs a median separately on the red, green, and blue channels of an image using a $3 \times 3$ neighborhood (which corresponds to radius = 1.414, but is sometimes confusingly described as radius = 1). A bubble sort is adequate for a small list of values. As the neighborhood size grows, substitution of other more efficient sorting methods (Quicksort, Shellsort, etc.) can be made without affecting the basic structure of the algorithm. As with the convolution procedure, this can be generalized to larger neighborhoods, and it is possible to make the routine faster by sorting portions of the neighborhood separately, keeping the sorted portions as the $x$ position is advanced, and merging them to obtain the final median, but doing so makes the algorithm underlying the procedure less clear.

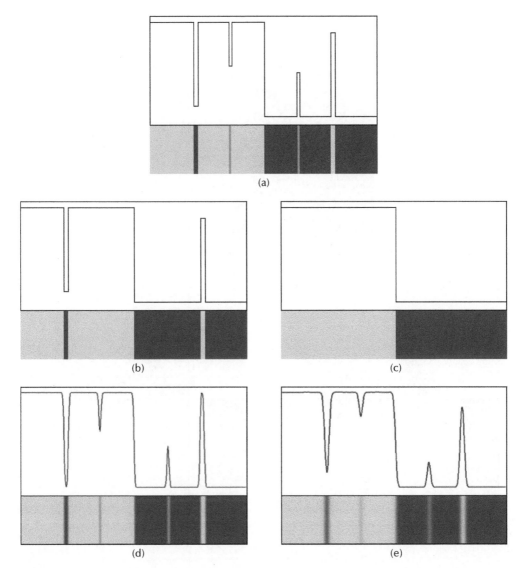

**FIGURE 2.28**
Example of median filtering with different width neighborhoods: (a) original image containing a step and lines with widths of 4 and 8 pixels; (b) application of a median with a 4-pixel radius neighborhood removes the narrow line but not the wide line; (c) application of a median with a 8-pixel radius neighborhood removes both narrow and wide lines — the step is not blurred or shifted; (d) blurring produced by a Gaussian convolution with standard deviation of 1.5 pixels; (e) blurring produced by a Gaussian convolution with standard deviation of 3 pixels — the steps and lines are blurred and not removed.

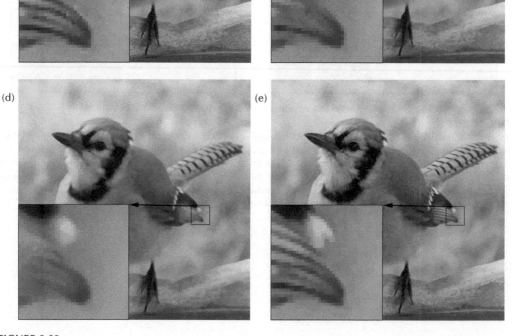

**FIGURE 2.29**
Application of a median filter to an image containing random noise and fine details: (a) original image, L channel [bird.tif], (b) radius = 1 pixel, (c) radius = 2 pixels, (d) radius = 3 pixels, (e) radius = 1 pixel, applied three times.

(a)  (b)

**FIGURE 2.30**
Application of a median filter to an image containing shot noise: (a) original image [girl.tif] with 400 of the pixels set to black or white, (b) after application of a 3 x 3 median filter.

```
// Code Fragment 2.11 — RGB median filter
// ... define variables and array, create line pointers
for (y = 0; y < height; y++)
{  // Read 3 lines for neighborhood
   ReadOriginalLine(y-1, LineA);
   ReadOriginalLine(y  , LineB);
   ReadOriginalLine(y+1, LineC);
   for (x = 0; x < width; x++)
   {
      Array[0] = LineA[x-1].red;
      Array[1] = LineA[x  ].red;
      Array[2] = LineA[x+1].red;
      Array[3] = LineB[x-1].red;
      Array[4] = LineB[x  ].red;
      Array[5] = LineB[x+1].red;
      Array[6] = LineC[x-1].red;
      Array[7] = LineC[x  ].red;
      Array[8] = LineC[x+1].red;
      Sort(Array, 9);             // sort array into order
      Result[x].red = Array[4]; // median value in array
      // ... repeat for .green and .blue
   } // for x
   WriteResultLine(y, Result);
} // for y
// ... dispose pointers

void Sort(float *Array, long N) // shell sort array of length N
{
```

```
    long  jump, i, j;
    for (jump = N / 2; jump > 0; jump /= 2)
       for (i = jump; i < N; i++)
          for (j = i-jump; (j > 0) && (array[j] > array[j+jump]); j -= jump)
          {
              float swap = array[j];  // SWAP array[j] <-> array[j+jump]
              array[j] = array[j+jump];
              array[j+jump] = swap;
          }
} // Sort
```

### 2.2.2   Color Issues (Again)

There are some circumstances in which operating on the individual RGB color channels is acceptable. For instance, a digital image acquired in dim light may have different amounts of noise in each channel, which is best removed within the channel itself. Also, single-chip color sensors use a grid of colored filters so that individual sensors respond to a specific colors, and some algorithms for demosaicing the data use median filters as part of the interpolation process.

However, in most cases the independent processing of the RGB color channels can lead to the introduction of false colors, particularly near feature edges. The problem is that the values selected by the median filter for the RGB values at each pixel may be taken from quite different locations, on different sides of the discontinuity. When combined, they produce erroneous and highly visible color artifacts.

The most common solution is to use the intensity values for sorting, as was done for the convolution examples presented above. The same code for converting from RGB values for each pixel to some other space, as shown in Chapter 1 and used in Code Fragment 2.2, can be used. Placing the intensity or luminance values into the list for sorting and selecting the median is not enough, however. Merging the new intensity value with the original color values is not appropriate. Instead, the full set of color values from the neighborhood pixel that was selected by the median filter must be copied to the target location. That requires carrying along the color values in the sorting process, as shown in Code Fragment 2.12.

```
// Code Fragment 2.12 – Median brightness filter
// ... define variables and arrays, create line pointers
for (y = 0; y < height; y++)
{  // Read 3 lines for neighborhood
   ReadOriginalLine(y-1 , RGBLine); // ... requires usual boundary protection
   ConvertRGBtoLAB(width, RGBLine, LineA); // convert to LAB space as above
   ReadOriginalLine(y   , RGBLine);
   ConvertRGBtoLAB(width, RGBLine, LineB);
   ReadOriginalLine(y+1 , RGBLine);
   ConvertRGBtoLAB(width, RGBLine, LineC);
   for (x = 0; x < width; x++)
   { // ... make three separate arrays.
      // sort on L but keep all three color values
```

```
      ArrayL[0] = LineA[x-1].L; ArrayA[0] = LineA[x-1].A;
          ArrayB[0] = LineA[x-1].B;
      ArrayL[1] = LineA[x  ].L; ArrayA[0] = LineA[x  ].A;
          ArrayB[0] = LineA[x  ].B;
      ArrayL[2] = LineA[x+1].L; ArrayA[0] = LineA[x+1].A;
          ArrayB[0] = LineA[x+1].B;
      ArrayL[3] = LineB[x-1].L; ArrayA[0] = LineB[x-1].A;
          ArrayB[0] = LineB[x-1].B;
      ArrayL[4] = LineB[x  ].L; ArrayA[0] = LineB[x  ].A;
          ArrayB[0] = LineB[x  ].B;
      ArrayL[5] = LineB[x+1].L; ArrayA[0] = LineB[x+1].A;
          ArrayB[0] = LineB[x+1].B;
      ArrayL[6] = LineC[x-1].L; ArrayA[0] = LineC[x-1].A;
          ArrayB[0] = LineC[x-1].B;
      ArrayL[7] = LineC[x  ].L; ArrayA[0] = LineC[x  ].A;
          ArrayB[0] = LineC[x  ].B;
      ArrayL[8] = LineC[x+1].L; ArrayA[0] = LineC[x+1].A;
          ArrayB[0] = LineC[x+1].B;
      Sort3(ArrayL, ArrayA, ArrayB, 9);// sort array into order
      Result[x].L = ArrayL[4]; // median value in array
      Result[x].A = ArrayA[4];
      Result[x].B = ArrayB[4];
    } // for x
    ConvertLABtoRGB(width, Result, RGBLine);// convert back to RGB
    WriteResultLine(y, RGBLine);
} // for y
// ... dispose pointers

void Sort3(float *array1, float *array2, float *array3, long N)
// sort array1 by value, carry along values in other arrays
{
    Boolean  done;
    long     jump, i, j;
    float    temp;
    jump = n;
    while (jump > 0)
    {
        jump /= 2;
        do // while !done
        {
            done = TRUE;
            for (j = 0; j <= n-jump; j++)
            {
                i = j + jump;
                if (array1[j] > array1[i])
                { // sort on array1 but carry 2 & 3 values along
                    temp = array1[j];            // SWAP array1[i] <-> array1[j]
                    array1[j] = array1[i];
```

```
                array1[i] = temp;
                temp = array2[j];           // SWAP array2[i] <-> array2[j]
                array2[j] = array2[i];
                array2[i] = temp;
                temp = array3[j];           // SWAP array3[i] <-> array3[j]
                array3[j] = array3[i];
                array3[i] = temp;
                done = FALSE;
             } // if a[j] > a[i]
          } // for j
       } while (!done);
    } // jump > 0
} // Sort3
```

Many median filters work in this way, using an intensity value for the ranking operation and bringing along the color information. For this purpose, any of the various color spaces (LAB, YIQ, HSI, etc.) can be used and will give generally similar results. Figure 2.31a shows an example. As can be seen in the illustration, the fact that pixels along the original boundary between the red and green areas differ in color but are similar in brightness produces visual artifacts when the pixel intensity alone is used for the ranking. Consequently, in such cases it is more appropriate to use all of the channel values in the median process.

Many scientific imaging applications produce more than three channels, such as satellite imagery with its six or nine bands, and medical imaging in which some channels represent quite different signals and contain very different types of information. It is appropriate to question what the notion of a median value means in the case of such multidimensional

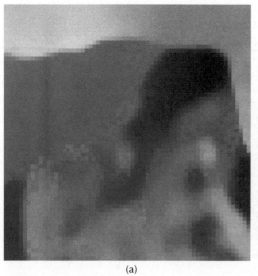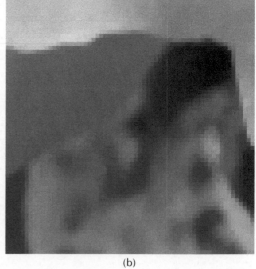

(a)                                                        (b)

**FIGURE 2.31**
(See **Color insert following page 172.**) Enlarged detail from the [rose.tif] image: (a) median calculated using the brightness of each pixel and carrying the color values along, (b) median calculated using the vector median in RGB coordinates. Compare the pixels near the edge of the red petal.

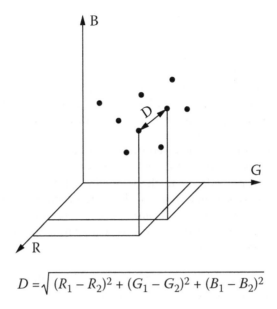

$$D = \sqrt{(R_1 - R_2)^2 + (G_1 - G_2)^2 + (B_1 - B_2)^2}$$

**FIGURE 2.32**
Calculating the distance between the coordinates of nine pixel values in color space.

data. In many cases there is no obvious way to combine the channels to produce a single composite value for ranking. However, there is another way to apply the median filter that can be used with any multichannel image (Astola, Haavisto, and Neuvo 1990; Oistämö and Neuvo 1990; Russ 1995).

The most frequently used definition of a *vector median* in this situation is illustrated in Figure 2.32. For a neighborhood of pixels, the RGB values for each pixel can be treated as coordinates in a basic orthogonal set of axes. The distance between any two of these points can be calculated by the Pythagorean theorem as shown.

The calculation of the vector median filter can be used with other sets of color coordinates such as LAB or YIQ as well. It is more difficult to adapt this method to hue–saturation–intensity color spaces because the hue value is in polar coordinates, and because the various axes are neither orthogonal nor straight. The relative weighting of distances along the various axial directions can also be varied, which is particularly important for spaces such as LAB.

The vector median of the nine points is defined as that point (set of color coordinates) for which the sum of the distances to its eight neighbors is smallest. That corresponds to the most "central" location in the cluster of points. Figure 2.31b shows the results obtained with a vector median. There are alternative definitions for the vector median that use the angles between the vectors to achieve the same result, but the distances are generally the easiest to calculate. Note that this method requires calculating the distances between all pairs of points for each neighborhood of pixels to decide which sum is the smallest. Even with clever programming to keep the values for some pixel pairs as the neighborhood advances, or to prune the list to eliminate outliers, this is an inherently slow operation that becomes rapidly slower as the neighborhood size increases (Huang, Yang, and Tang 1979).

### 2.2.3 Neighborhood Size and Shape

In fact, everything about the median filter calculation is very sensitive to the neighborhood size. As the size increases, ranking values into order takes more time. Ideally, the neighborhood should not be a square, as was used for convolutions, but an approximation to a circle. The use of a square neighborhood simplifies coding and is therefore used in some applications, but can produce visible artifacts and distortions in the image. A circle is ideally isotropic and can be approximated in a pixel grid as indicated in Figure 2.33.

The use of a square neighborhood for convolutions does not have this problem. The weight values in the corners of a symmetrical kernel, such as the Gaussian, are either very small or zero. The additional effort needed to fetch the pixel value and perform the multiplication and addition is small (and simpler than the logic required to limit the operation to a circular region), and those neighbor pixels are effectively ignored. However, for a ranking operation, they must be excluded, for example, by calculating the distance from the center or by including logical tests in the code.

In spite of these practical considerations, the great utility of the median filter makes it a very important tool for image processing (Davies 1988).

In some cases, the use of a neighborhood that is not circular but has a shape adapted to solve a specific known problem may be useful. The most common example has to do with scratches on film and the lines that result when the images are digitized. The scratches often run in the direction of film travel and result from handling. Knowing the orientation of the scratches makes it possible to design a median filter to remove them. The same method can be used to remove other linear artifacts, such as power lines in photographs or scan lines in surveillance video. As shown in Figure 2.34, a median filter using a neighborhood consisting of a row of pixels perpendicular to the lines and at least twice as long as their width will remove them.

Adaptations of the median filter to locating scratches and dust on scanned images, and automatically adapting a median filter neighborhood to remove them from the images, are now incorporated in some commercial film scanners.

Although the median filter is well known for its ability to preserve the magnitude and position of edge and step discontinuities in an image (Yang and Huang 1981), it does round sharp corners. This fact, combined with the removal of lines that are narrower than

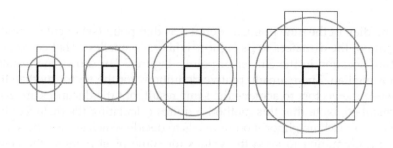

**FIGURE 2.33**
Examples of approximating a circular neighborhood in a pixel grid. From left to right, the neighborhoods include pixels whose centers lie within a radius of 1, $\sqrt{2} = 1.414$, $\sqrt{5} = 2.236$, and $\sqrt{10} = 3.162$ pixels from the center.

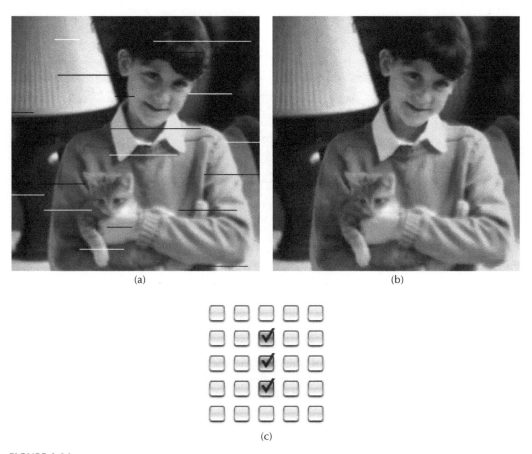

(a)                                           (b)

(c)

**FIGURE 2.34**
Removal of linear defects using a median filter with a custom neighborhood: (a) original image [girl.tif] with black and white horizontal lines added, (b) after application of a median filter with a neighborhood (c) consisting of the central pixel and its neighbors above and below.

the neighborhood radius, imposes limitations on the use of the technique. Various conditional or adaptive modifications of the median filter have been devised (Mastin 1985; Hwang and Haddad 1995). One modification or extension of the median method is particularly suitable for addressing these problems because it does not involve any arbitrary thresholds. Usually called a ___hybrid median___, the method consists of a two-stage approach. In the first stage, several different neighborhoods with specifically chosen shapes are used to select median values. In the second stage, these medians are again ranked, along with the original central pixel, to select a final median that is assigned to the original pixel location.

Figure 2.35 diagrams this process for the case of a 3 × 3 neighborhood. There are two partial neighborhoods used in the first stage, corresponding to the "+" and "×" patterns surrounding the original pixel. The median values from each of these neighborhoods, and the original pixel, are then ranked again in the second stage, to produce a final result. For larger radii, there are more orientations of the cross-shaped neighborhood that can be ranked in the first stage. The results for each of these $N$ orientations, plus $(N-1)$ repetitions of the central pixel, are then ranked in the second stage.

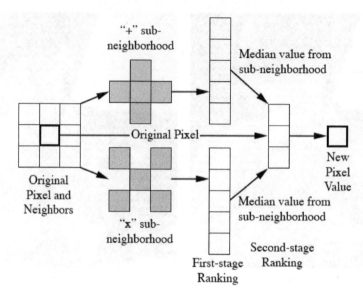

**FIGURE 2.35**
Diagram of the hybrid median procedure as described in the text.

Figure 2.36 shows the results of applying a hybrid median with a radius of 3 pixels to an image in which there are narrow lines and sharp corners, and compares the results to a conventional median of the same size and to three repetitions of a radius = 1 conventional neighborhood median.

### 2.2.4   Noise

Two quite different types of noise have been described in previous sections of this chapter: random or speckle noise, and shot or salt-and-pepper noise. Sometimes, it is useful to generate noise to add to an image. One purpose for doing this is to create test images that can then be utilized to evaluate various noise removal techniques.

A second reason is to superimpose noise on an image to mimic the effects of detector behavior or film grain. This is most often done when images have been manipulated (e.g., by drawing or pasting in information that was not originally there, either to add something to the picture or, more often, to cover up something that was originally there). It is (fortunately) difficult to perform this type of image "editing" without leaving tell-tale signs of the manipulation. Adding noise may hide the differences between the original noise signatures of the original image and the added information, although the presence of high levels of noise may in itself be considered evidence of probable tampering.

Noise can be added to an image in several different ways, depending on the purpose and the type of detector characteristics it is intended to model. Shot noise, corresponding to drop-out, locked, or dead pixels in the detector (or sometimes to dirt on a piece of scanned film), results in individual pixels being set to either black or white (or sometimes, depending on the detector type, to a primary color). In real detectors, these defects are almost always present but in small numbers, and are either corrected in the electronics of the readout process (by methods similar to the median filter described above) or are overlooked by the casual viewer. In most cases they occur at random locations, although some

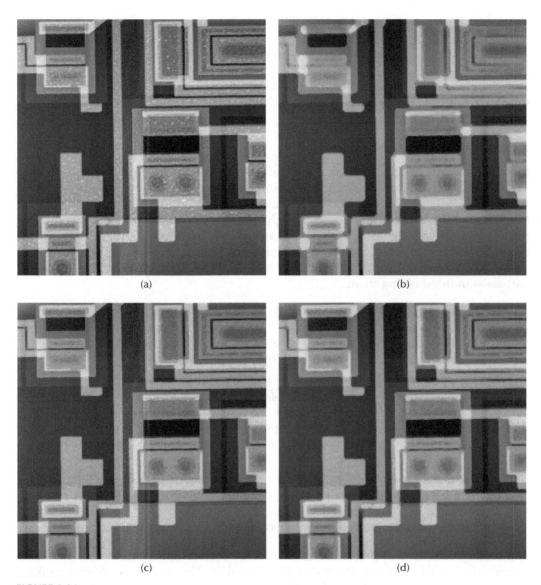

(a)  (b)

(c)  (d)

**FIGURE 2.36**
Comparison of a hybrid and conventional median: (a) original image [chip.tif] with textural noise, sharp corners, and narrow lines; (b) conventional median with radius 3, which rounds corners and fills in lines; (c) hybrid median with radius 3; (d) conventional median with radius 1 repeated three times.

types of defects are more likely to occur in rows or columns (which also makes them more objectionable).

Simulating shot noise at random locations can be accomplished as follows: A random number in the range 0.0 to 1.0 is generated for each pixel and the value compared to the probability that a pixel is defective. If the value is below that threshold, the pixel is set to black or white. For some purposes, if the detector being modeled has transistors that are filtered to detect red, green, or blue, it is also possible to use the same logic to set just the red, green, or blue value to zero or maximum. An alternative method for performing the simulation is to generate random numbers for location addresses and then modify those pixels accordingly.

Random speckle noise is also simulated with a random number generator. In this case the noise may be either uniform or Gaussian in distribution, and may be either added to or multiplied by the original pixel values, depending on the source of the noise being modeled. It is also possible to either apply the noise individually to the red, green, and blue channels (and to apply different amounts of it in each channel) or to apply the noise to just the luminance, leaving the colors unchanged.

### 2.2.5   Ranking and Morphology

Ranking the pixels in a neighborhood into order is not used only to find the median value. The minimum and maximum values can also be used in a variety of ways (Sussner and Ritter 1997; Wilburn 1998). For example, the features in Figure 2.37a are everywhere brighter than the local background, although because of nonuniform illumination of the field of view in the microscope, there is considerable variation in brightness from the edges toward the center. A ranking operation, combined with an arithmetic procedure, can be used to correct this variation so that the features are of uniform brightness, which will assist in thresholding them.

The steps in the procedure are illustrated in the figure:

1. Create a new image in which each pixel is replaced by its darkest neighbor (this is called an ___erosion___). Repeat this until the features are removed.
2. Create a new image from the result in step 1 in which each pixel is replaced by its brightest neighbor (this is called a ___dilation___). Repeat this the same number of times as used for the erosion in step 1. The combination of erosion followed by dilation is called an ___opening___. The purpose is to restore the brightness values of the pixels to their original levels except where features have been entirely removed by the erosion.
3. Create a final image by subtracting the result from step 2 pixel-by-pixel from the original image. The result from step 2 is a background image that has the non-uniform lighting characteristics of the original, but with the features missing. The difference between this image and the original is just the features.

The requirement for this procedure to work is that the bright features must be small enough in width to disappear after the erosion procedure. The erosion and dilation ranking can be carried out iteratively as described, or in a single pass using a large enough neighborhood (i.e., one with a diameter greater than half the width of the widest feature). In most cases, unless the features have some known shape and alignment, the neighborhood should approximate a circle to minimize artifacts in the resulting image.

If the features are dark on a light background, the same procedure is used except that the dilation is applied first, and the erosion afterward. The sequence of dilation followed by erosion is called a ___closing___. The processes of erosion, dilation, opening, and closing are collectively called ___morphological operations___. These techniques were originally developed (and named) for application to binary (black and white) images, and will be encountered again with slight variations in Chapter 4.

A note of caution: The sense of the erosion and dilation used here is that bright pixels are foreground (features) and dark ones are background, so eroding the features (making them smaller) is accomplished by replacing each pixel with its darkest neighbor. However, you are just as likely to encounter systems in which the sense of bright and dark are reversed

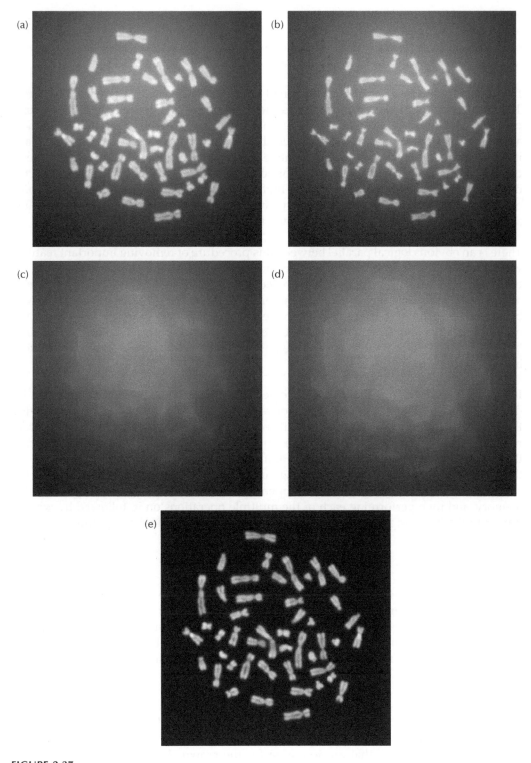

**FIGURE 2.37**
Erosion and dilation used to level nonuniform brightness in an image: (a) original image [chromosome.tif], (b) erosion (replacing each pixel with its darkest neighbor, radius 1), (c) after seven iterations of erosion (features removed), (d) after seven iterations of dilation (background brightness values restored), (e) result of subtracting image (d) from the original.

and features are treated as dark pixels on a white background. In that case the terms erosion and dilation may be reversed. As with the median filter, application of these morphological operations to color images must either choose to operate on the brightness values only, or to work directly in a color space (Comer and Delp 1999; Sartor and Weeks 2001).

It is not obvious from a visual examination of Figure 2.37 that the dilation step is required. That is because the variation in brightness across the image is gradual, and visual comparisons of brightness are difficult. In the example of Figure 2.38 the variation of brightness across the face of the moon, and the sudden change at the edge, produces a more dramatic illustration. When erosion is used to remove the bright markings, it enlarges the size of the dark maria and shrinks the moon. Subtracting this image without first performing the dilation produces a difference that fails to highlight (or even reverses the contrast of) some of the lunar features and also results in an obvious artifact at the limb of the moon. When an opening is used, these problems are avoided.

There is an obvious logical parallel between this procedure of removing important features with a rank-based neighborhood opening or closing, followed by subtraction from the original image, with the unsharp mask procedure described in Section 2.1.5. That method also uses the approach of removing the important features (by a Gaussian blur), followed by subtraction. The rank-based parallel to the unsharp mask is less used, but offers many of the same advantages, with the added benefit that it does not produce bright and dark "halos" around edges in the image, as the unsharp mask can. Figure 2.39 shows the step-by-step process.

Important note: Many of these morphological operations, such as erosion followed by dilation, or several iterations of erosion, or repeating a median filter, etc., require saving the intermediate images as they are produced. As pointed out before, the host program receives the modified version from processing when it is written back, but reading from the host always obtains the original unmodified values. Consequently, it is necessary for the program to allocate a temporary image in memory (using the tools provided on the companion CD and explained in the Appendix). Copying the original image into this memory, and then performing each of the multiple operations on it, followed by writing it back to the host program, makes it possible to perform multiple procedures.

### 2.2.6   Top Hat Filter

It is sometimes useful to perform the ranking operation in two different neighborhoods, a central one and a surrounding annulus. This is usually called a *top hat* filter, because (as shown in Figure 2.40a) the central neighborhood can be visualized as the crown of a top hat and the annular outer region can be thought of as the hat's brim. Of course, the circular shape of both regions is an idealized one, and in actual practice each consists of discrete pixels.

Comparing the brightest pixel in the inner region to the brightest pixel in the outer region, and comparing the difference to a threshold (which is the height of the hat's crown) can be used to find features that are bright and small enough to fit inside the crown, as shown in Figure 2.40b. Features that are too large for the crown (Figure 2.41a), or long lines (Figure 2.41b), or features that are closer together than the width of the brim (Figure 2.41c) will not produce a large brightness difference between the inner and outer regions, and will be ignored.

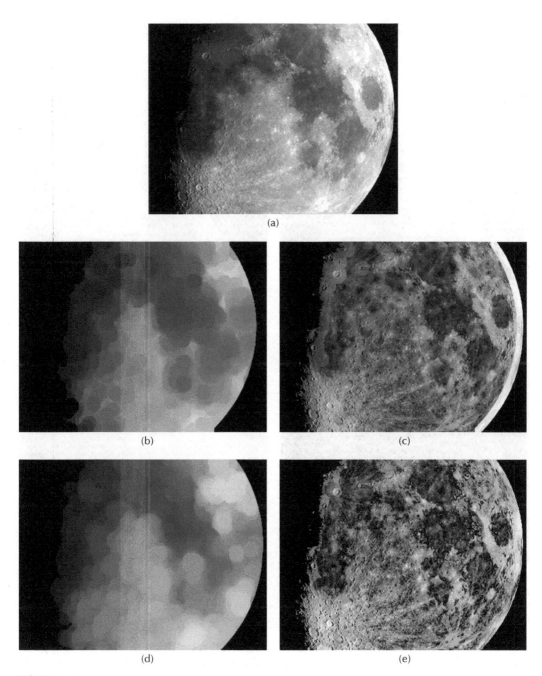

**FIGURE 2.38**

Comparison of erosion and opening: (a) original image [moon.tif], (b) after erosion of the bright features in the original, (c) result of subtracting image (b) from the original, (d) after erosion and dilation (opening) of the original, (e) result of subtracting image (d) from the original.

This logic can be reversed to find the darkest pixel value in each neighborhood for the purpose of trying to isolate dark features on a light background. Figure 2.42 illustrates a simplified example of the latter situation. The image consists of two different sizes of features. Applying a top hat filter with a crown size just larger than the small features and smaller than the large ones erases the large features but retains the smaller ones (Figure 2.42b).

**FIGURE 2.39**
Sharpening an image with rank-based morphology: (a) original image [rose.tif], (b) after applying a closing to remove detail, (c) result of subtracting (b) from (a), (d) result of adding (c) to the original.

In practice, the top hat filter is not usually the most appropriate tool for this type of procedure. Cross-correlation, which was introduced above and is discussed further in Chapter 3 (because it is most commonly applied using Fourier transforms) is a superior tool for most situations in which features of a particular size, shape, and brightness pattern are to be located in a complex image. However, the top hat filter is often applied to a slightly different problem: removing dirt or scratches from an image. In this version it is often called a ***rolling ball*** filter.

The logic of the rolling ball filter is the same as for the top hat: the brightest (or darkest) pixel values in a central region and in a surrounding annulus are found by ranking, and

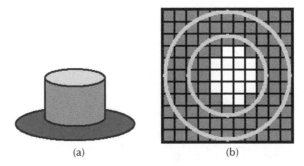

(a)            (b)

**FIGURE 2.40**
The top hat filter: (a) schematic diagram of the brim and crown, (b) comparison of the brightest pixel value in the inner (crown) and outer (brim) regions detects the presence of a feature.

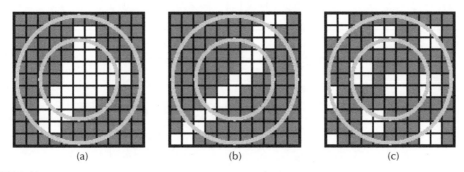

(a)         (b)         (c)

**FIGURE 2.41**
Features that are ignored by the top hat filter: (a) features too large to fit inside the crown, (b) lines that cross the crown and brim, (c) features closer together than the width of the brim.

the two values are compared. If the difference exceeds some threshold value, the rolling ball filter replaces the brightness of the central pixel by either the median or the mean of the pixels in the outer annular (brim) region. This erases bright (or dark) features that are smaller than the inner (crown) region, as shown in Figure 2.42c.

This procedure is often quite useful for removing dirt or dust from an image. Altering the shape of the inner neighborhood to be a line in specific direction, and the outer neighborhood to consist of two flanking lines on either side, allows the same method to remove scratches.

Another important application of the top hat filter, which will be used again in Chapter 3, is locating "spikes" in the Fourier transform of a image that shows a periodic structure. Figure 2.43 shows an example. The original image is a half-toned printed picture from a newspaper, showing the typical coarse dot pattern. The Fourier transform power spectrum of this image, which will be explained in detail in the next chapter, has a series of bright spots that represent large magnitudes for the periodic sinusoidal functions that correspond to the spacing and orientation of the halftone dots in the image. Finding those spikes is a necessary first step to processing the image to remove or mitigate the halftone pattern. As shown in the figure, the top hat filter locates them.

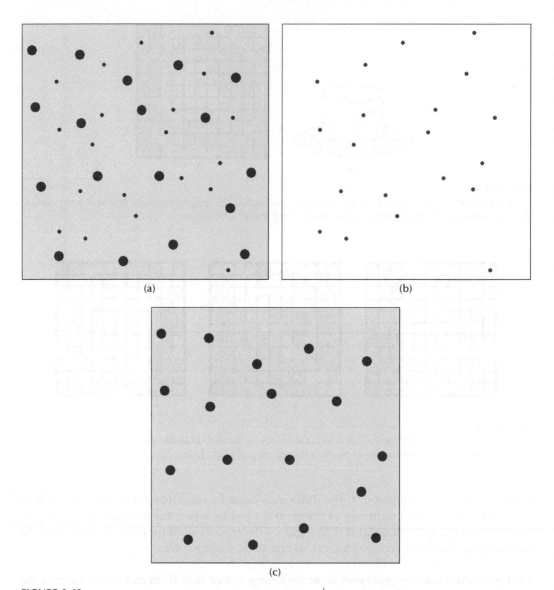

(a)

(b)

(c)

**FIGURE 2.42**
Example of a top hat filter: (a) original image containing two different size features [dots.tif], (b) features kept by a top hat filter as described in the text, (c) removal of small features by a rolling ball filter as described in the text.

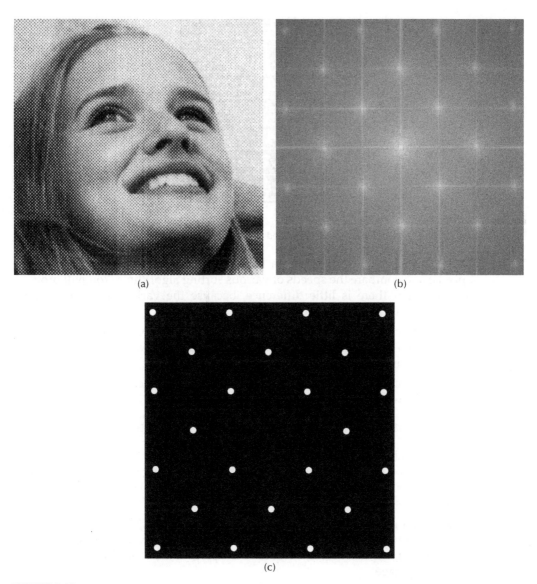

(a)      (b)

(c)

**FIGURE 2.43**
Use of a top hat filter with Fourier transforms: (a) original image [print.tif] showing a half-tone printing pattern; (b) Fourier transform power spectrum, showing bright "spikes" whose location represents the spacing and orientation of the halftone dots; (c) location of the spikes with a top hat filter to produce a mask. This example is shown with full details in Chapter 3, Section 3.2.

## 2.2.7 Problems

2.2.7.1#. By entering pixel values into the list to be sorted in a sequence that starts at the center and works outward, it is possible to vary the size of the neighborhood used in a rank filter just by changing the value of $N$ used in the sort. Figure 2.44 illustrates the procedure. Sorting just the first 5 values corresponds to radius 1, the first 9 values to radius $1.414 = \sqrt{2}$, the first 21 values to radius $2.236 = \sqrt{5}$, the first 37 to radius $3.162 = \sqrt{10}$, and so on. Implement this procedure.

|   | 45 | 37 | 26 | 30 | 38 |   |
|----|----|----|----|----|----|----|
| 44 | 25 | 21 | 10 | 14 | 22 | 39 |
| 36 | 20 | 9  | 2  | 6  | 15 | 31 |
| 29 | 13 | 5  | 1  | 3  | 11 | 27 |
| 35 | 19 | 8  | 4  | 7  | 16 | 32 |
| 43 | 24 | 18 | 12 | 17 | 23 | 40 |
|   | 42 | 34 | 28 | 33 | 41 |   |

**FIGURE 2.44**
Initially entering pixel values into the array to be sorted starting at the center and working radially outward.

2.2.7.2.  Implement a median filter using a large neighborhood (at least a radius of 4 pixels) and compare the speeds of various sorting algorithms. You will probably find that there is little difference, because the various pixel accesses account for much of the processing time.

2.2.7.3.  Implement a color vector median as described in the text. Compare the results to those from an intensity-based median, particularly near edges, lines, and other discontinuities.

2.2.7.4.  Implement a color vector median filter in which the distance between points is perceptually weighted, for example,

    D = sqrt(w1*(R1-R2)^2 + w2*(G1-G2)^2 + w3*(B1-B2)^2),

where the weights for red, green, and blue can be set to w1 = 0.25, w2 = 0.65, w3 = 0.10 as discussed in Chapter 1. Weighting can also be applied to other color spaces, such as LAB, and is equivalent to changing the scales of the various axes.

2.2.7.5.  Implement a median filter that operates on individual R, G, and B channels, and compare the result to one that uses the intensity values to select the neighboring pixel whose color values are copied to the central pixel location in the new image.

2.2.7.6.  Implement a gated median: IF AND ONLY IF the center pixel is the extreme value (rank 0 or $N - 1$ in the ranked list) THEN replace it in the new image with the median. This can also be done with the vector median filter. This method is particularly effective at removing shot noise.

2.2.7.7.  Implement an iterated median filter: Repeat the application of a small median filter (e.g., radius = 1.5 pixels, a $3 \times 3$ neighborhood of 9 pixels) until there are no further changes in the image. Note that this requires allocating a memory buffer to hold the image as discussed in the text.

2.2.7.8#.  Implement a hybrid median filter with neighborhood radius of 2 pixels. Compare the results to a conventional median filter of the same size.

2.2.7.9.  Implement a conditional or adaptive median in which pixels within a radius of 3 are added to the list to be sorted only if the difference between their brightness values and those of the original central pixel does not exceed a preset threshold value (e.g., 50 brightness levels). Compare the result to a conventional median. Note that you will need to count the number of pixels in the list each time, as this will vary. In the event of an even number of pixels in the list, the median value is usually taken (incorrectly) as the average of the two central values in the sorted list. The preferred method is to either randomly or alternately choose between the two central values.

2.2.7.10.  Using the [cat.tif] image, compare the ability of Gaussian smoothing with a conventional, hybrid, and conditional median filter to retain fine detail (e.g., the whiskers) while suppressing random speckle noise.

2.2.7.11#.  Implement a morphological filter that performs $N$ iterations of erosion followed by $N$ iterations of dilation (an opening), or the reverse (closing), reading a text file to determine which procedure to use and the value of $N$. Note that this requires allocating a memory buffer to hold the image as discussed in the text.

2.2.7.12.  Implement a rolling ball filter that removes dark dust particles from an image. Assume that the particles will be no more than 3 pixels in radius and will be separated by more than 2 pixels. Replace the central pixel by either the mean or median of the pixel values in the annular outer region and compare the results.

2.2.7.13#.  Implement a routine that adds a specified amount of random noise to an image. Compare the effects of adding values independently to the red, green, and blue channels against adding them to the luminance only. Use the resulting images to compare various random noise removal methods.

2.2.7.14.  Implement a routine that creates shot noise in an image by setting a preset number or percentage of the pixels at random locations to black and/or white. Use the resulting image to compare various shot noise removal methods.

## 2.3   Statistical Operations

### 2.3.1   The Variance Filter

Another class of neighborhood operations calculates various statistical properties of the pixel values in a neighborhood. Some of these share portions of the procedures outlined above, including pixel access and boundary protection, but are collected here to emphasize their basic similarities. A variety of statistical properties are available, but the most basic is the mean value. This can be obtained using the convolution approach, for example, by using a kernel like the one in Table 2.9. Using a divisor of 21 (the number of pixels added together), this calculates the average value in an approximately circular neighborhood with radius equal to $2.236 = \sqrt{5}$ pixels. The average is inferior to a Gaussian smooth for reducing random noise, and is seldom used (it was employed historically when multiplication was slow compared to simple addition).

One of the most used and useful statistical neighborhood calculations is the *variance filter*. The variance of a series of values is defined as shown in Equation 2.4 using the sum of squares of differences from the mean. Because the variance sums the squares of the differences, the range of values can be quite large. Consequently, in many cases, the square root of the variance (the standard deviation) is actually used for the new pixel value, as the resulting image is more easily viewed. But the filter is still generally called a variance operator.

$$Variance = \frac{\sum_{i=1}^{N} (x_i - mean)^2}{N}$$

(2.4)

Whether the variance or standard deviation is actually used, the range of values produced by the filter does not fit well into the 0 to 255 range. Either an arbitrary scaling value must be introduced, with the usual protection to clip values that would exceed the legal range, or the resulting values may be autoscaled, which requires a second pass through the data as described previously.

The calculation of the variance is not performed as Equation 2.4 might suggest, by first determining the mean. Code Fragment 2.13 shows a more direct method (based on Equation 2.5), which counts the number of pixels and builds sums of the values and their

**TABLE 2.9**

A Convolution Kernel That Calculates the Mean
or Average Value in a Neighborhood

| 0 | 1 | 1 | 1 | 0 |
|---|---|---|---|---|
| 1 | 1 | 1 | 1 | 1 |
| 1 | 1 | 1 | 1 | 1 |
| 1 | 1 | 1 | 1 | 1 |
| 0 | 1 | 1 | 1 | 0 |

squares in a single pass through the list. For a neighborhood filter that moves through the image, even more efficiency can be achieved by adding or removing values from the sums as the moving neighborhood reaches or leaves various pixels, but for clarity such modifications are avoided here.

$$Variance = \frac{\sum_{i=1}^{N} x_i^2}{N} - \left(\frac{\sum_{i=1}^{N} x_i}{N}\right)^2 \tag{2.5}$$

```
//Code Fragment 2.13 — Calculate the variance of N Values
sum1 = sum2 = 0;
for (k = 0; k < N; k++)
{
   sum1 += Value[k];
   sum2 += Value[k] * Value[k];
} // for k
mean = sum1 / N;
variance = (sum2 / N) - (mean * mean);
stddev = sqrt(variance);
```

The calculation shown in Code Fragment 2.13 is easily extended with additional sums to calculate the higher moments of a distribution, such as skew and kurtosis, but these are rarely useful for image pixel values.

The variance filter has two different types of response and, hence, two different areas of application, depending on the size of the neighborhood used. As illustrated in Figure 2.45b, when a small neighborhood is used the filter highlights edges and discontinuities. Similar to the Sobel gradient technique, it is often used for this purpose. But when a large-radius neighborhood is used, the filter responds to a different property of images, which for want of a better term is often called *texture*.

Figure 2.46a shows an image that contains five visually distinguishable regions. All have the same average brightness, but differ in the amount of local variation in pixel brightness. This can take many forms, including the magnitude of the variation, its spatial scale (sometimes referred to as the coarseness or granularity), any regularity or periodicity that may be present, and any directionality or anisotropy. In addition, the variation may be in pixel brightness and/or in color. A variety of image processing tools may be used to convert these variations to brightness differences that can be used to threshold the regions.

In the example of Figure 2.46b, a variance filter is applied with a radius (3.5 pixels) large enough to encompass the spatial scale or granularity of the variation. It converts the original image to one in which the various regions have distinguishable brightness values. There are other filters illustrated in the figure that accomplish this task also; they are discussed later in the text but are shown together for comparison.

The original image in Figure 2.46a is obviously an artificial construct. Figure 2.47a shows a real example, a microscope image of stained liver tissue. Distinguishing the various components in complex structures such as this is something that people are quite good

**FIGURE 2.45**
Application of statistical filters with a small (radius 1.5 pixel) neighborhood: (a) original image [saturn.tif],
(b) application of a variance filter, (c) application of a range filter (discussed later in the text).

at, but which can be difficult for computer algorithms to accomplish. In this case, the
visually "smooth" regions (mitochondria) do not have a unique value of brightness compared to the other areas of cytoplasm, but they do have a very different visual texture.
The variance filter (Figure 2.47b) converts the original image to one in which the brightness
values of the regions are distinct. Again, the results of other filters not yet discussed are
shown for comparison.

Because it is relatively straightforward to calculate, and has a variety of purposes, the
variance filter is rather widely used. It is by no means the only tool useful for isolating
texture in images. For example, periodic structures are best analyzed and processed using
Fourier transforms, which are considered in the next chapter. For color images, it is often
necessary to isolate the important variation (such as hue or saturation, or in some cases
red, green, or blue) in a single channel before texture processing.

Figure 2.48 shows another artificial image (which is representative of many types of fabric,
muscle tissue, plowing patterns in agriculture seen from the air, and other situations). The
statistics of the brightness patterns in the visually distinguishable regions are identical,
and the differences lie in the orientations of the patterns. Applying the Sobel gradient

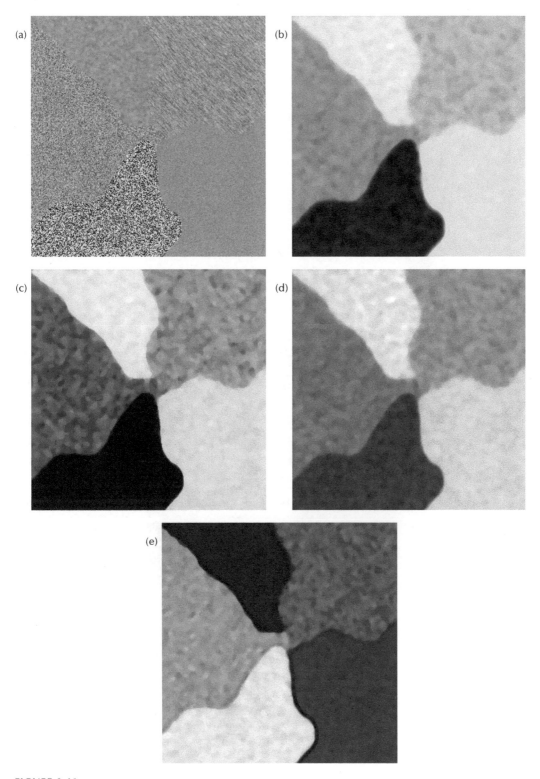

**FIGURE 2.46**
Application of statistical filters with a large (radius 3.5 pixel) neighborhood: (a) original image [textures.tif], (b) application of a variance filter, (c) application of a range filter, (d) application of an entropy filter, (e) application of a fractal dimension filter (c, d, and e are discussed later in the text).

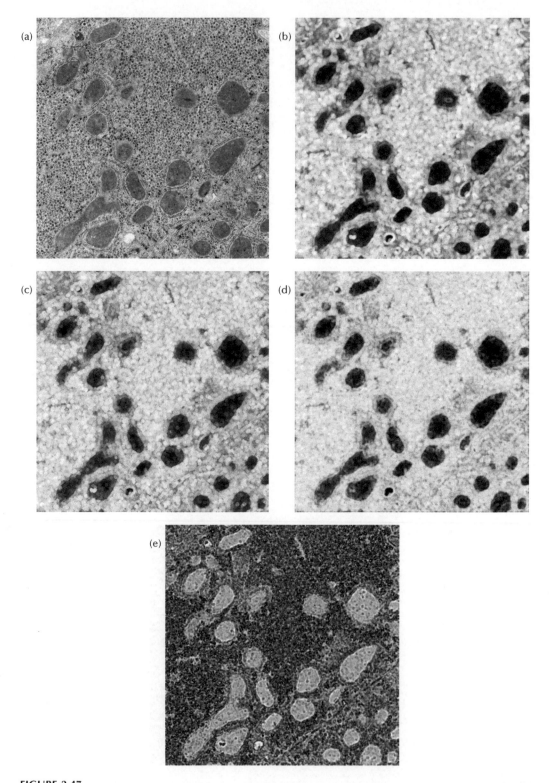

**FIGURE 2.47**
Application of statistical filters with a large (radius 4.5 pixel) neighborhood: (a) original image [liver.tif], (b) application of a variance filter, (c) application of a range filter, (d) application of an entropy filter, (e) application of a fractal dimension filter (c, d, and e are discussed later in the text).

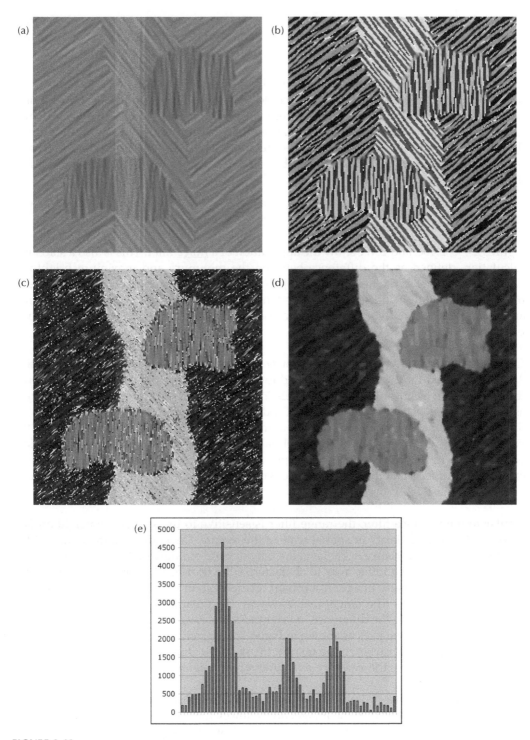

**FIGURE 2.48**
Conversion of texture orientation to brightness differences: (a) original image [directions.tif]; (b) application of the Sobel orientation filter with angle values from 0 to 360° represented by 0 to 255 pixel brightness values; (c) application of the Sobel orientation filter with angle values from 0 to 180° and from 180 to 360° represented by 0 to 255 pixel brightness values; (d) application of a median filter (radius = 2.5 pixels) to image (c); (e) histogram of values in image (d) showing peaks on the 0 to 180° scale corresponding to angles of about 36, 90, and 130 degrees which represent the orientation of texture in the regions of the original image.

filter and using the vector angle to form a new image reveals the differences best when the angles are mapped to pixel brightness values in such a way that the range from 0 to 180° is represented by the 0 to 255 range of pixel values, and the range from 180 to 360° repeats the 0 to 255 pixel values as shown. As is often the case for orientation images, the result is still somewhat noisy. In the example, a median filter was used to remove the noise, but a Gaussian smoothing filter could also have been applied. A conventional neighborhood convolution or median filter experiences problems with images in which grayscale represents orientation angle because of the discontinuity at 0°, and thus requires dealing with special cases. This example illustrates not only the conversion of texture orientation to unique brightness values, but also the need to apply multiple processing operations in many instances.

## 2.3.2  Other Texture Filters

Figures 2.45c, 2.46c, and 2.47c show the application of a *range* filter. This is also a statistical calculation, defined as the difference between the minimum and maximum brightness value for any pixel in the neighborhood. Just as with the variance filter, this can serve to highlight edges, lines, and discontinuities when used with a small neighborhood radius, and to convert texture to brightness when used with a larger neighborhood that encompasses the scale of the texture.

Calculating the range value can be performed with the same code discussed in Section 2.2 that was used for the median filter and for grayscale morphological processing. Once the values have been ranked for the pixels in the neighborhood, the range is calculated as the difference between the minimum and maximum values. Because it depends on these two extreme values, the range filter is much more sensitive to image noise than the variance filter, which uses all of the pixel values in the neighborhood in its calculation. Sometimes, with larger neighborhood sizes, a practical way to reduce the sensitivity to extreme values is to take the difference between the average of a few values at each end of the ranked list.

Similar to the variance filter, the range filter is sensitive to the size of the neighborhood used and to the granularity or spatial scale of the texture. Depending on the nature of the texture present, the variance or range value may depend strongly on the radius used. This dependence offers yet another way to characterize the texture. One kind of texture that is naturally present in many images is a *fractal* relationship in which the difference between pixel brightness values (the range) increases with the separation between the pixels in a particular way. Plotting the difference in brightness (the range) against the diameter of the neighborhood on log-log axes as shown in Figure 2.49, produces a straight-line plot whose slope and intercept characterize the texture (Russ 1990). Qualitatively similar results can be obtained by plotting the variance as a function of neighborhood size.

It is not intended here to analyze the underlying physical connections between this relationship and the objects in the image, nor to imply that all images and structures have this characteristic behavior. However, it is present often enough that using it as a basis for distinguishing textures in different structures and regions is often worthwhile. This also serves as an example of a fairly complicated neighborhood filter, which is still fast enough with modern computers to be practical for routine use. Figures 2.46e and 2.47e show representative results. The procedure for performing this function is outlined in Code Fragment 2.14.

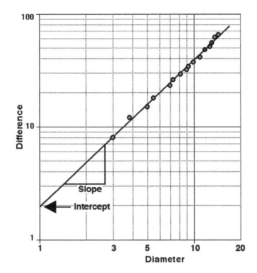

**FIGURE 2.49**
Example of a fractal relationship between extreme pixel brightness differences (the range) and neighborhood size.

```
// Code Fragment 2.14 — Calculate fractal filter
radius[16] = {1.0, 1.414, 2, 2.235, 3.0, 3.162, 3.606, 4.0, 4.472, 5.0,
              5.385, 5.831, 6.0, 6.325, 6.633, 7.0};
  // radii of quasi-circular neighborhoods
sumX = sumY = sumN = sumXX = sumXY = 0; // for linear regression
for (k = 0; k < 16; k++)
{ // ... find minimum[k] and maximum[k] by ranking each neighborhood
  range[k] = maximum[k] - minimum[k];
  // alternatively, use the standard deviation for each neighborhood
  Xvalue = log(radius[k]);
  Yvalue = log(range[k]);
  sumN  += 1;
  sumX  += Xvalue;
  sumY  += Yvalue;
  sumXX += Xvalue * Xvalue;
  sumXY += Xvalue * Yvalue;
} // for k
// classic linear regression for best fit line through the points
numerator   = sumN * sumXY - sumX * sumY;
denominator = sumN * sumXX - sumX * sumX;
if (denominator > 0) slope = numerator/denominator;
else                 slope = 0;
intercept = (sumY - slope * sumX) / sumN;
```

The procedure shown in the code fragment uses the range (difference between maximum and minimum pixel brightness values) in a series of approximately circular neighborhoods of increasing size, with radii from 1 to 7 pixels. This is a somewhat arbitrary limit, and either a smaller or larger limit on neighborhood size may be useful in some cases. Because each neighborhood includes all of the smaller ones, these range values increase monotonically. Classic linear regression is performed to fit a straight line through the data points on log–log axes. The slope or intercept of this line may then be scaled as appropriate and used as the pixel brightness value in the new image. This entire procedure is then repeated for each pixel in the image.

Another way to distinguish the textures present in different areas of an image, so that they can be separated by thresholding, is the ***entropy filter***. Entropy is a measure of the amount of variation in pixel brightness values. A formal definition is shown in Equation 2.6, and the corresponding calculation procedure is outlined in Code Fragment 2.15.

$$Entropy = \sum_{k=0}^{255} -\left(\frac{Histo_k}{N}\right) \cdot \log\left(\frac{Histo_k}{N}\right) \tag{2.6}$$

```
// Code Fragment 2.15 — Calculate neighborhood entropy
// ... N is the neighborhood size (number of pixels)
// ... Value[] is the array of neighborhood values
long histogram[256];                    // histogram of neighborhood values
float Entropy = 0;                      // accumulate Entropy here
for (k = 0; k < 256; k++)
   histogram[k] = 0;                    // clear out histogram
for (k = 0; k <= N; k++)
   histogram[(long)Value[k]]++;         // build the histogram
for (k = 0; k < 256; k++)               // accumulate Entropy
   if (histogram[k] > 0)                // skipping the empty bins
      Entropy -= (histogram[k] / N) * log(histogram[k] / N);
```

Since the procedure depends on the number of pixels that have each possible brightness value, the first step is to build a histogram for the pixels in the neighborhood. The code fragment shows the construction of a classic 256-bin histogram, but when working with small neighborhoods (in which every pixel may have slightly different values), a smaller array of 8, 16, or 32 bins may be used. The choice of number of bins depends on the precision of the original image data and on the number of pixels in the neighborhood. Too many bins will produce too few counts in each, whereas too few bins will hide local variations in pixel values. In some applications to color images, it is useful to employ a histogram of hue values to examine the variation in color rather than brightness.

The considerable variety of texture-sensitive filters reflects the many different types of texture that may be present in images. Distinguishing fields planted in wheat, corn, or soybeans in aerial photographs, or different types of biological tissues in a microscope slide, or subtle variations in soil granularity in sedimentary profiles, or different cloth-weaving patterns in a tapestry, or different wear patterns on prehistoric stone tools, are only a few of the situations in which humans are able to recognize textural variations. Learning to be more analytical observers who not only recognize the variation but also learn to identify the factors used to make the distinction is the first step toward choosing a texture filter that will enable a computer to perform the same step.

### 2.3.3 Enhancing Local Contrast

The entropy filter introduced the idea of constructing a histogram for the pixels in a neighborhood centered around each pixel in the image. This opens the door to applying

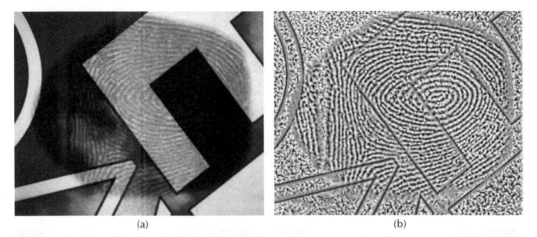

(a)                                          (b)

**FIGURE 2.50**
Example of local equalization: (a) original image of a fingerprint on a printed page, (b) local equalization using a neighborhood with radius = 4 pixels.

the histogram manipulation procedures that in Chapter 1 were applied to the entire image. The most widely used of these techniques is called local equalization.

In Chapter 1, histogram equalization was shown to be a method that spreads the pixel values out so that equal areas of the image are displayed with each of the available brightness levels. When it is applied to a local neighborhood, only the new pixel value for the central pixel in the neighborhood is used, to create a new derived image. As shown in Figure 2.50, the result of this procedure is to suppress long-range brightness variations across the image, while increasing local contrast. Pixels that are slightly brighter than their surroundings are made brighter still, while ones that are slightly darker than their neighbors are made darker. The neighborhood size must be large enough to encompass the scale of the structures whose visibility is to be enhanced.

The *local equalization* procedure is best understood in terms of the local neighborhood histogram (Zhu, Chan, and Lam 1999), but for actual calculation it can also be carried out efficiently using the same neighborhood ranking method used for median filtering and other operations described above. If there are $N$ pixels in the neighborhood, and the central pixel is has a position $R$ in the ranked list, then the new pixel brightness is just $255 * R/N$. A useful shortcut to determining the ranked position, which does not require performing a sorting operation, is simply to count the number of pixels that are darker and add to it half the number that are equal in brightness.

The local equalization method often produces so much increase in local contrast and so much suppression of overall contrast that the result is added back in some proportion to the original image, to produce a more visually pleasing final image. There are other approaches to local contrast enhancement that are more consistent with the way that human vision responds to image detail. One of these compares the brightness of a pixel to the mean brightness of the surrounding neighborhood, and uses the ratio of these values to adjust the pixel brightness up or down. The purpose is the same as that of local equalization, namely to make pixels darker than their surroundings darker still, and vice versa.

Code Fragment 2.16 outlines the typical procedure. At first glance this may seem to be the same as the unsharp mask described above, but it differs in several important ways.

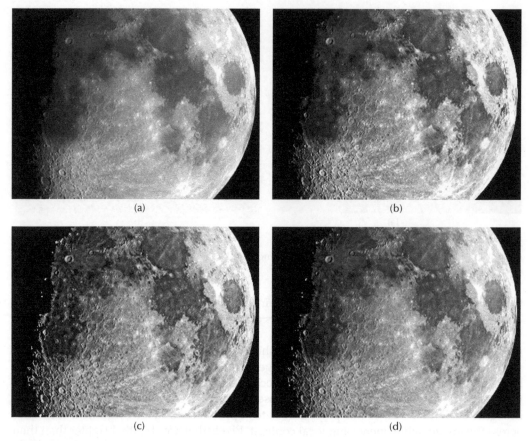

(a)                                             (b)

(c)                                             (d)

**FIGURE 2.51**
Local contrast enhancement: (a) original image [moon.tif], (b) adding the difference between the pixel and its neighborhood average, (c) result of the algorithm described in the text based on the square of the ratio between the pixel and its neighborhood average, (d) result of modifying the algorithm to use the ratio without squaring it.

First, the neighborhood is averaged rather than being a Gaussian-weighted convolution. In fact, this is a minor difference, and a Gaussian-weighted mean value can be substituted in the algorithm without significantly altering the results. The second, more important difference is the use of the ratio rather than the difference. Because human vision detects variations in an image based on the ratio of brightness values rather than the brightness difference, this algorithm produces superior results. Figure 2.51 compares the results of using the brightness ratio with the brightness difference. The ratio method brings out details in both bright and dark regions of the image much better (note, in particular, the details in the dark maria and in the shadowed regions near the terminator).

```
// Code Fragment 2.16 — Enhance local contrast
// process the intensity, leave colors unchanged
for (y = 0; y < height; y++)
   for (x = 0; x < width; x++)
   {  // center = value of pixel;
      // average = mean value of pixels in neighborhood;
      // ratio = average / center;
      // exponent = ratio * ratio;
      // normalized_pixel_value = center / 255;
```

```
// new_value = 255 * exp(exponent * log(normalized_pixel_value));
// equivalent to "new_value = 255 * (center/255)^exponent"
} // for x
```

To understand this difference in behavior, consider two pixels that differ in brightness by 20 gray levels (out of 255). If the actual values are 40 and 60, this corresponds to a 50% change in brightness. On the other hand, if the actual values are 200 and 220, it corresponds to only a 10% change in brightness. The latter is far less noticeable than the former. The ratio method used in the algorithm correctly represents the perceived change in brightness.

The third difference between the algorithm in Code Fragment 2.16 and an unsharp mask is the use of a power, in this case the square (but sometimes the power is made into an adjustable parameter) to vary the effect based on the ratio. The power causes small ratios to produce a proportionately greater change than large ones. Figure 2.51d also shows the result of using the ratio without squaring it.

A different approach to enhancement of feature sharpness is intended to make gradual transitions in brightness at edges and discontinuities into abrupt ones, which appear sharper visually and also can simplify the thresholding operation to accurately delineate the size of features for subsequent measurement. Human vision detects contrast at details and steps based on two factors, the brightness ratio and the distance over which the change occurs. The previous methods increased the amount of the change. This technique reduces the distance, which also increases the visual perception of the contrast as well as making the steps and edges easier to locate.

The technique, sometimes called a ***maximum likelihood*** filter or a ***Kuwahara*** filter, compares the variances calculated for all of the smaller neighborhoods that include the central pixel but (unlike the neighborhoods used in the preceding examples) are not symmetrically placed around it, as shown in Figure 2.52. The subregion that has the smallest variance

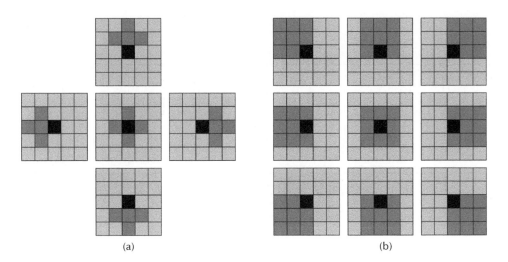

(a)                                             (b)

**FIGURE 2.52**
Schematic diagram of a Kuwahara filter. The shaded regions that include the dark central pixel are compared and the mean value of the one with the smallest variance is assigned to the central pixel location in the new image. (a) Five 5-pixel neighborhoods, (b) nine 9-pixel neighborhoods.

value is chosen as the one that the central pixel is most like, and the pixel is assigned the mean value of that subregion (Kuwahara et al. 1976). The figure illustrates the cases in which 5-pixel or 9-pixel subregions are used, but the same idea may be directly extended to larger neighborhoods as dictated by the breadth of transitions and edges in the original image.

As shown in Figure 2.53, the result is to replace pixels along smoothed or blurred edges with the value from the region on one side or the other. This procedure replaces gradual brightness variations at steps and discontinuities with more abrupt ones. Figure 2.54 shows that it also produces stair-stepping, or aliasing, along boundaries, and contouring, or posterization, in regions of the image that originally consisted of smooth or gradual variations in brightness, which can be visually distracting in photos but is often helpful for measurements.

For comparison purposes, application of a median filter also results in narrower discontinuities at edges and steps and can result in stair-stepping, or aliasing, of edges. The unsharp mask method of sharpening can produce somewhat narrower discontinuities, but also produces halos of brighter and darker pixels adjacent to the edge that are avoided by either the median or Kuwahara methods. As it does not result in contouring or aliasing of edges, the unsharp mask is usually the preferred technique for purely visual enhancement.

The Kuwahara filter is especially useful as a precursor to thresholding. Pixels along edges and discontinuities in an image often have values that are intermediate between those of the regions on either side of the step. This can result from image enlargement or from recording an image that is slightly out of focus. But in many cases it is just the consequence of the finite size of the detector, which averages the incoming light from both sides of a boundary. The Kuwahara filter assigns those pixels to whichever of the neighboring regions they are statistically most like, producing a reproducible and usually accurate delineation of the location of the step. As these edges are then used for measurement of features in images, it improves the reproducibility and accuracy of those measurements as well.

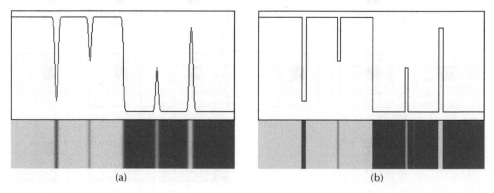

(a)                                                          (b)

**FIGURE 2.53**
Operation of a Kuwahara filter: (a) original image with blurred step and lines, (b) after application of the filter.

**FIGURE 2.54**
Example of a Kuwahara filter: (a) original image [girl.tif], (b) processed result, (c) enlargement of a portion of
(a) showing gradual brightness transitions at edges, (d) enlargement of the same portion of (b) showing sharp
transitions.

### 2.3.4 Problems

2.3.4.1#. Implement a variance filter with a fixed size, radius = 2. Compare the results
of using the variance and the standard deviation. Use autoscaling to adjust
the final values to fit the 0 to 255 range.

2.3.4.2. Implement a variance filter with a variable size, determined by reading in the
radius value from a text file.

2.3.4.3. Implement a range filter with a variable size. Compare the results, particularly
the sensitivity to image noise, with the variance filter in Problem 2.3.4.2. Make

this comparison for both small neighborhoods (for edges) and large neighborhoods (for texture).

2.3.4.4. Implement an entropy texture filter. Use autoscaling to adjust the final values to fit the 0 to 255 range.

2.3.4.5. Implement a fractal texture filter with maximum neighborhood radius of 5 pixels. Use autoscaling to adjust the final values to fit the 0 to 255 range. Compare the use of slope and intercept on various images.

2.3.4.6#. Implement a local equalization filter. Compare the results with large and small neighborhoods.

2.3.4.7. Implement a local enhancement filter using the ratio of the pixel value to the mean of a 6-pixel radius neighborhood, as outlined in Code Fragment 2.16. Compare the results to using a Gaussian-weighted convolution with a standard deviation of 3 pixels instead of a mean. Compare the results to using other powers between 1 and 3 instead of the square of the ratio.

2.3.4.8#. Implement a Kuwahara filter using nine $3 \times 3$ neighborhoods. Compare the results to a Laplacian sharpening convolution and to a median filter.

# 3

## Image Processing in the Fourier Domain

Treating images as a grid of pixels and processing them in the ___spatial domain___ or ___pixel domain___, as discussed in the previous chapter, is straightforward in implementation, generally conceptually clear, and can provide solutions to many problems. Converting an image to the ___Fourier domain___ offers additional capabilities that are very powerful, but requires learning some new ways of representing and interpreting data (except for people with a strong background in electrical engineering or acoustics, for whom thinking in terms of frequencies has become natural). This chapter presents the basics of the underlying math, but moves on quickly to the practical use of the Fourier domain for processing of images.

### 3.1 The Fourier Transform

Joseph Fourier's theorem, which students typically encounter in undergraduate mathematics classes, states that any function can be represented by a summation of sinusoids (Bracewell 2000). There are three parameters that control a sinusoid: the amplitude (or magnitude), frequency, and phase, as shown in Figure 3.1. The Fourier theorem imposes a few constraints on the functions to which it applies, but for our purposes the only important ones are that the function exist, that it be single-valued over a finite interval, and that beyond that interval it repeat endlessly in both directions. Equation 3.1 shows the relationship between the function in the spatial domain $f(x)$ and the Fourier transform $F(u)$, where $x$ represents position (or, in many cases of signal processing, time) and $u$ represents frequency. The exponential notation used in the equation is compact and elegant, but is more easily understood for many purposes when it is expressed in terms of sine and cosine functions.

$$F(u) = \int_{-\infty}^{+\infty} f(x) e^{-2\pi i u x} dx$$

$$f(x) = \int_{-\infty}^{+\infty} F(u) e^{+2\pi i u x} du \tag{3.1}$$

$$e^{-2\pi i u x} = \cos(2\pi u x) - i \cdot \sin(2\pi u x)$$

The integrals shown in Equation 3.1 are infinite, but in practice a "good enough" fit to any arbitrary function can be obtained with a modest number of terms because the

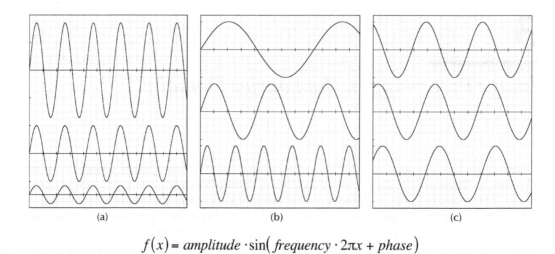

$$f(x) = amplitude \cdot \sin(frequency \cdot 2\pi x + phase)$$

**FIGURE 3.1**
Sine functions showing the effect of varying each controlling parameter: (a) amplitude, (b) frequency, (c) phase.

amplitudes of the lower frequency terms dominate the series, and it converges quickly. Many mathematics textbooks that introduce the Fourier theorem illustrate it by showing how a series of terms converges to model a step function, which is a particularly difficult case because of its extreme shape. This is shown in Figure 3.2 for 1, 2, 3, 5, 10, 20, and 40 terms in the series.

The ability to model a step function is not just a mathematical curiosity, as the edges and detail in images are also abrupt changes in brightness that must be modeled by Fourier

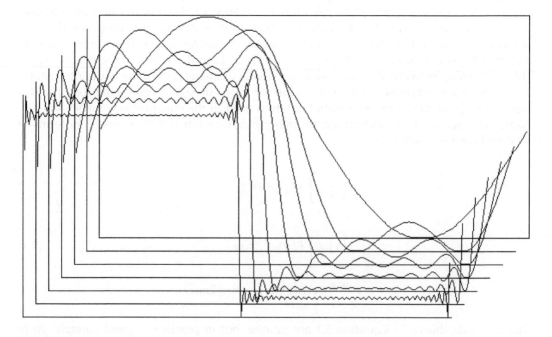

**FIGURE 3.2**
A series of plots showing the Fourier series approximation to a step function. From back to front, the plots show 1, 2, 3, 5, 10, 20, and 40 terms in the series.

series. Notice in particular that as the number of terms used to model the step increases, the sharpness of the step improves. Also note the oscillations that remain near the step, which are often described as "ringing." Both characteristics will also be found in the case of images.

The terms in the *__Fourier transform__* are complex, meaning that they have real and imaginary parts as shown in Equation 3.2. The calculation of the transform, which will be shown below, produces the complex values in this form, but for interpreting the results it is usually more convenient to express them in terms of the magnitude and phase. The relationship is the usual one in complex coordinates, in which the vector magnitude and phase are just a polar coordinate representation of the orthogonal real and imaginary coordinates. It is also useful for a variety of reasons to define the power of each term as the square of its magnitude.

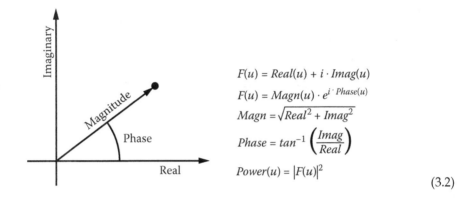

$$F(u) = Real(u) + i \cdot Imag(u)$$
$$F(u) = Magn(u) \cdot e^{i \cdot Phase(u)}$$
$$Magn = \sqrt{Real^2 + Imag^2}$$
$$Phase = tan^{-1}\left(\frac{Imag}{Real}\right)$$
$$Power(u) = |F(u)|^2$$

(3.2)

### 3.1.1 The Fourier Transform of an Image

For a function that is not continuous, but consists of values that are sampled at regular intervals (e.g., an image consisting of pixels), the infinite integrals in Equation 3.1 become the finite summations in Equation 3.3. Note the slight differences between the forward and reverse transforms. $N$ is the number of points. The highest frequency, determined by the spacing of the points, is the *__Nyquist frequency__*. A minimum of two points is required to specify the peak and trough of a sinusoid. This is another way of stating the limit of resolution in an image, which is the distance corresponding to two pixels. In other words, the smallest feature separation that can be resolved in an image is two pixels, corresponding to the distance between the centers of two bright pixels separated by one dark pixel, or vice versa. Consequently, this defines the resolution limit of a digitized image. (The ability to actually see detail at this limit also depends on the amount of contrast present.)

$$F(u) = \frac{1}{N}\sum_{x=0}^{N-1} f(x) \cdot e^{-2\pi iux/N}$$

(3.3)

$$f(x) = \sum_{u=0}^{N-1} F(u) \cdot e^{+2\pi iux/N}$$

Of course, images are two-dimensional representations of intensity as a function of position. Equation 3.4 shows the extension of the Fourier summations to two dimensions. The implementation of the two-dimensional calculation of the transform can be performed as a series of one-dimensional procedures. The transform is calculated on each row, and then on each column (or vice versa). Indeed, this is true for any number of dimensions, and some applications such as three-dimensional computerized tomography, which rely on the Fourier transform, make use of this fact.

$$F(u,v) = \frac{1}{N^2} \sum_{x=0}^{N-1} \sum_{y=0}^{N-1} f(x,y) \cdot e^{\left(\frac{-2\pi i(ux+vy)}{N}\right)}$$

$$f(x,y) = \sum_{u=0}^{N-1} \sum_{v=0}^{N-1} F(u,v) \cdot e^{\left(\frac{+2\pi i(ux+vy)}{N}\right)}$$

(3.4)

In order to perform a Fourier transform, it is necessary to create a temporary array of complex values. This is shown in Code Fragment 3.1, which reads the image from the host program and writes the brightness values to the temporary array. Initially, the imaginary parts of the complex values are zeros.

```
// Code Fragment 3.1 — Create temporary complex image array
//note: these structs are defined in the glue (PhotoshopShell) files
typedef struct
{
    double    real, imag;
} FFTelem;

typedef struct
{
    float red, green, blue;
} RGBPixel;


{
    long        Err, ID, x,
                width, height;
    float       brightness;
    RGBPixel *OrigLine;
    FFTelem    *TempLine;

    GetOriginalDimensions(&width, &height);              // image dimensions
    Err = MakeTemporaryImage(width, height, sizeof(FFTelem), ID)
    OrigLine = CreateAPointer(width, sizeof(RGBPixel)); // for original image
    TempLine = CreateAPointer(width, sizeof(FFTelem));  // for temp image
    for (x = 0; x < width; x++)
        TempLine[x].imag = 0; // zero the imaginary part
    // read in the original
    for (y = 0; y < height; y++)
    {
        ReadOriginalLine(y, OrigLine);
        for (x = 0; x < width; x++)
        {
```

```
        brightness = (OrigLine[x].red+OrigLine[x].green+OrigLine[x].blue)/3;
        TempLine[x].real = brightness;
    } // for x
    WriteTempImageLine(ID, y, TempLine);
  } // for y
  DisposeAPointer(OrigLine);
  DisposeAPointer(TempLine);
}
```

Code Fragment 3.2 shows the implementation of the discrete Fourier transform of a one-dimensional array. This is a straightforward illustration of the equations. It can be applied to a line array of any size, but as the size becomes large this method becomes very slow. The time required is proportional to the square of the number of points in the series. A much faster algorithm was developed more than 40 years ago called the FFT (***Fast Fourier Transform***) and is shown below. The computing time for the FFT is proportional to $N \cdot \text{Log}_2(N)$ so, for example, a transform on 1024 points can be as much as 100 times faster with the FFT.

```
// Code Fragment 3.2 - Discrete Fourier Transform (DFT)
// The Line arrays contain m points, each of type FFTelem.
DFT(int dir,int m, FFTelem *Line1)     // dir=+1 (forward) dir=-1 (inverse)
{
  long      i, k;
  double    arg;
  double    cosarg, sinarg;
  FFTelem   *Line2;

  Line2 = CreateAPointer(m, sizeof(FFTElem));   //do transform in new array
  for (i = 0; i < m; i++)
  {
    Line2[i].real = Line2[i].imag = 0;
    arg = - dir * 2.0 * 3.141592654 * (double)i / (double)m;
    for (k=0;k<m;k++)
    {
      cosarg = cos(k * arg);
      sinarg = sin(k * arg);
      Line2[i].real += (Line1[k].real * cosarg - Line1[k].imag * sinarg);
      Line2[i].imag += (Line1[k].real * sinarg + Line1[k].imag * cosarg);
    }
  }
  // Copy the data back to old array
  if (dir == 1)
    for (i = 0; i < m; i++)
    {
      Line1[i].real = Line2[i].real / (double)m;
      Line1[i].imag = Line2[i].imag / (double)m;
    }
  else
    for (i = 0; i < m; i++)
```

```
        {
            Line1[i].real = Line2[i].real;
            Line1[i].imag = Line2[i].imag;
        }
    DisposeAPointer(Line2);
}
```

The FFT technique for calculating the Fourier transform applies a "divide-and-conquer" strategy. The most common, and certainly the clearest, version of the method is the Cooley–Tukey algorithm (Cooley and Tukey 1965), which performs a repeated halving and doubling applied to the array. The method is fundamentally recursive although it is not usually programmed that way. Starting with the relationship shown in Equation 3.3, it is useful to define the function $w_K$ as the $K$th complex root of one, $w_K = \exp(-2\pi i / K)$. Then, for an initial array of $N$ points ($N = 2K$), Equation 3.3 can be written as shown in Equation 3.5, subdividing the array into two halves.

$$F(u) = \frac{1}{N} \left( \sum_{x=0}^{N/2-1} f(2x) w_N^{(2x)u} + \sum_{x=0}^{N/2-1} f(2x+1) w_N^{(2x+1)u} \right) \tag{3.5}$$

But the square of a $2K$th root of 1 is also a $K$th root of 1, as shown in Equation 3.6.

$$w_N^{(2x)u} = w_{(N/2)}^{xu} \tag{3.6}$$

So the expression can be rewritten as shown in Equation 3.7.

$$F(u) = \frac{1}{N} \left( \sum_{x=0}^{(N/2)-1} f(2x) \cdot w_{N/2}^{xu} + \sum_{x=0}^{(N/2)-1} f(2x+1) \cdot w_{N/2}^{xu} \cdot w_N^u \right)$$

$$= \frac{1}{2} \left( F_{even}(u) + F_{odd}(u) \cdot w_N^u \right) \tag{3.7}$$

The two sums in Equation 3.7 are called the even and odd terms, respectively. But because of Equation 3.8, there is also the relationship shown in Equation 3.9, where the even and odd terms are recombined.

$$w_K^{K+u} = w_K^u$$

$$w_{2K}^{K+u} = -w_{2K}^u \tag{3.8}$$

$$F(u+K) = \frac{1}{2} \left( F_{even}(u) - F_{odd}(u) \cdot w_{2K}^u \right) \tag{3.9}$$

Hence, it is possible to compute an $N$-point Fourier transform by dividing it into two parts, the first half using Equation 3.7 and the second half by using the same terms again as in Equation 3.9. If the number of data points is restricted to an exact power of two, the halving can be repeated down to the trivial case of a transform of a single value. The

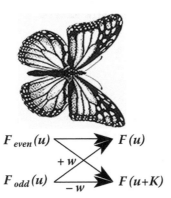

$$F_{even}(u) \longrightarrow F(u)$$
$$+w$$
$$F_{odd}(u) \xrightarrow{-w} F(u+K)$$

**FIGURE 3.3**
The butterfly relationship.

relationships in these two equations are generally referred to as a butterfly, because of the perceived similarity of the diagram showing the combinations of terms to the insect, as shown in Figure 3.3.

The usual way of performing this procedure is to rearrange the ordering of the data points as shown in Code Fragment 3.3. The example is for a one-dimensional array, which can be used for a two-dimensional image by processing the rows and then the columns as noted above. The shuffling of the data points is performed according to a table (almost universally called the twiddle matrix) that for the simplest version of the algorithm can be derived by writing the addresses of the values in binary form and reversing the bit order, as shown in Table 3.1 for the case of 16 points.

Note in the example that four of the values (positions 0, 6, 9, and 15) do not move, and the remaining moves are paired so that there are only six actual swaps of values that are needed. Creating a list of the pairs of positions to be swapped for each specific power-of-two size (16, 32, 64, 128, 256, ...) is a clear and efficient way to program this, although it is also possible to calculate the pairs on the fly.

**TABLE 3.1**

Twiddle Matrix Used to Shuffle 16 Input Values

| Initial | Binary | Reversed | Destination |
|---------|--------|----------|-------------|
| 0 | 0000 | 0000 | 0 |
| 1 | 0001 | 1000 | 8 |
| 2 | 0010 | 0100 | 4 |
| 3 | 0011 | 1100 | 12 |
| 4 | 0100 | 0010 | 2 |
| 5 | 0101 | 1010 | 10 |
| 6 | 0110 | 0110 | 6 |
| 7 | 0111 | 1110 | 14 |
| 8 | 1000 | 0001 | 1 |
| 9 | 1001 | 1001 | 9 |
| 10 | 1010 | 0101 | 5 |
| 11 | 1011 | 1101 | 13 |
| 12 | 1100 | 0011 | 3 |
| 13 | 1101 | 1011 | 11 |
| 14 | 1110 | 0111 | 7 |
| 15 | 1111 | 1111 | 15 |

```
// Code Fragment 3.3 - Applying a twiddle matrix
// rearrange values in the complex array of values X[k]
// each element of which consists of .real and .imag parts.
if (dimension == 16)
{
   swaplist[13]={6,1,8,2,4,3,12,5,10,7,14,11,13};
   // 0th value indicates six pairs of indices
   for (j = 0; j < swaplist[0]; j++)
   {
      float    temp;
      int    srci = swaplist[j*2+1];
      int    dsti = swaplist[j*2+2];
      temp = X[srci].real; X[srci].real = X[dsti].real; X[dsti].real = temp;
      temp = X[srci].imag; X[srci].imag = X[dsti].imag; X[dsti].imag = temp;
   }
}
// ... and so on for other powers of 2
```

Once the reordering of values has been performed, the butterfly calculation can be applied. This is done (as shown in Code Fragment 3.4) by stepping through the power-of-two levels. Some languages, such as Fortran and C++, have a predefined complex data type and arithmetic operators that function with those values. In the example, as in the previous ones, a data type has been defined as a **struct** with real and imaginary parts, and the calculations are performed as shown. The variable **step** is the distance between the two inputs to the butterfly, starting with 1 and doubling at each **level**. The W array (**W[][].real and W[][].imag**) contains the precalculated complex roots of unity for each level.

```
// Code Fragment 3.4 - Butterfly FFT calculation
// step - 2^(level-1), stride = 2^level
// Dimension = power of 2; LogDim = base 2 log of the Dimension
// i.e., Dimension = 2^LogDim
// ... the W[][] array (FFTelem) contains precalculated complex roots of
//      unity for each level and step position
long  step = 1;
for (level = 1; level <= LogDim; level++)
{
   long   stride = step * 2;
   for (j = 0; j < step; j++)
   {
      FFTelem U, T;
      U.real = W[level, j].real;
      U.imag = W[level, j].imag;
      // Complex U = exp(-2*π*i/(2^level)
      for (k = j; k < Dimension; k += stride)
      {  // in-place butterfly
         T = U;
         T.real = X[k+step].real * T.real - X[k+step].imag * T.imag;
         T.imag = X[k+step].real * T.imag + T.real * X[k+step].imag;
```

```
        X[k+step].real = X[k].real - T.real;
        X[k+step].imag = X[k].imag - T.imag;
        X[k].real      += T.real;
        X[k].imag      += T.imag;
    } // for k
  } // for j
} // for level
```

The only requirement of the most straightforward implementation of this algorithm is that the number of points in the series be a power of 2 (e.g., 128, 256, 512, 1024). In terms of image processing, that means the image must be a square of that size. For images that do not have those dimensions, a common solution is to pad the image out to the next larger power of 2, filling the surrounding space with a medium gray value (theoretically, the mean value of the pixels in the image is best, but in practice a value of 128 usually works fine, and even black or white are often used).

The student should be aware that there are other implementations of the FFT that do not have the power-of-two restriction on image size. For example, there is a downloadable C language library available at <http://www.fftw.org/> that provides very fast and flexible results, but it is much more complicated and less suitable for learning the principles involved. With non-power-of-two dimensions, the arrangement of magnitude and phase data within the transform changes and is more difficult to interpret.

One way to perform the padding operation is to create a temporary image of the appropriate size (with complex values) and copy the image into it before performing the Fourier transform. Alternatively, most host programs allow you either to change the canvas size directly, or to copy and paste the image into a new one of the appropriate dimensions, which is another way of accomplishing the padding. Because of the inherent assumption in the Fourier transform that the image repeats vertically and horizontally, it does not matter whether the original image is centered in the padding (equal amounts on all sides), or whether the padding is placed on two sides, usually the bottom and right, as shown in Figure 3.4. (The statement that the location of the padding does not matter is not strictly true: there is no effect on the power spectrum or on any of the filtering and other operations described below, but the phase values for the sinusoids are shifted by the offset.) For convenience, many of the images referred to in this section have dimensions that are exact powers of two.

When the Fourier transform of an image is calculated, the result (as noted above) is a set of complex values that specify the amplitude and phase of every sinusoid of increasing frequency that contributes to the original image. Adding together all of those sinusoids, each one with its corresponding amplitude and shifted by its phase value, will exactly reproduce the original image. In the example code fragments that follow, this is done for color images using the mean intensity (average of red, green, and blue values). However, the procedure can also be applied to the individual color channels, whether they are the original R, G, and B, or any of the other color spaces discussed in previous chapters.

Code Fragment 3.5 shows the procedure for calculating the FFT of a one-dimensional array. This is used as a subroutine in Code Fragment 3.6 to perform the procedure on a two-dimensional array (an image), by first applying the one-dimensional procedure to each horizontal row of values in the image, and then repeating the operation on each

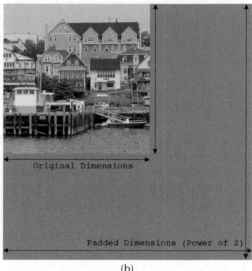

(a)                                                      (b)

**FIGURE 3.4**
Padding an image to increase the dimensions to a power of 2 to allow using the Fast Fourier Transform:
(a) padding added on all sides of the original image, (b) padding added on two sides. Because of the assumption
that the data repeat vertically and horizontally, either method can be used.

vertical column of values. A final step in the process performs a ***block-swap*** of the four
quadrants of the values, as indicated in Figure 3.5. This is done so that the initial term in
the summation, the constant value that represents the average intensity of the image, is
shifted to the center of the resulting array. By convention and for convenience in interpreting the data, this is the origin of the plots shown below.

```
// Code Fragment 3.5 — One-dimensional FFT
// This computes an in-place Cooley-Tukey, Radix 2 complex-to-complex FFT
// The array Line contains 2^m points of complex type FFTelem.
// dir = 1 gives forward transform, = -1 gives reverse transform.
void FFT(short dir, long m, FFTelem *Line)
{
    long     n, i, i1, j, k, i2, l, l1, l2;
    double   c1, c2, tx, ty, t1, t2, u1, u2, z;

    n = 1 << m; // Calculate the number of points (2^m)

    // bit reversal (DECIMATION IN TIME)
    i2 = n >> 1;
    j = 0;
    for (i = 0; i < n-1; i++)
    {
        if (i < j)
        {   // complex swap Line[i] with Line[j]
            float tx = Line[i].real;
            float ty = Line[i].complex;
            Line[i].real = Line[j].real;
```

```
         Line[i].imag = Line[j].imag;
         Line[j].real = tx;
         Line[j].imag = ty;
      }
      k = i2;
      while (k <= j)
      {
         j -= k;
         k >>= 1;
      }
      j += k;
} // for i
// Compute FFT (DIVIDE AND CONQUER)
c1 = -1.0;
c2 = 0.0;
l2 = 1;
for (l = 0; l < m; l++)
{
      l1 = l2;
      l2 <<= 1;
      u1 = 1.0;
      u2 = 0.0;
      for (j = 0; j < l1; j++)
      {
         for (i = j; i < n; i += l2)
         {
            i1 = i + l1;
            t1 = u1 * Line[i1].real - u2 * Line[i1].imag;
            t2 = u1 * Line[i1].imag + u2 * Line[i1].real;
            Line[i1].real = Line[i].real - t1;
            Line[i1].imag = Line[i].imag - t2;
            Line[i].real += t1;
            Line[i].imag += t2;
         }
         z = u1 * c1 - u2 * c2;
         u2 = u1 * c2 + u2 * c1;
         u1 = z;
      }
      c2 = sqrt((1.0 - c1) / 2.0);
      if (dir == 1)
         c2 = -c2;
      c1 = sqrt((1.0 + c1) / 2.0);
}
// Scaling for forward transform
if (dir == 1)
   for (i = 0; i < n; i++)
   {
```

```
         Line[i].real /= n;
         Line[i].imag /= n;
      } // for i
}


// Code Fragment 3.6 — FFT of an image and display of the power spectrum
// ... read image, verify ^2 dimensions, create temporary complex array
// ... for each line, perform 1-D FFT
// ... for each column, perform 1-D FFT
// swap quadrants (block swap)
for (y = 0; y < height/2; y++)
{
   ReadTempImageLine(ID, y, L1);
   ReadTempImageLine(ID, y+height/2, L2);
   for (x = 0; x < width/2; x++)
   { // swap values
      long   x2=x+width/2;
      temp = L1[x].real; L1[x].real = L2[x2].real; L2[x2].real = temp;
      temp = L1[x2].real; L1[x2].real = L2[x].real; L2[x].real = temp;
      temp = L1[x].imag; L1[x].real = L2[x2].imag; L2[x2].imag = temp;
      temp = L1[x2].imag; L1[x2].real = L2[x].imag; L2[x].imag = temp;
   }
   WriteTempImageLine(ID, y, L1);
   WriteTempImageLine (ID, y+height/2, L2);
} // for y
// ... read each line, compute power (linear or log), find max
// power = L1[x].real * L1[x].real + L1[x].imag * L1[x].imag
// power = log(power); if (power > max) max = power;
// ... write autoscaled result image
```

## 3.1.2   Displaying the Transform Information

The Fourier transform is usually depicted onscreen for visual interpretation and subsequent processing by showing just the ***power spectrum***. As noted above, the power of each term in the Fourier transform is the square of the amplitude. These values cover a large range, and to make better use of the display, they are often shown as the logarithm of the actual power; that is the convention used in the examples that are shown, and corresponds to the outline in Code Fragment 3.6. The display is autoscaled so that the maximum value (usually but not always the constant term at the origin) is set to white.

To understand the organization of the information in the power spectrum display, consider the simple image shown in Figure 3.6, which consists of just three superimposed sinusoidal patterns. Each one has a different frequency and orientation. The Fourier transform power spectrum consists of three ***spikes***, bright points that represent large values of the power at the corresponding frequencies, on a background of much smaller values that correspond to all of the other minor terms that result from finite noise and other imperfections in the image.

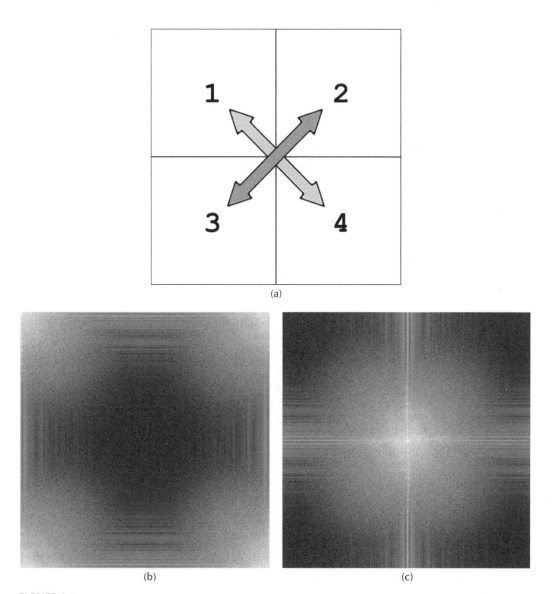

(a)

(b)                                    (c)

**FIGURE 3.5**
Block-swapping the Fourier transform values interchanges quadrant 1 with quadrant 4 and quadrant 2 with quadrant 3, as shown diagrammatically in (a). This places the low-frequency terms (which generally have greater magnitude) near the center of the display as shown in (c) rather than at the four corners as in (b). The example is the power spectrum from the [manuscript.tif] image from Figure 3.8.

Figure 3.6c provides a guide to the correspondence between the location of the spikes and the corresponding sinusoids. The top half of the power spectrum plot and the bottom half are identical, with 180° rotational symmetry about the origin. It would not actually be necessary to show both halves of the power spectrum, but by convention this is the way the plot is usually displayed. The locations of the three points are characterized by the radius from the center, which is the frequency of the corresponding sinusoid, and an angle, which is the orientation. High frequencies are shown at large radii, and low frequencies are at small radii. The angle from the origin to the point in the transform is perpendicular to the orientation of the lines for the corresponding sinusoid in the original image. The

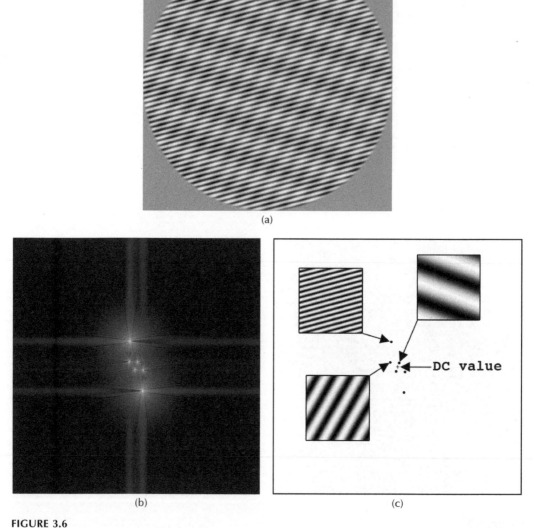

(a)

(b)                                                    (c)

**FIGURE 3.6**
Fourier transform power spectrum display: (a) original image [sinusoids.tif], (b) power spectrum display show-ing three spikes (each appears twice because of the display symmetry), (c) index to the display showing the individual sinusoid patterns responsible for each spike.

point right at the origin, in the center of the image, is called the **_DC value_** and has a value in the Fourier transform that represents the mean brightness of the entire image.

For a typical image, the power spectrum is more complex than the one shown in Figure 3.6 and contains power at all frequencies. Usually, the power drops off gradually at higher frequencies (as noted above, this is a characteristic of the Fourier transform because the series converges rapidly). There may be a visible cross pattern, as in Figure 3.7, which results from the mismatch between the left and right boundaries of the image and the top and bottom of the image. Remember that the Fourier transform method relies on the assumption that the data in the interval modeled by the series (in this case, the data in the image) repeat endlessly in both directions, which is equivalent to assuming that the

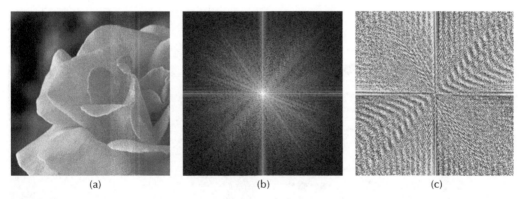

|     |     |     |
| --- | --- | --- |
| (a) | (b) | (c) |

**FIGURE 3.7**
(See **Color insert following page 172**.) Fourier transform displays: (a) original image, $256 \times 256$ pixel portion of [rose.tif], (b) power spectrum (log scale) showing prominent cross pattern resulting from boundary mismatches, (c) phase values shown as hue.

image repeats endlessly, similar to a series of tiles. Figure 2.5 (Chapter 2) illustrated the wrap-around mating of the left-right and top-bottom image boundaries.

If the boundaries of the image do not match, which in general they will not, then just as for the step function shown in Figure 3.2, many terms extending to high frequencies are needed to produce an exact match. The bright vertical line in the center of the power spectrum in Figure 3.7b shows the terms needed for the top and bottom boundaries of the image, and the bright horizontal line shows those needed for the left and right boundaries. Other prominent radial lines in the power spectrum result from the steps at the edges of petals in the flower, but do not extend to the very highest frequencies.

The phase information is less commonly displayed, because it is usually more difficult to visually interpret. One way to show the phase values, which can vary from 0 to $2\pi$, is to represent them by a hue value, which also conveniently varies from 0 to 360 degrees around the color wheel. In Figure 3.7c this use of color is illustrated.

### 3.1.3 Low-Pass Filters

After the block-swap procedure, the power spectrum plot of the Fourier transform result is arranged with the low-frequency terms in the summation near the center origin, and higher-frequency terms at larger radii. If all of the terms are used in the inverse transform, the original image is reconstructed exactly. However, it is possible to perform the reconstruction using only some of the terms instead of the entire series. As indicated in the one-dimensional example of Figure 3.2, if just the first few terms are kept, the result is an approximation of the original image in which sharp edges (the step in the profile) are somewhat rounded off, and other differences such as ringing appear as well. Similar results are obtained in the two-dimensional case.

Figure 3.8 shows an image with its FFT power spectrum. Keeping the lower-frequency (small-radius) terms in the transform and reducing the amplitude of higher-frequency (large-radius) terms to zero produces a reconstructed result when the inverse FFT is performed that smoothes the image. The mask used for this is shown in Figure 3.8c. It has a Gaussian cross section, and the magnitude of the values in the mask are multiplied

**FIGURE 3.8**
Low-pass filtering: (a) original image [manuscript.tif], (b) FFT power spectrum of (a), (c) Gaussian mask that keeps low frequencies and eliminates high frequencies, (d) reconstructed result showing smoothed image.

by the amplitudes of the corresponding terms in the Fourier transform, which are located at the same positions in the complex array.

There is potential for confusion between the terms "filter" and "mask." Some texts use the terms filter and mask interchangeably. Although the operation of low-pass, high-pass and other frequency selection operations are commonly called filters, for consistency in this chapter the term _**mask**_ is preferred for operations that apply an image array as a set of scalar constants that are multiplied by the amplitude of the Fourier terms. The term _**filter**_ will be principally reserved for cases in which the array of multiplied values is itself a Fourier transform that contains complex values, as will be introduced in Section 3.3.

However, because of historical usage, the term filter is also applied to some masking operations.

$$G(u,v) = H(u,v) \cdot F(u,v) \qquad (3.10)$$

Mathematically, this masking procedure is written as shown in Equation 3.10. $F(u,v)$ is the transform of the original image and consists of complex values, $H(u,v)$ is the mask and consists of real values, and $G(u,v)$ is the array of complex values that results from the multiplication. The inverse transform of $G$ is the reconstructed, processed image. If $F$ is treated as real and imaginary values, both the real and imaginary terms must be multiplied by $H$. If $F$ is treated as phase and magnitude values, only the magnitude terms are multiplied by $H$ and the phase is unchanged.

Code Fragment 3.7 outlines the procedure for applying a mask array to the stored Fourier transform. The mask array has the same dimensions as the Fourier transform (and the original image from which the transform is derived). It is often convenient to create the mask as an image and store it in the reference image file introduced in Chapter 1, Section 1.5, for applying mathematical combinations of images. The code fragment uses that reference image as the mask, interpreting the pixel values from 0 to 255 as mask values ranging from 0 (total attenuation) to 1 (no attenuation). The images used as masks are shown in the figures that appear below.

Remember that the complex values in the temporary array were block-swapped to generate the power spectrum display, and must be swapped back before the inverse transform is computed. Alternately, if no display of the power spectrum is produced, the block swapping of the temporary array can be skipped. In that case the mask values must be block-swapped as they are read, to correctly align with the corresponding frequencies.

```
// Code Fragment 3.7 - Application of a mask to the FFT image
// ... Read original image, copy into temporary complex image
// ... Perform FFT on the temporary complex image
for (y = 0; y < height; y++)
{
   // ... Read a line from the reference image (ref_l)
   // ... Read a line from the temporary complex image (fftline)
   for (x = 0; x < width; x++)
   {
      // calculate weight = mask value[x] / 255
      float mask_lum = (ref_l[x].red + ref_l[x].green + ref_l[x].blue)/3;
      float mask_weight = mask_lum / 255;
      // multiply the real and imaginary value times the weight
      fftline[x].real *= mask_weight;
      fftline[x].imag *= mask_weight;
   }
   // ... Write the line back to the temporary complex image (fftline)
}
// ...Perform inverse FFT on the temporary complex image
// ...Display the filtered result
```

There are several points about the procedure shown in Figure 3.8 that are important to understand, and which explain many of the reasons that processing in Fourier space is used as an important tool. First, the result (Figure 3.8d) is a smoothed image that is identical to the result that could be obtained by performing a Gaussian smoothing operation in the pixel domain, as shown in Chapter 2. It can be shown mathematically that the two operations are exactly the same. But to apply this procedure in the pixel domain would require a very large number of multiplications for each pixel (a 23 × 23 kernel of weights, which would be needed to produce the amount of smoothing shown, requires more than 500 multiplications and additions for each pixel). Only a single multiplication of the mask value times the magnitude of the corresponding sinusoid is required in the Fourier transform implementation, plus of course the overhead in performing the forward and reverse FFT procedures. Some of the other procedures shown in this chapter would require a spatial domain filter as large as the image itself.

It must be emphasized that although Fourier domain processing is mathematically equivalent to the filters described in Chapter 2, Section 2.1, that can be applied with multiplicative kernel operations in the spatial domain, the statistical and rank-based neighborhood operations described in the remainder of Chapter 2 do not have Fourier domain equivalents.

Although issues of efficiency are not unimportant, there are more fundamental advantages to working in the Fourier domain. One is the ability to choose specific frequencies and orientations to be selectively removed from the image, as will be shown below. Another is the ability to perform several types of measurements that would be very difficult to accomplish in the pixel or spatial domain, but which become quite easy in the Fourier domain. Also, as will be shown in Section 3.4, deconvolution can be used to remove blur from some images. Perhaps most important of all is the rearrangement of the data in the original image that allows the user (after some practice to develop the necessary experience and intuition) to see and interact with the image information in a new way.

The mask applied in Figure 3.8 is an example of a ***low-pass filter***, so called because it passes, or leaves unchanged, the low frequencies, while attenuating or removing the high frequencies in the image. The filtering in the example is performed with a mask that has a specific shape, namely, a Gaussian cross section, which has the same optimal properties for this purpose that the Gaussian kernel has in the spatial domain. (In fact, there is a relationship between the two: as will be shown below, the Fourier transform of any spatial domain kernel, which is after all just an array of values that can be treated similar to any other image, generates the corresponding filter that can be applied in the Fourier domain. The FFT of a Gaussian is another Gaussian.)

Many texts start out with a simpler mask, one that simply keeps all frequencies below a certain value and reduces to zero the amplitude of all higher frequencies. This would correspond to a mask shaped like a circle or disk. Figure 3.9 shows the effect of using such a mask. The image is certainly smoothed by the removal of high frequencies, but there are artifacts present that appear as ripples around all of the edges, and obscure much of the detail. This phenomenon is called ***ringing*** (or sometimes, Gibbs ringing). It can be reduced or eliminated by using masks that do not cut off abruptly at specific frequencies. As shown in Figure 3.10, simply smoothing the edges of the mask reduces the ringing. In the various examples that are shown below, the mask arrays that cut off selected frequencies are always smoothed to soften the edges of the mask. This can be done by applying a small averaging or smoothing filter (in the spatial domain) to the mask itself.

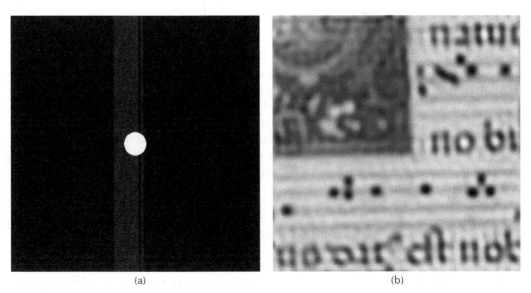

(a)                                          (b)

**FIGURE 3.9**
Effect of mask shape: (a) mask with sharp frequency cutoff, (b) reconstructed result showing ringing.

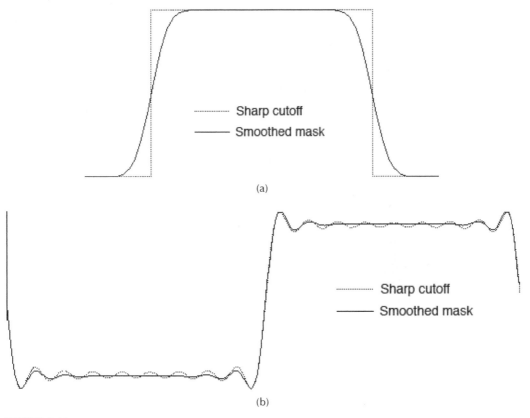

(a)

(b)

**FIGURE 3.10**
Effect of mask shape in one dimension: (a) smoothing the edge of the mask, (b) reconstructed result for the step function in Figure 3.2.

### 3.1.4  High-Pass and Band-Pass Filters

Low-pass filters, as shown in the previous section, keep (pass) the low frequencies shown in the Fourier transform power spectrum with small radii, and attenuate or block the high frequencies. Of course, it is possible to do the opposite. A **_high-pass filter_** keeps (passes) the high frequencies while attenuating low frequencies. It can be constructed as a sharp-edged mask similar to that shown in Figure 3.9 (by reversing the values in the mask), but much better results are obtained with other profiles. The **_ideal inverse filter_** is a ramp that keeps the highest frequencies (at the maximum radius or Nyquist limit) and progressively attenuates lower frequencies, decreasing to zero amplitude at the center. As shown in Figure 3.11, the result is very similar to the Laplacian and unsharp mask operations performed in the pixel or spatial domain.

The examples of low-pass and a high-pass filter masks shown above are not the only profiles of mask amplitude as a function of frequency that may be used. In particular, the Butterworth filters (masks) constitute a family of such profiles, as shown in Figure 3.12. The filters are radially symmetrical, and as shown in the equation and plots are high-pass filter masks that attenuate low frequencies and preserve high frequencies. The frequency $f_0$ at which the amplitude reaches 50% is a user-defined parameter, corresponding to the radius of a mask with a sharp cutoff. Reversing or complementing the mask produces the corresponding Butterworth low-pass filter mask. Figure 3.13 illustrates the result of second-order ($n = 2$) Butterworth low- and high-pass filter masks with a 50% attenuation point set to one-quarter the maximum (Nyquist) frequency.

A high-pass filter emphasizes the highest frequencies, which are likely in many images to be pixel-to-pixel variations that result from random noise. Of course, this was also a problem with the Laplacian and unsharp mask methods applied in the spatial domain. It is possible to attenuate the highest frequencies as well, creating a **_band-pass_** filter mask as shown in Figure 3.14. This is similar in purpose and concept to the Difference-of-Gaussians, or DoG, filter used in the spatial domain. The advantage to working in the

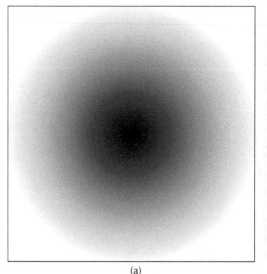

(a)

(b)

**FIGURE 3.11**
High-pass filter: (a) an ideal inverse filter mask (the values increase linearly with radius), (b) the reconstructed result.

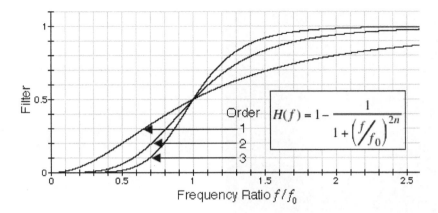

**FIGURE 3.12**
Butterworth filter profiles for orders $n$ = 1, 2, and 3.

Fourier domain is the ability to directly visualize the mask in terms of frequencies, and to examine the FFT power spectrum display to see the magnitude of the various frequency ranges. This often indicates where the real image data end (the highest frequency at which there is any real information defines the image resolution) and the magnitude of the random noise.

Other types of masks can be constructed and applied to keep selected frequencies and to attenuate others. All of the foregoing examples shown above are symmetrical isotropic masks that affect all orientations uniformly. It is similarly possible to create masks that selectively remove certain orientations, as shown in Figure 3.15. This method is sometimes useful for processing video images, particularly from surveillance cameras that record images from a single video field having half of the horizontal scan lines of full video. In can also be useful for scratch removal. This type of mask is sometimes called a ***notch filter***.

(a)

(b)

(c)

(d)

**FIGURE 3.13**
Application of Butterworth second-order filter masks to the Fourier transform of the image in Figure 3.8a. The $f_0$ frequency is set to 25% of the maximum. (a) Low-pass filter mask, (b) low-pass result, (c) high-pass filter mask (reverse of a), (d) high-pass result.

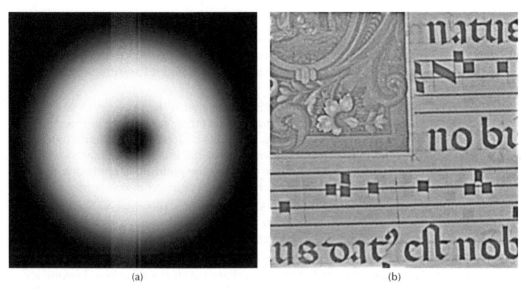

(a)  (b)

**FIGURE 3.14**
Band-pass filter: (a) an arbitrary mask (both high and low frequencies are attenuated, and intermediate frequencies kept), (b) the reconstructed result.

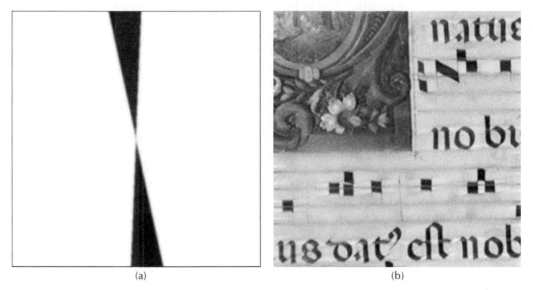

(a)  (b)

**FIGURE 3.15**
Directional filter: (a) mask that removes frequencies with orientations near the vertical, (b) the reconstructed result in which horizontal lines are suppressed.

### 3.1.5    Problems

3.1.5.1#.  Implement a two-dimensional Fourier transform for an image. Test the image
first to ensure that it is square and has dimensions that are an exact power of
two. Display the power spectrum using a linear and a logarithmic scale and
compare them.

3.1.5.2.  Extend the program from Problem 3.1.5.1 so that if the image does not have
exact dimensions, it is padded out to the next larger proper size as it is copied
into the temporary array in memory, filling the padded area with intermediate
gray or with the mean brightness value from the image. Note that as shown
in Figure 3.4, it makes no difference whether the padding is applied with the
original image centered in the larger one, or positioned in one corner.

3.1.5.3.  Modify the program from Problem 3.1.5.1 to display the phase values for each
term, either scaled to fit into the 0 to 255 intensity range or as hue values
superimposed on the power spectrum.

3.1.5.4.  Implement a program to perform a Fourier transform followed by the application of a high-pass filter that applies a sharp cutoff at a preselected frequency
value, and then retransforms the data to show the filtered image. Modify this
to apply a gradual cutoff such as a Butterworth filter. Add a logical variable
that allows the program to function as either a high-pass or a low-pass filter.
Note that for convenience, both the cutoff frequency, the choice of a sharp
cutoff or shaped mask, and the selection of low- or high-pass filtering can be
read from a text file.

3.1.5.5.  Implement a program that applies Gaussian smoothing to an image by performing a Fourier transform and multiplying by a Gaussian low-pass filter,
and retransforming. Compare the results to a comparable spatial domain
Gaussian filter. (The standard deviation of the Gaussian filter in the Fourier
domain is inversely proportional to that in the spatial domain. As explained
later in this chapter, radius in the Fourier domain is a frequency proportional
to the width of the original image divided by a distance in the spatial domain.)

3.1.5.6#.  Implement a program that performs a Fourier transform on an image, uses a
previously stored reference image as a mask to reduce or erase selected frequency terms, and then retransforms the data to display the resulting image.
Use this with a variety of manually created masks, both with and without
blurring abrupt edges. Note that the stored reference image has values in the
range 0 to 255, and these must be divided by 255 to normalize them to the
0.0 to 1.0 range of values in the mask that are multiplied by the amplitudes in
the FFT array.

3.1.5.7.  Implement a program that creates a Butterworth filter as an image that can be
stored in the reference image to be used as a mask. Read the cutoff frequency
and the choice of high- or low-pass filtering from a text file.

## 3.2    Removing Periodic Noise

Filter masks that remove or reduce the magnitude of terms in the Fourier transform can
also be used to selectively remove frequencies that correspond to periodic noise in images.

Electrical interference, the flickering of fluorescent lights, vibration, detector characteristics such as the color filter arrays used in many digital cameras and the interlaced scan in video cameras, some image compression methods, printing technologies such as half-toning, and other causes, can superimpose patterns onto images that have regular spacings. The corresponding frequencies have large amplitudes, and show up in the FFT power spectrum as spikes, or bright points. Selectively removing them removes the pattern, while having minimal effects on the rest of the image.

### 3.2.1 Masks for Selected Frequencies

Figure 3.16 illustrates the process of removing selected frequencies, using an image that was printed with the common half-tone method used in practically all newspapers, books (including this one), and magazines. The size of the individual black half-tone dots is varied to create the visual effect of a range of gray values, but the spacing of the half-tone dots is regular. Consequently, the power spectrum shows a regular array of prominent spikes. The spikes closest to the center (lowest frequency) define the dot spacing, and the regularly spaced higher frequency spikes (called ***harmonics***) define the shape of the dots.

Creating a mask that selectively removes the spikes but leaves all of the other terms in the Fourier transform unchanged, and then performing the inverse transform, reconstructs the image with the half-tone pattern suppressed. Because the half-tone printing in this book of an image that was previously half-toned can generate moiré interference patterns, enlarged fragments of the original and reconstructed images are included in the figure to better show the effect of processing. The recovery of fine details is evident, such as the eyelashes that are present in the original but obscured by the periodic noise pattern.

When this procedure is applied to printed color images, it is generally best to convert the image to a set of CMYK channels to match the inks used for printing. The periodic noise pattern in each channel will be distinct, corresponding to the orientation of the half-tone grid used for that color. The procedure outlined below can be applied individually to each channel, with a different mask for each one.

Creation of the mask can be accomplished in several ways. It is important to remember that the display of the power spectrum is not the actual Fourier transform of the image, which consists of complex floating point numbers and is not easily manipulated directly. The routine outlined in Code Fragment 3.6 (and programmed in Problem 3.1.5.6) can be used to apply a mask to a Fourier transform and reconstruct the result. The display of the power spectrum with the visible spikes is an important starting point for the creation of the mask. Manual marking of the spikes (and erasure of the intervening space) can be used to create a mask that can then be applied in a subsequent calculation. In some cases this manual method is quick and easy, but in general it is preferred to use algorithmic procedures that can be programmed to perform the operation.

Two principle approaches are typically used to generate a mask image from the image of the FFT power spectrum. Both use the spatial domain processing methods described in the previous chapters. The first employs the top hat filter introduced in Chapter 2. Because the spikes are usually small and brighter than their surroundings, the top hat method can locate them. The second method uses thresholding (discussed in Chapter 4) to select bright pixels and create a mask, but must first deal with the fact that the power spectrum is not uniform in brightness. The power spectrum display is generally brighter (representing

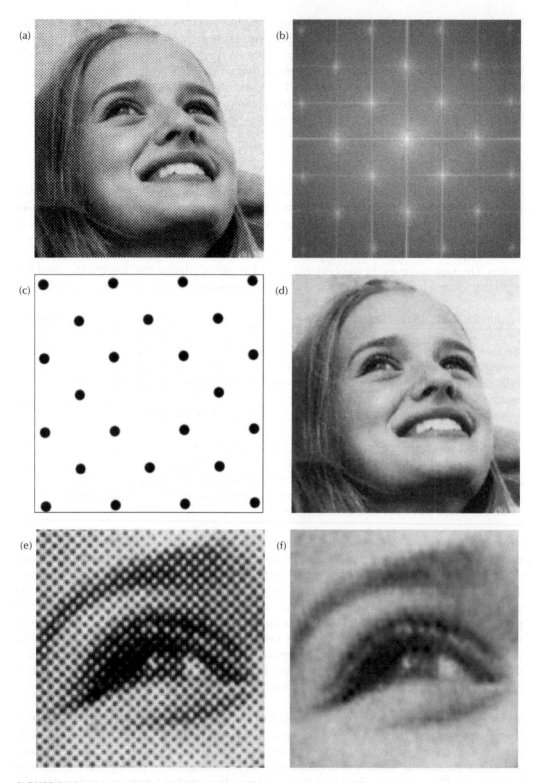

**FIGURE 3.16**
Removal of periodic noise: (a) original half-tone-printed image [print.tif], (b) FFT power spectrum showing spikes, (c) mask to eliminate the spikes, (d) reconstructed image with spikes removed, (e) enlarged detail of (a), (f) enlarged detail of (d).

larger magnitudes) near the center (low frequencies) than at the edges (high frequencies). An approximate background may be generated by applying either a large median filter, a large morphological opening (erosion followed by dilation), or a large Gaussian smooth to the power spectrum image to remove the spikes and produce a smoothly varying background. Subtracting this background levels the overall brightness as shown in Chapter 1 and makes the thresholding operation possible.

With either of these methods, the mask will typically locate the spikes but will also mark the points at the center of the power spectrum, where the Fourier transform holds the constant term that represents the average image brightness and a few low-frequency terms that represent the gradual variation of brightness in the original image. This central hole in the mask must be filled to preserve this information in the reconstructed result. Note that in Figure 3.16c the regular array of dark circles in the mask does not remove terms at the origin. The mask was created using the top hat method, and then the central disk filled.

The example in Figure 3.17 is used to illustrate the thresholding approach, and also to show that sometimes it is the periodic information in the image that should be kept and all of the nonperiodic terms that should be eliminated. The original image is cloth with a herringbone weave pattern. The original image is a color image, and the processing is performed only on the brightness values. The reconstructed result is combined with the

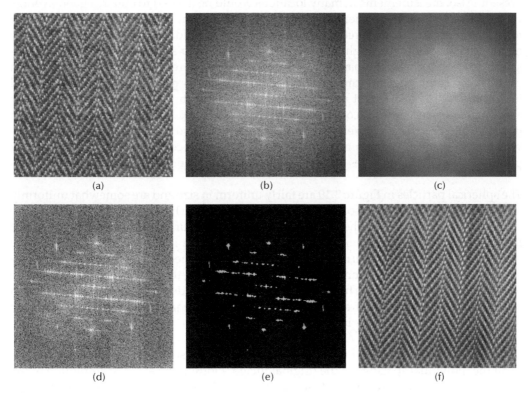

(a)      (b)      (c)

(d)      (e)      (f)

**FIGURE 3.17**
(See **Color insert following page 172**.) Reconstruction of periodic information: (a) original image [cloth.tif], (b) FFT power spectrum display, (c) background generated by applying a median filter to image (b), (d) leveled result produced by subtracting (c) from (b), (e) thresholded mask that preserves the spikes and removes other terms, (f) result obtained by inverse Fourier transform of masked terms and recombining the reconstructed brightness with original color values.

original color values for each pixel (as shown in the previous chapter) to produce the final result. Note that in this case the mask is reversed so that the spikes, which represent the periodic information in the original image, are preserved while the other terms, which correspond to the nonperiodic or random noise in the original image, are removed. In this case, of course, the terms at the center of the transform are kept as well.

### 3.2.2 Measurements

Some measurements that are difficult to make on the spatial domain image become much simpler in the Fourier domain. Figure 3.6 showed that sinusoids with a large spacing (low frequency) produce spikes in the power spectrum display that are positioned close to the origin, and vice versa. The radius in the power spectrum represents frequency, which is the inverse of spacing following the relationship shown in Equation 3.11.

$$Radius = \frac{Image\ Width}{Spacing} \tag{3.11}$$

This relationship can be used to simplify measurements in images in which spacings are only approximately regular. Figure 3.18 shows an example. The spacing of the Z-bands in the muscle is not easily measured because of the variations in contrast and the noise present. Also, measurements in many locations would be desired to obtain a good average value. However, in the Fourier transform power spectrum, the distance from the center to the first peak can be measured with good accuracy. The additional peaks at progressively higher frequencies are the higher-order terms in the Fourier series, called higher harmonics, and occur because the bands in the original image are not perfectly sinusoidal in contrast. By combining information from all of the bands present in the entire original image, the radius to the first peak in the power spectrum (19.24 pixels) represents the average frequency. The width of the original image is 512 pixels, and the spacing calculated from Equation 3.11 is 26.6 pixels. The width of the peak also provides information on the variation in spacing. Using a mask that preserves just the central peak and the first few principal harmonics generates an image that shows the band structure.

The spherical particles in Figure 3.19 are fairly uniform in size and are somewhat uniformly arranged, but not perfectly so. The FFT power spectrum shows a pattern of rings. The ring pattern indicates that the repetition of features occurs in all directions. (Anyone familiar with diffraction patterns will recognize that these images are identical to those patterns, and in fact the Fourier transform describes mathematically the physics of diffraction.) The radius of the innermost ring corresponds to the nearest-neighbor distance (outer rings correspond to the higher frequencies needed to describe the brightness pattern of the features, which is not a simple sinusoid and so requires higher-order harmonic terms in the Fourier series as in the previous example).

Measurement of the radius of the inner ring in the power spectrum provides an average of all the individual spacings in the original image. Plotting the average amplitude of the Fourier transform as a function of radius (Figure 3.19c) makes this measurement even easier. In this example, the location of the peak of brightness lies at a radius of 24.5 pixels. Using Equation 3.11, the image width is 256 pixels and the calculated spacing of the particles in the original image is 10.5 pixels. Because they are touching, this is also the particle diameter.

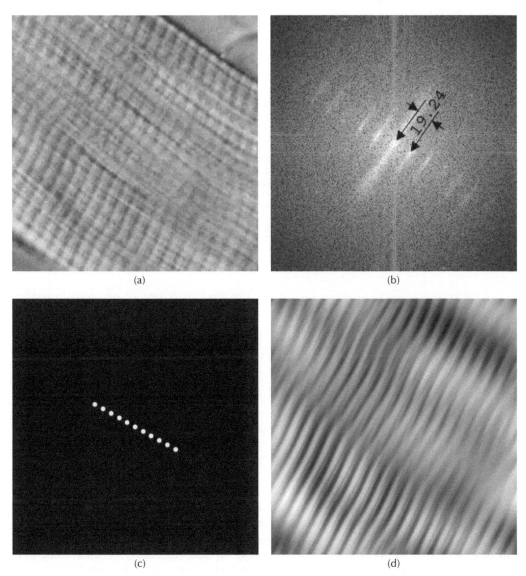

(a)

(b)

(c)

(d)

**FIGURE 3.18**
Measurement of spacing: (a) original image [muscle.tif] showing regularly spaced bands (the lines running at a slight angle to the vertical); (b) enlarged central portion of the FFT power spectrum showing the radius from the center to the first peak, which corresponds to the spacing of the bands; (c) mask that preserves the central peak and the first few principal harmonics; (d) reconstructed image showing the band structure.

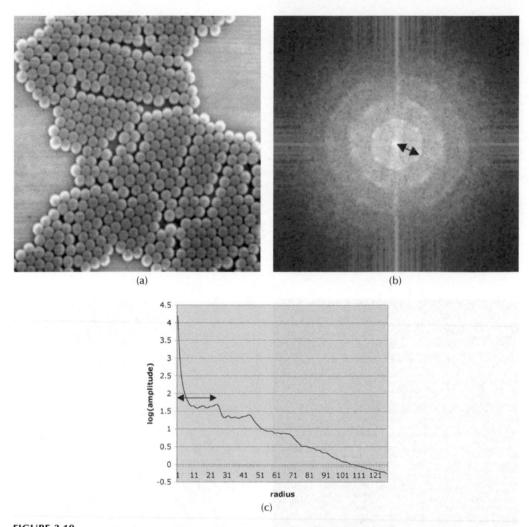

**FIGURE 3.19**
Measurement: (a) original image [particles.tif], (b) FFT power spectrum showing ring patterns with arrow indicating the radius of the innermost ring, (c) plot of average amplitude (log scale) of the Fourier transform as a function of radius with arrow indicating the location of the peak corresponding to the innermost ring.

### 3.2.3  Problems

3.2.3.1.  Use the program from Problem 3.1.5.6 to apply masks and reconstruct an image in which periodic information is either removed or preserved.

3.2.3.2.  Process the image of an FFT power spectrum by subtracting a background produced by smoothing or median filtering so that the spikes can be selected by a threshold. Use this mask to selectively remove the frequencies corresponding to the spikes, or alternatively to keep those frequencies and eliminate everything else.

3.2.3.3.  Process the image of an FFT power spectrum using a top hat filter to create a mask that corresponds to the spikes. Use this mask to selectively remove the frequencies corresponding to the spikes, or alternatively, to keep those frequencies and eliminate everything else.

3.2.3.4. Implement a program that averages the amplitude of terms in the Fourier transform as a function of frequency (radius from the origin). Write the data to a text file so that it can be plotted, for example, by using a spreadsheet program.

## 3.3 Convolution and Correlation

### 3.3.1 Convolution in the Fourier Domain

The low-pass filter applied to the Fourier transform to smooth the image in Figure 3.8 has a Gaussian profile and produces the same result as a Gaussian spatial domain filter. That is because the Fourier transform of the Gaussian spatial domain filter also has a Gaussian profile, as shown in Figure 3.20. The Gaussian is one of very few functions whose Fourier transform contains no imaginary terms, which is why the Gaussian convolution can be treated as a scalar mask.

In the general case, the Fourier transform of any spatial domain filter can be obtained, but the result will be an array of complex values with varying amplitude and phase. Multiplying this transform, point by point, by the Fourier transform of an image, and then performing an inverse transform, will produce a resulting spatial domain image that is identical to the result that would be obtained by the application of the original filter in the spatial or pixel domain. The procedure of applying a filter to an image is called *convolution*, whether it is performed in the spatial or the Fourier domains. Equation 3.12 summarizes the convolution relationship. Unlike the application of a mask in Equation 3.10, in which $H$ is a scalar (real) value, for convolution the values of both $F$ and $H$ are complex and, hence, the multiplication is complex. Convolution of an image $f(x,y)$ with a filter $h(x,y)$, indicated by the $*$ symbol can be performed by multiplying the complex Fourier transforms of each, $F(u,v)$ and $H(u,v)$, respectively.

$$f(x,y)*h(x,y) = \frac{1}{N^2}\sum_{n=0}^{N-1}\sum_{n=0}^{N-1} f(m,n)h(x-m,y-n)$$

$$f(x,y)*h(x,y) \Leftrightarrow F(u,v)H(u,v)$$

(3.12)

The advantage of performing convolution in the Fourier domain, as noted before, is that only a single multiplication is needed for each point in the transform, instead of the dozens or hundreds of multiplications per pixel that would be required to perform the equivalent operation in the spatial domain. The disadvantage is the overhead of the forward and inverse Fourier transform calculations. Code Fragment 3.8 outlines the procedure. Except for very small images and filters, the advantages outweigh the disadvantages. Note that the filter (shown as $h(x,y)$ in Equation 3.12) must be padded with zero values to be the same size as the image, so that its transform $H(u,v)$ will be the same size as the transform of the image.

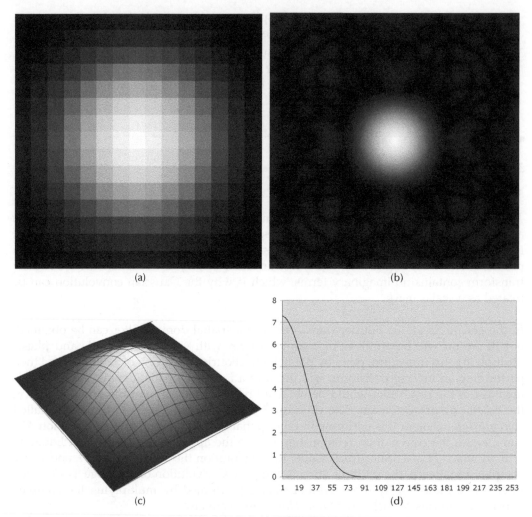

(a)

(b)

(c)

(d)

**FIGURE 3.20**
Fourier transform of a spatial domain filter: (a) values for a Gaussian filter with standard deviation of 3 pixels, represented as grayscale values (the image shows an enlargement of the individual pixels at the center of a 512-pixel-wide image); (b) amplitude (linear scale) of the FFT power spectrum of (a); (c) isometric plot of the pixel values in (a) showing Gaussian shape; (d) radial plot of the amplitude of Fourier transform terms in (b) showing Gaussian shape.

```
// Code Fragment 3.8 — Convolution
{  // ... Verify that the original and reference images are the same size
   //      and a power of two in dimension (and square)
   // ... Allocate two FFTelem lines (FL1 & FL2)
   // ... Read the original image and create a temporary complex image (#1)
   // ... Perform the forward FFT calculation on the image (#1)
   // ... Read the reference image and create a temporary complex image (#2)
   // ... Perform the forward FFT calculation on the image (#2)
   for (y = 0, y < fsize; y++)
   {
```

```
    ReadTempImageLine(tempimage1, y, FL1);
    ReadTempImageLine(tempimage2, y, FL2);
    for (x = 0; x < fsize; x++)   // convolution is complex multiplication
    {
        // complex multiplication = (a + bi) * (c + di)
        //                        = ac - bd + i(bc + ad)
        float tr = FL1[x].real * FL2[x].real - FL1[x].imag * FL2[x].imag;
        float ti = FL1[x].imag * FL2[x].real + FL1[x].real * FL2[x].imag;
        FL1[x].real = tr;
        FL1[x].imag = ti;
    } // for x
    WriteTempImageLine(tempimage1, y, FL1);
  } // for y
  // ... Perform the inverse FFT calculation on temporary complex image (#1)
  // ... Write the processed image back to the host program
}
```

Figure 3.21 illustrates convolution performed using the FFT method for the case of a directional derivative. The original kernel values are shown plotted as a grayscale image; because the kernel contains both positive and negative numbers, zero is represented as medium gray. The result of the convolution (Figure 3.21b) is identical to that produced by applying the $7 \times 7$ kernel to the spatial domain image using the procedures in Chapter 2.

### 3.3.2 Correlation

Chapter 2 also introduced the idea of correlation, showing it to be the result of performing the same operation as convolution but with the spatial target rotated by 180 degrees. To perform the equivalent procedure in the Fourier domain, the phase of each term in the complex Fourier representation of the target must likewise be shifted by $\pi$. Referring back to the diagram in Equation 3.2 shows that this is equivalent to changing the sign of both the real and imaginary parts of the complex value. Equation 3.13 shows the correlation relationship of a target $h$ with an image $f$, and its similarity to the convolution expression shown in Equation 3.12 (note the changes in sign within the summation). $F^*$ and $f^*$ are the complex conjugates of $F$ and $f$, meaning that they have been phase shifted by $\pi$.

$$f(x,y) \circ h(x,y) = \frac{1}{N^2} \sum_{n=0}^{N-1} \sum_{n=0}^{N-1} f^*(m,n) h(x+m, y+n)$$

$$f(x,y) \circ h(x,y) \Leftrightarrow F^*(u,v) H(u,v)$$

(3.13)

Figure 2.9 (Chapter 2) showed an example of ***cross-correlation***. Performing the operation in either the spatial or Fourier domains will produce identical results. Figure 3.22 shows another example, in which the target image is selected as one feature or region in the image itself, which is then padded to square power-of-two dimensions that match those of the image (also padded in this example). The Fourier transforms are then multiplied (with the phase shift) point by point and the result retransformed. The resulting image shows peaks wherever the original image contains features similar to the selected target.

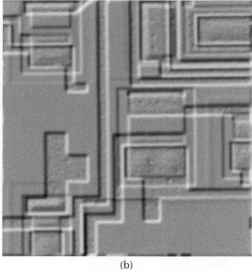

(a)

(b)

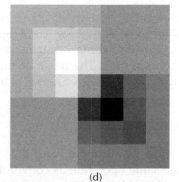

| 0.350 | 0.320 | 0.280 | 0.250 | 0.000 | 0.000 | 0.000 |
|-------|-------|-------|-------|-------|-------|-------|
| 0.320 | 0.707 | 0.625 | 0.500 | 0.000 | 0.000 | 0.000 |
| 0.280 | 0.625 | 1.414 | 1.000 | 0.000 | 0.000 | 0.000 |
| 0.250 | 0.500 | 1.000 | 0.000 | -1.000 | -0.500 | -0.250 |
| 0.000 | 0.000 | 0.000 | -1.000 | -1.414 | -0.625 | -0.280 |
| 0.000 | 0.000 | 0.000 | -0.500 | -0.625 | -0.707 | -0.320 |
| 0.000 | 0.000 | 0.000 | -0.250 | -0.280 | -0.320 | -0.350 |

(c)

(d)

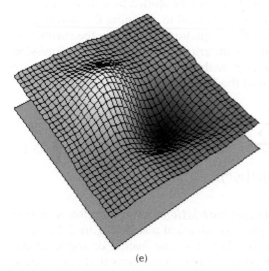

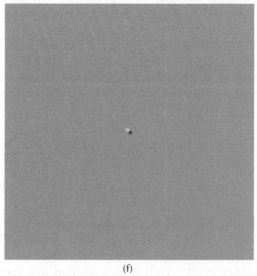

(e)

(f)

**FIGURE 3.21**
Convolution with a kernel: (a) original image [chip.tif], (b) result after application of the filter, (c) numeric values
for a 7 × 7 directional derivative kernel, (d) the values from (c) plotted as grayscale pixels, (e) graphical
presentation of the kernel values showing the positive and negative regions, (f) the kernel values padded with
intermediate gray to the same size as the original image.

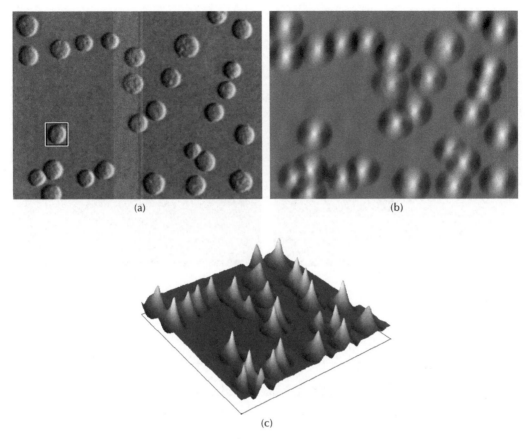

(a)

(b)

(c)

**FIGURE 3.22**

Cross-correlation: (a) original image [bubbles.tif] showing features that are illuminated from one side, and vary somewhat in size and texture; the box indicates the single bubble selected as a target; (b) result of cross-correlation, showing peaks at the location of each bubble; (c) rendered version of image (b), showing peaks whose height measures the similarity between each feature and the target.

This procedure is capable of finding features that are partially obscured, for example, by camouflage or random noise. Each peak value is a measure of the degree of match between the feature and the target. A top hat filter is often a suitable tool for locating the peaks.

If correlation is performed using an entire image as its own target (called *__autocorrelation__*) the result is similar to the example shown in Figure 3.23. One way to understand this is to imagine printing a copy of the image as a transparency and sliding the copy (rotated by 180°) over the original. When the shift amount and direction move features off themselves, a value representing the degree of match between the two drops. If the image contains similar features in a regular array, shifting the image so that one feature aligns with another produces another peak. The result of the autocorrelation is an image in which bright peaks show the predominant directions and distances of the neighbors around each particle in the original image.

Note that the cross-correlation and autocorrelation images are spatial domain results, so that locations and distances are directly in pixels rather than frequency, as was the case for the measurements on the Fourier domain power spectrum shown above. A plot (shown in Figure 3.23c) of the mean intensity as a function of radius in this autocorrelation result has peaks that identify the mean neighbor distances of first, second, third, fourth, and

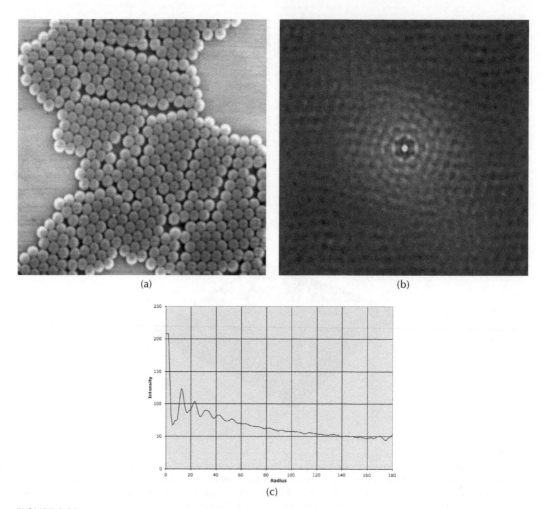

(a)                                                      (b)

(c)

**FIGURE 3.23**
Autocorrelation: (a) original image [particles.tif]; (b) autocorrelation result, an image showing the spacing and direction of neighbors; (c) plot of average intensity versus radius for the autocorrelation result in (b), with peaks showing the mean distance to first, second, third, and farther neighbors.

farther neighbors throughout the image. Analysis of this plot shows that the peaks become smaller and increase in width with neighbor distance, indicating a loss of regularity in the particle arrangement over greater distances.

### 3.3.3   Problems

3.3.3.1.   Implement a program to compute the FFT of a spatial domain filter, perform complex multiplication of the terms with those in the FFT of an image, and perform the inverse transform on the result. Read the spatial domain filter from a text file as in Chapter 2. Note that the filter array may need to be scaled to the 0 to 255 range and must be padded with an appropriate value (usually 0 or 127.5, depending on whether the filter has all positive, or positive and negative values) to have the same dimension as the image. Compare the results of applying spatial domain filters as shown in Chapter 2 with the equivalent procedure in the Fourier domain.

3.3.3.2#. Implement a program that performs convolution with the image stored as the reference image. Test the images first to verify that they are the same size and a square power of two in dimension.

3.3.3.3#. Implement a program that performs cross-correlation using the image stored as the reference image as the target. Test the images first to verify that they are the same size and a square power of two in dimension. By saving the image as the reference image before using this program, you can also perform autocorrelation.

## 3.4 Deconvolution

Equation 3.12 showed that convolution can be carried out by performing a complex multiplication of the terms in the Fourier transform of an image with the terms in a Fourier transform of a kernel, such as a smoothing function. The blurring of an image owing to imperfect optics, camera motion, or other imaging defects may also be considered to be a smoothing function, and can be similarly described as a convolution. There is an implicit assumption that the same blurring is applied to every point in the image.

This suggests that if a suitable kernel can be found that corresponds to the blurring, it might be removed from the blurred image by dividing the Fourier transform of the blurred image by that of the blur function. The inverse procedure is called ***deconvolution***. Equation 3.14 shows the relationship between convolution (top) and deconvolution (bottom). $F$ is the Fourier transform of the ideal image $f$, $H$ the transform of the smoothing or blur function $h$, and $G$ the transform of the blurred image $g$.

$$G(u,v) = H(u,v)F(u,v)$$

$$F(u,v) = \frac{G(u,v)}{H(u,v)}$$ 

(3.14)

### 3.4.1 Wiener Deconvolution

In most cases, the expression for deconvolution in Equation 3.14 is unrealistically optimistic. The actual blurred image $g(x,y)$ that is acquired always contains some random noise and, furthermore, consists of finite pixels (which average brightness over a small area rather than being an ideal point sample) that are measured with a finite precision. There are usually similar (or even more severe) problems with the blur function $h(x,y)$ if it can be measured (as discussed below). These factors limit the degree to which the ideal image $f(x,y)$ can be recovered as well as constraining the procedure for performing the deconvolution. In spite of these limitations, the basic deconvolution idea is valid and actually useful in many real situations.

The first problem in most situations is to determine the blur function $h(x,y)$. This is called the ***point spread function*** or ***PSF***, since it represents the spreading or blur that a single bright point in the ideal image would display in the blurred image. It is assumed to describe the blur that is convolved with every point in the ideal image (in other words, whatever blur is present is the same everywhere in the picture). In some cases, the PSF can actually be measured directly. For example, in astronomy, a single isolated star should be a single point in most images, but if the optics are imperfect or other image blurring occurs (for example, owing to atmospheric effects), it will appear as a feature of finite size. The image of the star is a direct measure of the PSF.

This problem of blurred images arose in the case of the Hubble telescope, whose primary mirror was fabricated with the wrong curvature and produced out-of-focus images. The problem was later partially corrected by replacing a secondary mirror with one made with a compensating curvature (which also greatly improved the light gathering capability of the instruments), but for several years the images that were obtained were blurred. During this period, deconvolution was used to produce sharp, corrected images. The PSF was calculated from measurement data on the primary mirror, which had been carefully recorded during manufacture. That is not the only example of deconvolution used in the space program: the high-resolution camera in the "Deep Impact" mission to the Tempel-1 comet in 2005 was badly out of focus, and the images were salvaged by deconvolution.

When the Hubble telescope was launched (1990), image deconvolution was still considered to be a high-end computer procedure usually carried out on mainframes or supercomputers. Advances in computer speed, incorporation of floating point arithmetic processors, availability of large memories, and improvements in displays have made these techniques practical on desktop and laptop computers. The version of Photoshop released in 2005 (CS2) includes a deconvolution procedure (called Smart Sharpening) as part of the standard package. It uses a disk of adjustable radius for the PSF, approximating the effect of a camera aperture.

Figure 3.24 shows the result of deconvolution. An image of a single star is selected as a point spread function, scaled to a maximum of 255, and padded with zero values (because the background in the image is black) to the same size as the blurred image of the galaxy. The Fourier transform of the *PSF* is then used to deconvolve the blurred image by performing complex division, and the result is inverse-Fourier-transformed to produce the reconstructed image. This image shows much more resolved detail than the blurred original.

However, the procedure used to produce Figure 3.24c is not quite that shown in Equation 3.14. In order to limit the effects of noise, and also to prevent numerical overflows that would result because of division by very small values in the $H(u,v)$ array, a ***Wiener deconvolution*** was performed. This method produces an estimate of the original image that has minimum differences from the theoretical ideal image in a least-squares sense. For the case of random additive noise in a blurred image, the Wiener method produces the best estimate of the uncorrupted, unblurred original, and it is also fairly easy to implement.

A scalar constant $K$ is added to the denominator as shown in Equation 3.15, in order to prevent numerical overflow and to limit the effects of noise on the reconstruction. The value of the constant depends on the amount of noise present in the image, but in practice this is almost never known and the value is instead treated as an adjustable parameter. It can sometimes be estimated as the standard deviation of the histogram peak measured on a portion of the image that should be uniform in brightness. Code Fragment 3.9 outlines the calculation procedure. If the magnitude of $K$ is too small, noise will dominate the reconstructed image, and if it is too large, the amount of resolution improvement will be

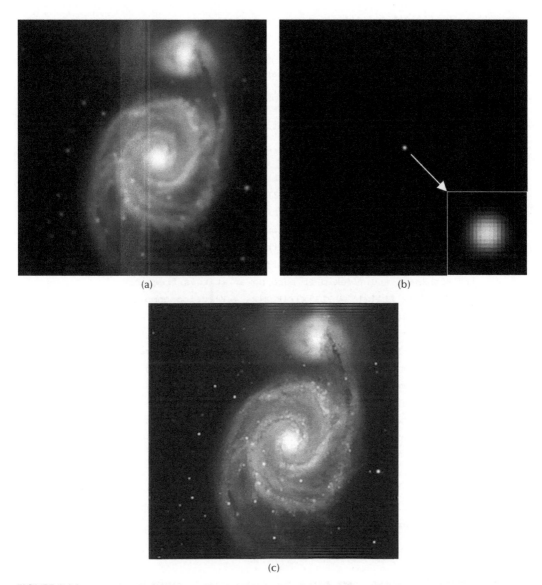

**FIGURE 3.24**
Deconvolution: (a) original out-of-focus image [galaxy.tif]; (b) image of a single star used as a point spread function [psf.tif], with enlargement to show pixel-level detail; (c) deconvolved result.

small, as illustrated in Figure 3.25. Note that the reconstruction $F$ is now shown with a hat to indicate that it is an estimate of the true function.

$$\hat{F}(u,v) = \left[ \frac{1}{H(u,v)} \cdot \frac{\left| H(u,v) \right|^2}{K + \left| H(u,v) \right|^2} \right] G(u,v) \tag{3.15}$$

```
// Code Fragment 3.9 — Wiener deconvolution
{  // ... Verify that the original and reference images are the same size
   //       and a power of two in dimension
```

```
// ... Read the Wiener constant from a text file
// ... Scale the constant according to the image area
// ... Allocate two FFTelem lines (FL1 & FL2)
// ... Read the original image and create a temporary complex image (#1)
// ... Perform the forward FFT calculation on the image (#1)
// ... Read the reference image and create a temporary complex image (#2)
// ... Perform the forward FFT calculation on the image (#2)
for (y = 0, y < fsize; y++)
{
    ReadTempImageLine(tempimage1, y, FL1);
    ReadTempImageLine(tempimage2, y, FL2);
    // complex divide with WEINER_K. (If K==0, standard complex divide)
    for (x = 0; x < fsize; x++)   // deconvolution is complex division
    {   //compute magnitude squared for the PSF frequency
        float fftL2Mag2 = FL2[x].real*FL2[x].real + FL2[x].imag*FL2[x].imag;
        if (fftL2Mag2 > 0) //skip if zero power at this frequency
        {   //compute FL1[x]' = FL1[x]/(FL2[x] + K) using complex division
            float tempImag = FL2[x].imag * ((fftL2Mag2+WIENER_K)/fftL2Mag2);
            float tempReal = FL2[x].real * ((fftL2Mag2+WIENER_K)/fftL2Mag2);
            float fftL2Mag2PlusK = (tempReal*tempReal + tempImag*tempImag);
            float tempr = FL1[x].real * tempReal + FL1[x].imag * tempImag;
            float tempi = FL1[x].imag * tempReal - FL1[x].real * tempImag;
            FL1[x].real = (tempr / fftL2Mag2PlusK);
            FL1[x].imag = (tempi / fftL2Mag2PlusK);
        } // if any power in PSF at this frequency
    } // for x
    WriteTempImageLine(tempimage1, y, FL1);
} // for y
// ... Perform the inverse FFT calculation on temporary complex image (#1)
// ... Write the processed image back to the host program
}
```

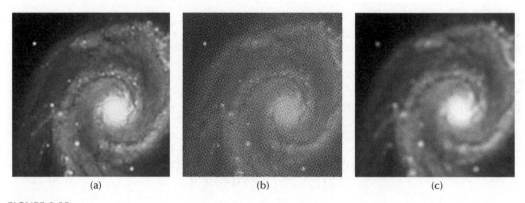

(a)                              (b)                              (c)

**FIGURE 3.25**
Effect of the Wiener constant (detail from the example in Figure 3.24): (a) good result; (b) constant too small, noise dominates result; (c) constant too large, little improvement.

## 3.4.2 The Point Spread Function

Direct measurement of the *PSF* is not always practical. In the case of the Hubble telescope, as mentioned above, the *PSF* was actually determined by calculation based on the measurements of the mirror curvature. Sometimes, capturing the image of a single point of light using the same optics can be accomplished. In other situations, an image of a known subject can be captured. If a test pattern with good contrast and sharply defined edges is imaged with very little noise, the *PSF* (or more specifically, the Fourier transform of the *PSF*, which is called the modulation transfer function) can be obtained by solving Equation 3.14 for $H(u,v)$. This can then be used in turn to deconvolve other blurred images acquired with the same optics. Because division is performed twice in this approach, the deleterious effects of any random noise are increased.

Motion blur results from the motion of either the camera or subject during exposure. Figure 3.26 shows an example, a picture taken from a low-flying airplane. In this case the *PSF* consists of a line corresponding to the motion vector of the plane. The vector's length and angle can be estimated rather closely by examining the blur pattern of small features in the image, or in principle can be calculated by knowing the plane's direction and speed and the exposure interval. A Wiener deconvolution using an image of the motion blur vector results in a great improvement in image sharpness. The figure also illustrates the effects of using too large or small a constant $K$.

It is also worth noting in Figure 3.26 that the reconstruction of the image is poorer at the top right and bottom left (the defect is even more apparent in the noisy result in Figure 3.26e). That is because in this instance the camera was also apparently rotating slightly during the exposure, and so the blur vector or PSF is not perfectly uniform across the entire image. The deconvolution technique relies on the assumption that the PSF is the same everywhere, and so this method cannot be applied in cases such as a blurred image of a running person, whose arms, legs, and torso all have different motion. It also cannot deal with out-of-focus images in which the amount of out-of-focus blur varies from place to place, for instance, due to depth-of-field limitations in a photograph. In some cases it may be possible to subdivide an image into many parts and perform deconvolution on each part separately.

Other deconvolution approaches exist besides the Wiener method. For example, the apodization technique avoids numeric overflows by eliminating (reducing to zero) the amplitude of any terms in the Fourier transform array that would become too large as a consequence of dividing by a small value in the transform of the *PSF*. This approach works well in most cases and is very easy to use. The Wiener method is the most widely used because of its good results and efficient implementation.

There are numerous approaches to modeling a shape for the *PSF* with equations having parameters that are iteratively adjusted as deconvolutions are performed. These iterative methods typically have some difficulty in determining when to stop the iteration, and in any event are rather time consuming. A comprehensive treatment of iterative deconvolution methods can be found in (Lagendijk and Biemond 1991; Campisi and Egiazarian 2007). Fortunately, many optical defocus problems (which are the greatest single cause of blurred images) can be modeled by a *PSF* that is approximately Gaussian in shape. With only a single parameter (the standard deviation) to adjust, the Gaussian is a relatively simple model to use, and works well enough in many cases to produce a significant improvement in the reconstructed image. As noted before, another of the advantages of

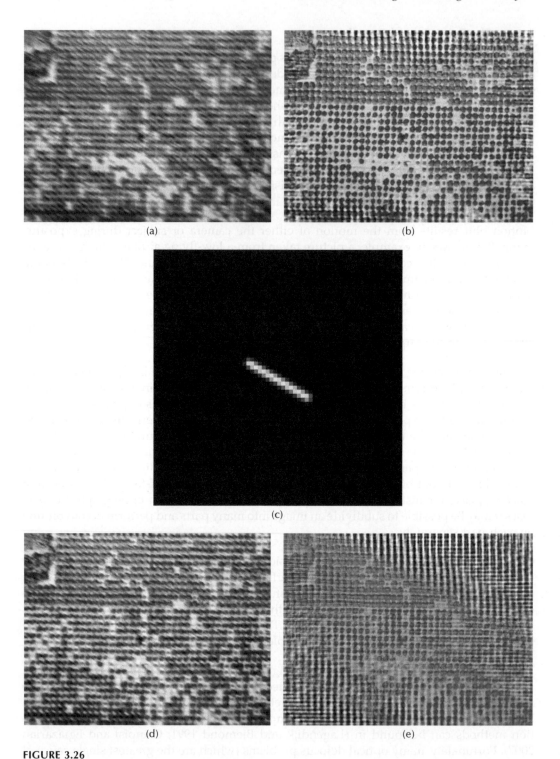

(a)

(b)

(c)

(d)

(e)

**FIGURE 3.26**

(See **Color insert following page 172.**) Deconvolution of motion blur: (a) original image [orchard.tif]; (b) result of Wiener deconvolution; (c) point spread function showing the motion vector [motion.tif], enlarged to show individual pixels; (d) deconvolution result using too large a $K$ value (image is still blurred); (e) deconvolution result using too small a $K$ value (noise dominates).

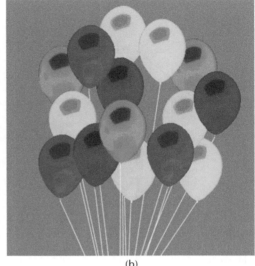

**FIGURE 0.3**
Reversing the values in a color image: (a) original [balloons.tif], (b) "inverse" or complement.

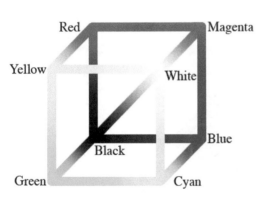
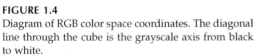

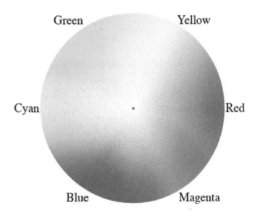

**FIGURE 1.4**
Diagram of RGB color space coordinates. The diagonal line through the cube is the grayscale axis from black to white.

**FIGURE 1.5**
The color wheel, with uniformly spaced R, G, and B.

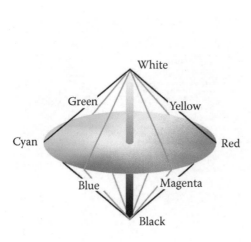

**FIGURE 1.6**
Diagram of bi-conical hue–saturation–intensity color space.

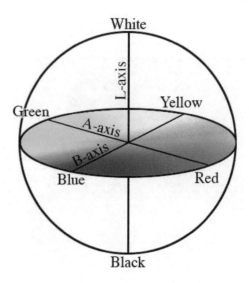

**FIGURE 1.8**
Diagram of a spherical LAB color space.

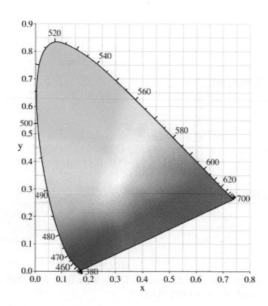

**FIGURE 1.10**
CIELab color. (The intensity axis is perpendicular to the plane shown.)

**FIGURE 1.11**
Maximizing image contrast: (a) original image [rose.tif], (b) stretching RGB histograms individually (note the yellow color cast), (c) using a single maximum and minimum for all channels, (d) stretching the luminance.

**FIGURE 1.12**
Stretching contrast with clipping: (a) original image [manuscript.tif], (b) result of stretching with clipping.

**FIGURE 1.18**
Setting the limits for each channel on the RGB values for the darkest and lightest pixels in the image: (a) original [anemone.tif], (b) result.

(a)           (b)

**FIGURE 1.22**
Application of tristimulus correction: (a) original image with color standards [sandpaint.tif], (b) corrected result.

(a)           (b)

**FIGURE 1.24**
Correcting vignetting: (a) original image [ocean.tif], (b) corrected result.

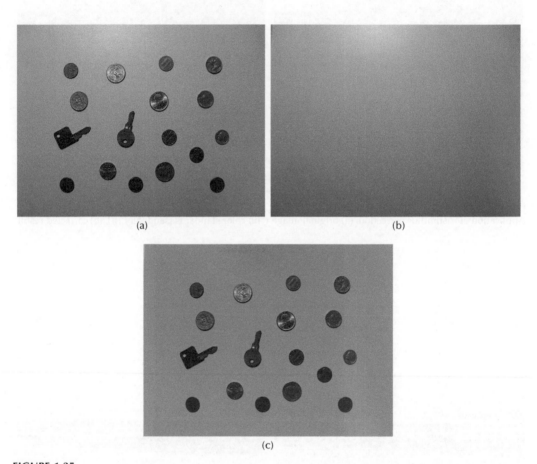

**FIGURE 1.25**
Correcting nonuniform lighting: (a) original [coins.tif], (b) background [background.tif], (c) leveled by subtraction.

(a)

(b)

(c)

**FIGURE 1.33**
Absolute difference: (a) reference image [clock1.tif], (b) original image [clock2.tif], (c) absolute difference.

(a)

(b)

(c)

(d)

**FIGURE 1.35**
Applying functions to pixel values: (a) original image [rose.tif], (b) conventional reverse applied to RGB channels,
(c) reversing the intensity only, (d) square root function applied to intensity values.

**FIGURE 2.1**
Example of smoothing random noise by applying the kernel in Table 2.1 to an image: original color image [bird.tif].

(a)                                                                                        (b)

**FIGURE 2.6**
Example of sharpening detail in the same image as Figure 2.1 using the Laplacian kernel shown in Table 2.2: (a) application to the L channel only, (b) application to the individual R, G, and B channels (note the introduction of extraneous colors for the pixels).

**FIGURE 2.21**
Displaying the vector orientation for the Sobel brightness gradient: hue values (with the brightness representing the vector magnitude).

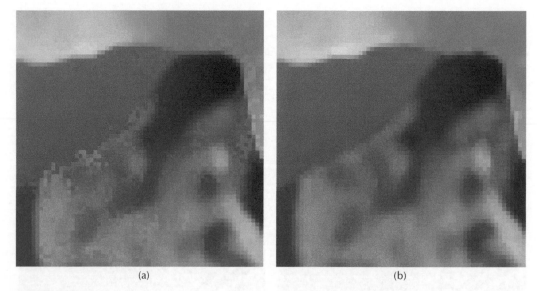

(a)                (b)

**FIGURE 2.31**
Enlarged detail from the [rose.tif] image: (a) median calculated using the brightness of each pixel and carrying the color values along, (b) median calculated using the vector median in RGB coordinates. Compare the pixels near the edge of the red petal.

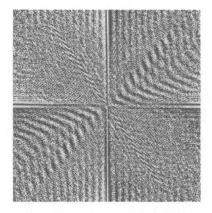

**FIGURE 3.7**
Fourier transform displays: (c) phase values shown as hue.

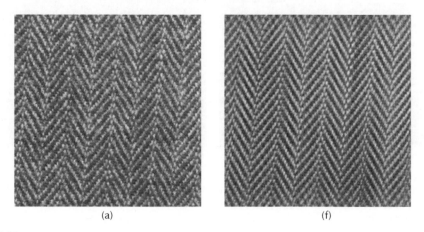

(a)　　　　　　　　　　　　　　　　　(f)

**FIGURE 3.17**
Reconstruction of periodic information: (a) original image [cloth.tif], (f) result obtained by inverse Fourier transform of masked terms and recombining the reconstructed brightness with original color values.

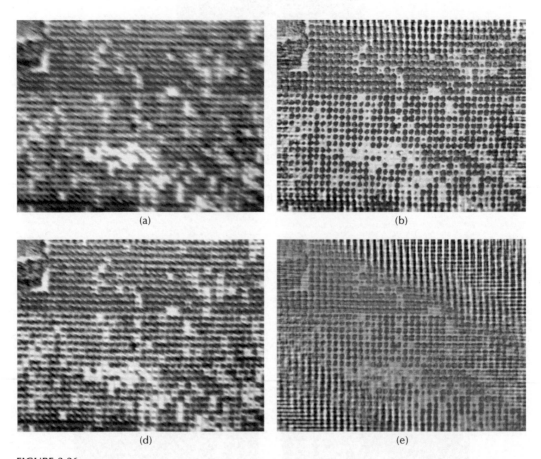

(a)

(b)

(d)

(e)

**FIGURE 3.26**
Deconvolution of motion blur: (a) original image [orchard.tif], (b) result of Wiener deconvolution, (d) deconvolution result using too large a *K* value (image is still blurred), (e) deconvolution result using too small a *K* value (noise dominates).

**FIGURE 3.32**
JPEG compression: (a) the [rose.tif] image after JPEG compression; (b) enlarged detail showing 8 × 8 pixel blocks; (c) changes in the image, produced by subtracting the original image from the JPEG version; (g) hue values in the JPEG version.

**FIGURE 4.10**
Thresholding a color image: (e) thresholded result in which background pixels are set to zero but feature pixels are left unchanged.

**FIGURE 4.11**
Multichannel thresholding: (a) original image [balloons.tif].

**FIGURE 4.37**
Measurement with a grid: (a) section through layers of paint [paint.tif].

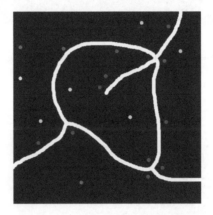

**FIGURE 4.42**
Distance measurement using the *EDM*: (c) result of combining the *EDM* values with a binary image of the malls in Figure 4.41. The brightness values are shown with false colors that vary from magenta to red, and the roads from image (a) have been superimposed, to assist in visual interpretation of the distances.

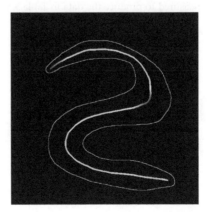

**FIGURE 4.43**
Measuring variable feature width: (e) the *EDM* values assigned to the skeleton. The skeleton has been dilated for visibility and the pixel grayscale values are shown in false color with the original feature outline superimposed to show the relationship between the feature width and the resulting values.

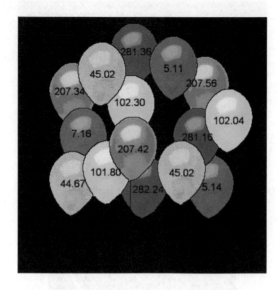

**FIGURE 5.32**
Measuring the mean hue in each feature [balloons.tif]: (b) combining the original image with the binary image. The numeric values correspond to an angle (0 to 360°) on the color wheel, and are shown superimposed on the original image for convenient reference.

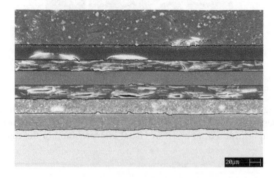

**FIGURE 5.33**
Layers in a cross section of paint, measured to produce the data in Table 5.5 [paint.tif].

**FIGURE 5.37**
Image of features for classification [nuts.tif].

a Gaussian model for the PSF is that the Fourier transform of a Gaussian is another Gaussian, with only real terms.

The use of a Gaussian *PSF* to model the blur function and perform deconvolution is illustrated in Figure 3.27. The result is nearly as good as that produced by using the measured *PSF* in this instance. Deconvolution with a Gaussian having too large a radial standard deviation produces a result showing ringing as illustrated in Figure 3.27c. The appearance of these rings can be used in an interactive or iterative procedure to estimate the best *PSF* model.

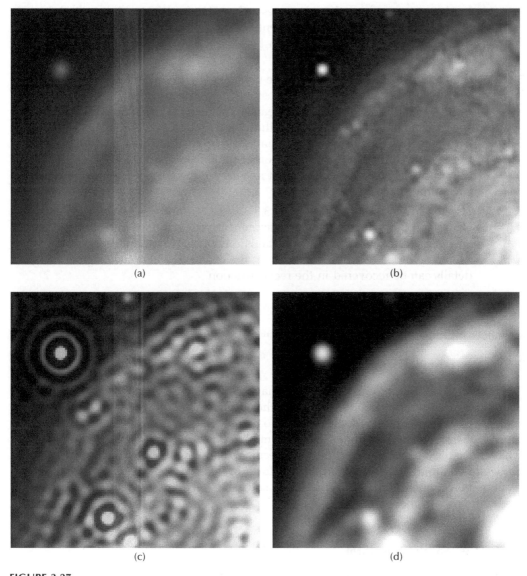

(a)

(b)

(c)

(d)

**FIGURE 3.27**
Deconvolution with a Gaussian PSF: (a) original (enlarged detail of Figure 3.24a), (b) deconvolved using a Gaussian PSF with $\sigma$ = 2.75 pixels, (c) deconvolved using a Gaussian PSF with $\sigma$ = 5 pixels (note ringing), (d) unsharp mask applied to original image.

It is instructive to compare the results of deconvolution with a Gaussian *PSF* with those from the unsharp mask operation shown in Chapter 2, which also uses a Gaussian function. The deconvolution procedure, which is accomplished by performing a complex division of the Fourier transforms, removes blur and recovers resolution of fine detail in the reconstructed image. The unsharp mask procedure (Figure 3.27d) subtracts the result of applying a Gaussian smooth in the spatial domain, and simply increases the contrast for the existing detail, without recovering any of the finer detail.

### 3.4.3   Problems

3.4.3.1   Implement a program that generates a Gaussian function with standard deviation value read from a text file, centered in the image. Scale the values to fit the 0 to 255 range. The result can be saved as the reference image for the following problems.

3.4.3.2#.  Implement a program that uses the stored reference image as a point spread function to deconvolve an image using Wiener deconvolution. Compare the results with different values of $K$ (read from a text file). $K$ is most conveniently specified as the approximate amplitude of the noise in the image, on the 0 to 255 scale. It can sometimes be estimated as the standard deviation of the peak in the histogram of pixels in a region of the image that should be uniform in brightness. Test to ensure that both images are the same size, with power-of-two dimensions.

3.4.3.3.  Use the programs from Problems 3.3.3.2, 3.4.3.1, and 3.4.3.2 with generated Gaussian point spread functions having different standard deviations to blur an image by convolution, add various amounts of random noise, and then deconvolve it. Note the limitations on the extent to which the original image details can be recovered in the reconstruction.

## 3.5   Other Transform Domains

The Fourier domain is used extensively in image processing, but is not the only alternative way to represent the image information. Other domains are used because they organize the image data in ways that make it easier to access, visualize, process, or measure some aspects of the information. Many possible mathematical transformations exist, some of which have been shown to be useful in specific situations. The wavelet transform has a particular advantage over the Fourier transform in that it does not make the assumption that the image repeats on all directions, and so is able to better process values right up to the boundaries of the image.

### 3.5.1   The Wavelet Transform

The Fourier transform constructs an arbitrary function (such as an image) as a summation of sinusoids, which continue indefinitely beyond the image. The sine functions are said to be the ***basis functions*** for the transform. The ***wavelet transform*** uses basis functions (there are many different choices) that are localized, so that the summation of the terms produces values only within the image boundaries. This limited area of coverage is described by mathematicians as "compact support" (the sinusoids used in the Fourier transform have "extended support").

Similar to the sine functions in the Fourier transform, the basis functions used in the wavelet transform constitute a series of terms with smaller and smaller horizontal scales (comparable to the increasing frequencies of the sines), but in addition to this series, because the basis functions are localized, there must also be a series covering all of the various positions of each function across the width of the image.

Consequently, even for a one-dimensional waveform, the wavelet summation is written as shown in Equation 3.16. The function $f(x)$ is expressed as a double summation over values of $s$ scalings and $t$ translations of amplitude coefficients $c$ times the scaled and translated copies of the mother wavelet $g(x, s, t)$. There are many possible functions that can be used as the basis function for the wavelet transform (Daubechies 1992). The selected *mother wavelet* $g(x,s,t)$ is scaled (reduced in lateral dimension) and translated (shifted to lateral positions).

$$f(x) = \sum_{s=0}^{S} \sum_{t=0}^{T} c_{s,t} \cdot g(x,s,t) \tag{3.16}$$

The simplest of the functions used as a mother wavelet is the Walsh function, also called the Haar wavelet, which satisfactorily demonstrates the basic characteristics of the procedure. As shown in Figure 3.28, it is simply a step function, and the most convenient way to scale this mother wavelet is by factors of 2. The extension of these functions to fit a two-dimensional image becomes a set of rectangular and square step functions in sizes that range downward from the full image area to the size of a single pixel, and the summation in Equation 3.16 must be extended to cover four indices, one each for scaling and translation in the $x$- and $y$-directions.

As with the Fourier transform, the wavelet transform is separable so that a one-dimensional transform can be applied to rows and to columns of an image, and there is a fast version of the wavelet transform (FWT) that requires the image to have power-of-two dimensions so that the factor-of-two scaling can be performed. As with the FFT, padding is used when necessary to expand an image to those dimensions.

As shown in Code Fragment 3.10, the one-dimensional Haar wavelet transform is calculated by repeatedly halving the number of values by averaging down the data, and finding the difference between the left and right halves. The forward and inverse transforms are symmetrical except for a change in sign. The procedure generates a one-dimensional transform in which the first value is the overall average of the original function, and the next two values are the differences of the left and right halves of the array from the average. Then, the doubling and differencing procedure is repeated until the individual pixel differences are reached. Because of the use of averages at each stage, many of the values in the transform are very small. Code Fragment 3.11 uses this one-dimensional transform as a subroutine to apply to a two-dimensional image.

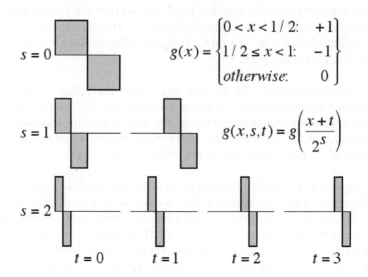

$$s = 0 \qquad g(x) = \begin{cases} 0 < x < 1/2: & +1 \\ 1/2 \le x < 1: & -1 \\ otherwise. & 0 \end{cases}$$

$$s = 1 \qquad g(x,s,t) = g\left(\frac{x+t}{2^s}\right)$$

$$s = 2$$

$$t = 0 \qquad t = 1 \qquad t = 2 \qquad t = 3$$

**FIGURE 3.28**
The Haar wavelet family. The mother wavelet (top line) is scaled (subscript *s*) and translated (subscript *t*) to form a complete set of basis functions, which when multiplied by appropriate amplitude coefficients and summed reconstruct the original signal being modeled.

```
// Code Fragment 3.10 — One-dimensional wavelet transform
void Haar_2 (float *data, long N, long direction)
//N is the number of points
{  // generate Haar coefficients for each iteration
   long   i, j, halfN;
   float  h0, h1;
   float  *tmp;              // temporary array for transform
   h0 = h1 = 0.5;
   tmp = CreateAPointer(N, sizeof(float));
   halfN = N / 2;
   if (direction == 1)      // direction = FORWARD
   {
      i=0;
      for (j = 0; j < N - 1; j += 2)
      {
         tmp[i]       = h0 * data[j] + h1 * data[j+1];
         tmp[i+halfN] = h0 * data[j] - h1 * data[j+1];
         i++;
      } // for j
   }
   else                     // direction = INVERSE
   {
      i = j = 0;
      h0 = h1 = 1.0;
      do
      {
```

```
         tmp[j]        = h0 * data[i] + h1 * data[i+halfN];
         tmp[j+1]      = h0 * data[i] - h1 * data[i+halfN];
         j += 2;
         i++;
      } while (i < halfN);
   }
   for (i=0; i<N; i++)         // copy data back to original array
      data[i] = tmp[i];
   DisposeAPtr(tmp);
} // Haar_2

void fwt_1d (float *data, long N, long direction)
{
   long  thisN, cptr;
   if (N > 2)
   {
      if (direction == 1)    // direction = FORWARD
      {
         thisN = N;
         while (thisN >= 2) // iterate through decreasing N's
         {
            Haar_2(data, thisN, direction);
            thisN /= 2;
         } // while
      } // FORWARD
      else                   // direction = INVERSE
      {
         thisN = 2;
         cptr = 0;
         while (thisN <= N)  // iterate through increasing N's
         {
            Haar_2(data, thisN, direction);
            thisN *= 2;
            cptr++;
         } // while
      } // INVERSE
   } // if N > 2
} // fwt_1d

// Code Fragment 3.11 — Two-dimensional wavelet transform
void fwt_2d (long direction) // 1 = FORWARD, 0 = INVERSE
{
   long  x,y;               // loop counters
   float *wline;            // one line or column
   long dimension;          // size of wavelet image (height & width)
   // ... read dimension = width of image (square, power of 2)
   // ... create temp image, read original,
```

```
// ... calculate brightness, write to temp
wline = CreateAPointer(dimension, sizeof(float));
// transform the rows
for (y = 0; y < dimension; y++)
{
   ReadTempImageLine(ID, y, dimension, wline);
   fwt_1d(wline, dimension, direction);
   WriteTempImageLine(ID, y, dimension, wline);
}
// transform the columns
for (x = 0; x < dimension; x++)
{
   ReadTempImageColumn(ID, x, dimension, wline);
   fwt_1d(wline, dimension, direction);
   WriteTempImageColumn(ID, x, dimension, wline);
}
DisposeAPointer(wline);
} // fwt_2d
```

One important difference from the Fourier transform is the fact that the wavelet transform produces real numbered values rather than complex ones. That makes it possible to directly display the result, by appropriately scaling the numbers so that they fit into the 0 to 255 range. Figure 3.29 shows an example. The diagram in Figure 3.30 indicates the arrangement of values in the transform. The multiple copies of the original image are reduced to different sizes, and each one contains just the differences in values at that scale from the copy that is reduced by a further factor of 2. The single point at the upper left

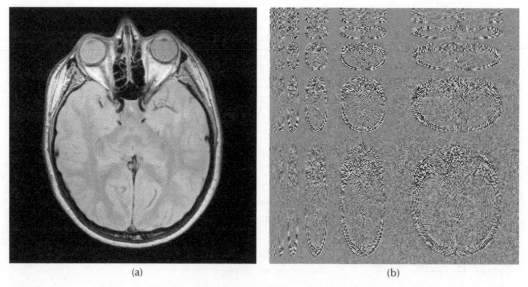

(a)　　　　　　　　　　　　　(b)

**FIGURE 3.29**
Wavelet transform: (a) original image [mri.tif], (b) transform (contrast increased to show details).

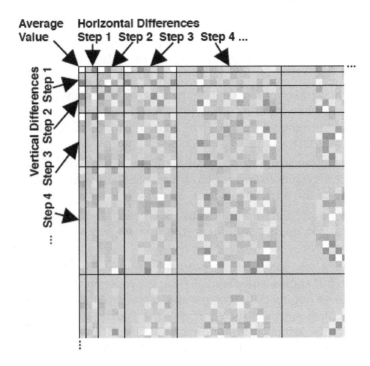

**FIGURE 3.30**
Diagram of the arrangement of values in the Haar wavelet transform.

contains the mean brightness value for the entire image. One way to think of this hierarchical representation of the data is as a pyramid of copies, each copy reduced by a factor of two progressing up the pyramid and each containing just the differences between the current copy and the one above it.

When the values are displayed as an image, it becomes possible to edit them, and to perform an inverse transform to recover a processed image. (Note that if the values have been scaled to produce a viewable display, the scaling factors must be preserved to enable the inverse to be performed.) If all of the terms in the transform are used, the inverse transform procedure shown in Code Fragment 3.11 will recover the original image exactly. Selectively removing or reducing terms accomplishes filtering, just as it does for the Fourier transform. The Haar mother wavelet is not optimum for visual filtering of images, but there are many others to choose from. For example, a popular mother wavelet is the Daubechies-4 function. Instead of the weight values that generate the Haar functions, shown in Code Fragment 3.10 with values of +0.5 and –0.5 applied to two pixels, the Daubechies-4 function treats a 4-pixel-wide region using weights of $\{(1+\sqrt{3})/4, (3+\sqrt{3})/4, (3-\sqrt{3})/4, (1-\sqrt{3})/4\}$.

One current use of the wavelet transform is in the identification of fingerprints. Because the spacing and width of the ridge markings that produce fingerprints is very consistent, only frequencies (spacings) in that narrow range are needed to produce useful images of the fingerprints. The national fingerprint database uses a wavelet transform to keep just the important range of terms, and discards the others. This requires much less storage space than keeping the full images of fingerprints. The next section considers the more general topic of image compression.

### 3.5.2   Problems

3.5.2.1#. Implement a program to calculate and display the two-dimensional Haar wavelet transform. You will have to scale the values to fit the 0 to 255 range of the display.

3.5.2.2. Implement a program to calculate the inverse Haar wavelet transform. Note that because the values of the terms were scaled to fit the 0 to 255 display range, your program will have to be able to read the scale factors to generate the reconstructed result.

3.5.2.3. Use the programs from Problems 3.5.2.1 and 3.5.2.2 with manual editing of the transform values to erase or modify selected terms and compare the results of filtering to the original image.

3.5.2.4#. Implement a program that performs the forward and inverse Haar wavelet transform that applies automatic filtering to remove (set to zero) small-amplitude terms.

3.5.2.5. Implement a program that performs the forward and inverse Haar wavelet transform but applies automatic filtering to proportionately reduce the amplitude of lower-frequency (larger-spacing) terms, by analogy to the ideal inverse filter described above in the context of the Fourier transform. Note that the location of the terms corresponding to large spacings/low frequencies is in the upper-left corner of the transform, rather than the center as in the case of the Fourier transform arrangement.

3.5.2.6 Modify the programs from the previous problems to utilize the forward and inverse Daubechies-4 wavelet transform. Note that because this method uses four pixel values rather than two, it must access values beyond the end of each row and column in the array. Use the same method described in Chapter 2 for repeating the end values in each line.

---

## 3.6   Compression

Images are large, and so are the files they produce. A 3000 × 4000 pixel image from a modern digital still camera, with R, G, B color values for each pixel each requiring 2 bytes (typical cameras use a 12-bit digital conversion because the sensors provide more than 8 bits of dynamic range) requires 72 MBytes. In order to reduce the size of these files, to save memory space in the camera and on the computer disk, and to make it easier to transmit copies of the image in e-mails and speedup downloads from Web pages, image compression is widely used. It is important to understand that all of the compression techniques that produce any significant reduction in file size work by discarding some information from the image. They are called "lossy" compression methods for that reason.

*Lossless compression* (such as that used in zip archives) finds repetitive patterns of values in the file and replaces them with codes. This works very well for compressing text files, which contain only a relatively small number of possible characters (letters, numbers, and punctuation) which furthermore occur in patterns that are often repeated in the file (common words and word fragments). Such methods produce compressed files that can be used to exactly reconstruct the original. But images contain many possible values and very few repetitive patterns, so that only slight reductions in file size can be achieved in this way, typically, less than 50% (2:1).

The goal of *lossy compression* methods is to achieve significantly greater amounts of compression by discarding some of the information. The intent is always to discard those details that are least important to human visual interpretation and recognition of the image contents. Of course, this works best for images of familiar subjects that are easily recognized. Lossy techniques must not be applied if the images are to be subsequently used for any measurements or other legal or analytical purposes. Compression factors of 10:1 and higher are common.

### 3.6.1 Lossless Compression

Black and white images or ones consisting of only a few discrete gray levels, such as typical fax transmissions, can be effectively compressed to require a much smaller amount of data than the original pixel array by using run-length encoding (Hunter and Robinson 1980). Instead of transmitting each pixel value, the image is represented as a series of horizontal line segments of uniform brightness (or color). A table of the starting position $(x, y)$, length, and value of each line segment can be used to exactly reconstruct the original. This corresponds to the way fax machines scan and print images. This *run-length encoding* is suitable for representing computer graphics and cartoons, which consist of large areas of uniform colors. It is also useful for thresholded binary images and will be illustrated again in Chapter 4. However, it does not work well for images containing many brightness or color values, because the table is often larger than the original pixel array.

A combination of delta coding and variable-length coding provides a useful illustration of a method that is more suitable for images. The first step (delta coding) replaces the value of each pixel (except for the first one in each line) with the difference between that pixel and the preceding one. These difference values may vary for an 8-bit image from +255 to 255, but most of the delta values will be small. By itself, delta coding does not reduce the size of an image file, but it makes the second step in the procedure more practical.

A situation in which a few values are relatively common and others occur only infrequently can be represented by a set of code values that use short code lengths for the common values and longer codes for the less common ones. Variable-length coding techniques, of which Huffman code is the earliest form (Huffman 1952), provide a mechanism to select the appropriate codes.

Perfect coding would produce a representation of an image in which the number of bits per pixel is just the entropy of the original image (defined previously in Chapter 2, Equation 2.6), a limit set by information theory. Variable-length code representation would only achieve the theoretical optimum in the specific case that the values being coded had the exact frequencies of occurrence of (1/2, 1/4, 1/8, ...). Figure 3.31 shows an example image, with the original histogram and the result after delta-coding. Although the original grayscale image uses 8 bits per pixel, the entropy of the delta-coded result indicates the theoretical possibility of encoding it with only 5.76 bits per pixel on average. The actual compression achieved for the image in the figure is 5.82 bits per pixel, a potentially useful reduction from the original 8 bits per pixel. Images from the Voyager spacecraft were reduced from the 8 bits per pixel recorded by the camera to an average of slightly more

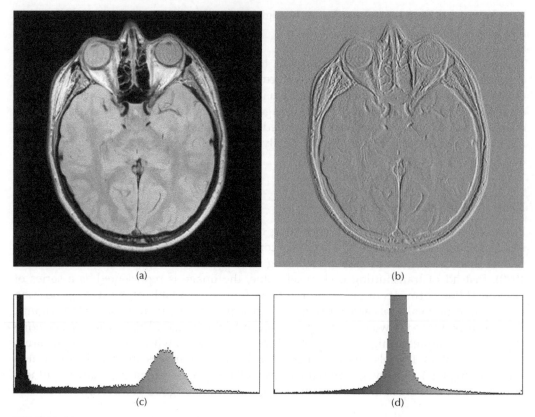

FIGURE 3.31
Delta-coding an image: (a) original [mri.tif], (b) after delta coding (display shifted to represent a zero difference as an intermediate gray), (c) histogram of image (a), (d) histogram of image (b).

than 3 bits per pixel for transmission back to earth by a combination of delta and variable-length coding.

The relative frequencies of the different pixel values (as summarized in the histogram) can be used to build a code table. Ideally, a table would be constructed for each image (which would then need to be stored or transmitted with it); in practice, after delta coding it is often adequate to use a single standardized table that is good enough. The actual code table is not unique, and many equivalent forms can be used. Table 3.2 illustrates the construction of a Huffman code for the simplified case of a 400-pixel image with 8 discrete pixel values. Representing the individual pixel values without compression would require 3 bits per pixel (**000, 001, ... , 110, 111**).

In the table, the eight groups of pixels correspond to a histogram that tallies the number of pixels with each value (in the example, these represent the delta code results, but the same procedure can be applied to the original pixel brightness values). In the first step, the two groups with the lowest frequency of occurrence are compared, and a **1** is assigned to whichever has the greater frequency and a **0** to the group with the lower frequency. Then, the counts of pixels in these groups are combined, reducing the number of groups by one. The process is repeated, comparing the two smallest groups, assigning **0** and **1** codes, and combining the counts, for a total of 7 steps.

**TABLE 3.2**

Construction of a Variable Length Code

| Pixel Value | -3 | -2 | -1 | 0 | +1 | +2 | +3 | +4 |
|---|---|---|---|---|---|---|---|---|
| Frequency | 0.065 | 0.1025 | 0.205 | 0.24 | 0.155 | 0.1325 | 0.0925 | 0.0075 |
| Number of Pixels | 26 | 41 | 82 | 96 | 62 | 53 | 37 | 3 |
| Step | | | | | | | | |
| 1 | 1 | | | | | | | 0 |
| | (29) | | | | | | | |
| 2 | 0 | | | | | | 1 | |
| | | | | | | | (66) | |
| 3 | | 0 | | | | 1 | | |
| | | | | | | (94) | | |
| 4 | | | | | 0 | | 1 | |
| | | | | | | | (128) | |
| 5 | | | 0 | | | 1 | | |
| | | | | | | (176) | | |
| 6 | | | | 0 | | | 1 | |
| | | | | | | | (224) | |
| 7 | | | | | | 0 | 1 | |
| | | | | | | | (400) | |
| Group Code | 10111 | 010 | 00 | 01 | 011 | 110 | 1111 | 00111 |

To construct the code for each group, begin at the top and follow a path to the bottom, producing a string of the successive digits. If and when the group is assigned a **0** and merged with another, the path follows down that group's column. Thus, the first column in the table (value = −3) begins with a **1** (step 1) and **0** (step 2) but then jumps to the value = +3 column and continues with **1** (step 4), **1** (step 6), **1** (step 7) for a final code sequence of **10111**. Note that when combined as a string of bits (e.g., for transmission), it is not possible to mistake any sequence of code values for another.

The overall number of bits per pixel required to represent the entire image is the summation of the length of each group code times the frequency of pixels in that group. For the example shown, this is 2.79 bits per pixel, compared to the 3 bits per pixel required to represent each of the original gray values. The theoretical limit calculated from the entropy is 2.73 bits per pixel. This is less than the 3 bits per pixel that would be required to represent each pixel value without compression. For an image with a larger number of grayscale values, the potential savings is greater, but so too is the effort required to construct the code table.

For the infrequent large delta values, a code will be needed that is much longer than the number of bits that would be required to represent the original pixel, but as these occur only rarely, the overall savings is significant. Modified forms of this coding use a unique reserved prefix code followed by the actual pixel value for the infrequent cases.

A somewhat different approach is taken by the most widely used lossless compression method, Lempel–Ziv–Welch (LZW) coding, which assigns fixed-length code words to variable-length sequences. This method is part of the TIFF and PDF file standards and is supported by almost all current programs that handle or display images (including web browsers), although licensing was required under U.S. Patent 4,558,302 (now expired).

LZW coding is often used for text compression. A standard code book or dictionary might contain, for example, 1024 entries (all of the possible 10 bit codes). The first 26 entries would be the letters of the alphabet, followed by another 26 for the capital letters, together with some additional entries for a space, the digits from 0 to 9, and commonly used symbols and punctuation marks. Further entries would include common two-, three- and four-letter sequences (an, wh, qu, the, ing, tion, ...) some of which are short words and some that occur as parts of words. For English text, such a code book can be constructed that efficiently replaces frequently encountered sequences of letters with a single code entry, and greatly reduces the total size of the file. Different code books would be optimized for different languages, because the common letter sequences differ. The use of a standardized code book means that the book does not need to be included with the individual compressed files.

For 8-bit grayscale images, a dictionary is constructed for each image. The codes begin with the 256 possible individual pixel values. The encoding process then scans the image line by line and adds symbols for gray-level sequences that are not already present, starting with 2-pixel sequences, then unique 3-pixel sequences, and continuing until the dictionary is filled. Any sequence that is already in the dictionary is replaced by the corresponding symbol. The length of the dictionary is a key parameter affecting the degree of overall compression that can be achieved. In this case, of course, the dictionary must be transmitted with the image.

When applied to color images, any of these coding processes operate on the individual color channels, but these may be RGB, LAB, or any other set as described in Chapter 1. They become less practical for images with a greater bit depth.

Color images are also frequently subjected to another compression procedure that is not loss free. By decreasing the number of individual color values that are present in the image (Braudaway 1987; Heckbert 1982), it is possible to construct an indexed color image in which each pixel is assigned a value (often a single-byte value from 0 to 255). The pixel value then selects a single entry in the color table, consisting of red, green, and blue intensities that are used to display the pixel. This method falls far short of true color representation, but is often adequate for image viewing and is widely used in the GIF and PNG image formats used for web display (the standard convention for Web browsers further restricts the color table to a fixed, predefined set of 216 entries).

Several algorithms exist for selecting a small number of colors to represent all of those in an image. Most are based on finding clusters of pixel values in color space. Programs such as Photoshop offer a choice of several techniques. Because this is a lossy method that is ill suited to any subsequent form of image analysis or to high-quality display or printing, it is not considered further here.

### 3.6.2  JPEG Compression

The most widely used image compression technique by far, JPEG compression is built into the firmware of many cameras (to reduce storage requirements) and was developed by professional photographers (JPEG stands for Joint Photographers Expert Group) (Pennebaker and Mitchell 1992). The lossy procedure consists of several steps:

1. The image is reduced if necessary to 8 bits per channel from the greater bit depth captured by the detector. This conversion is not linear, but approximates the

logarithmic response of human vision. Depending on the dynamic range of the original image this usually causes some information loss, particularly in the very bright and very dark regions.

2. The image is converted to a color space (YUV, similar to the YIQ space described in Chapter 1) that separates the luminance and color information. It is then subdivided into $8 \times 8$ pixel blocks. If the image is not an exact multiple of 8 pixels in width or height, it is temporarily padded out to that size. Usually, the color channels are then reduced in resolution by a factor of 4, because human vision tolerates a lack of sharpness in the color better than it does in brightness. This is done by averaging the color values for several pixels and is the next lossy step in the procedure.

3. Each $8 \times 8$ pixel block is processed using the discrete cosine transform (DCT). This is closely related to the Fourier transform, except that all of the values are real instead of complex numbers. The transform produces another $8 \times 8$ block of values for the frequency components. Although the original values are 1 byte = 8 bits (0 to 255), the transformed data are stored temporarily in 12 bits, giving 11 bits of precision plus a sign bit. Except for the possibility of round-off errors due to this finite representation, the DCT portion of the algorithm does not introduce any loss of data (i.e., the starting values can be exactly reconstructed from the transform by an inverse DCT). Nor does it produce any compression.

4. The 64 coefficients for each block are quantized to a lower precision by dividing by a fixed table of values that gives the least precision for high-frequency terms. Decreasing the "quality" factor in most implementations increases the division factors and reduces more terms to low precision or erases them altogether. This is the major lossy step in the compression. More precision is retained for the intensity or luminance than for the color data. This is because in the intended use of the compression method for human viewing of images, it is generally accepted that more fidelity is needed in image brightness than is needed in color, because human vision is very tolerant of and insensitive to color variations.

5. The first of the 64 coefficients for each block is the average brightness or "DC" term. It is represented as a difference from the same term for the preceding block in the image. The blocks are listed in raster-scan order through the image.

6. The remaining 63 coefficients for each block are scanned in a zigzag diagonal order that starts with the lowest frequencies and progresses to the highest. This sequence includes many zero values and can be compacted by substituting code values for repetitive patterns of bits, using the same coding procedures described above. This step is loss free.

Source code and compiled binary implementations of JPEG compression are widely available, from Intel <www.Intel.com>, the Independent JPEG Group <www.IJG.com>, and many other sites. These libraries are highly optimized for speed. Practically all computer programs that can load and display images support JPEG compression, and it is also used in most consumer digital cameras. Rather than trying to program the algorithm, it is useful for the student to try to understand the consequences that this type of compression has for images. This can be done by saving a copy of an image in JPEG format using the host program (e.g., Photoshop), and then opening that file and comparing the contents to the original image. Side-by-side comparison may not reveal many obvious differences, but subtracting one from the other will show that many pixel values have been changed.

(a)                             (b)

(c)                             (d)

**FIGURE 3.32**
(See **Color insert following page 172.**) JPEG compression: (a) the [rose.tif] image after JPEG compression; (b) enlarged detail showing 8 × 8 pixel blocks; (c) changes in the image, produced by subtracting the original image from the JPEG version; (d) difference in the intensity values between the original and compressed images.

Figure 3.32 shows the result of applying JPEG compression to the rose image used in previous chapters. The file was reduced from its original 583 KB to 115 KB, a factor of 5. This is not an extreme amount of compression; factors of 10 or 20 are not uncommon in Web downloads. Because of the use of 8 × 8 pixel blocks, the JPEG version shows a characteristic blocky appearance. Comparing the original to the compressed file by subtracting the original from the JPEG version reveals many differences. The intensity values are preserved with more fidelity than the color values, but there are still differences at edges where details have been eliminated or shifted in position. The color values, shown in the figure as saturation and hue, are compressed more than the intensity values, and in many cases a single color has been used for an entire 8 × 8 block.

**FIGURE 3.32 (continued)**
(e) The saturation values in the JPEG version; (f) difference in the saturation values between the original and compressed images; (g) hue values in the JPEG version; (h) difference in the hue values between the original and compressed images.

Considering the amount of information that has been discarded by this lossy compression method, it is interesting that the image is still easily visually recognizable. In an image containing much fine detail, some of the smaller features may be altered or lost. However, the assumption underlying the JPEG method is that a human viewer of the image would not have noticed those details anyway. Other lossy compression methods (such as fractal compression) use different algorithms but all depend on the same assumption. Some of them (e.g., fractal compression) introduce artificial texture and details as well as removing some information.

### 3.6.3  Fourier and Wavelet Compression

Using the Fourier and wavelet transform routines developed above, it is possible to explore directly the consequences of discarding specific values from the transforms in order to perform lossy image compression. Neither the Fourier nor wavelet transform reduces the number of values required to represent an image. Each transform has the same number of values, and because they are floating point values, they actually occupy more memory space than the original image may have required. Replacing some of these values by zero can reduce the size.

For example, Figure 3.33 shows the consequences of reconstructing an image after removing various fractions of the terms in a Fourier transform. One way that this can be accomplished is by thresholding the power spectrum display to construct a mask that retains only the desired fraction of the larger-magnitude terms. As expected, the quality of the image declines as compression increases. Because it is primarily the higher-frequency terms that have smaller amplitudes and are eliminated, the sharpness of edges is particularly affected. Other methods for removing selected terms that vary the selection threshold with frequency may improve the results. That is part of the strategy used in the JPEG method, which uses the closely related discrete cosine transform.

The wavelet transform can also be used for compression, in the same way. The most recent version of JPEG compression (called JPEG2000) has not been widely adopted, but allows the wavelet transform to be used as an alternative to the discrete cosine transform. Figure 3.34 shows the consequences of applying the same thresholding method to eliminate the smaller terms in the transform, followed by reconstructing the image. The same general trends can be noted, as well as the fact that the Daubechies-4 wavelet is somewhat superior to the simple Haar function for preserving visual quality in images.

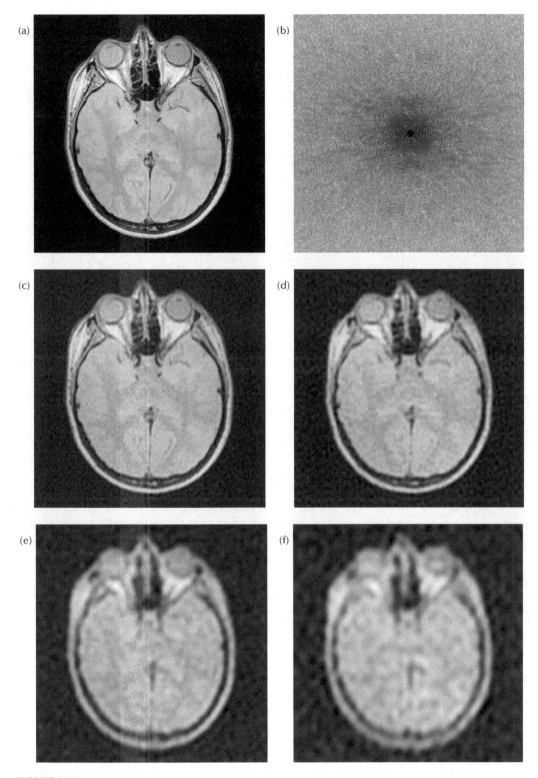

**FIGURE 3.33**
Fourier compression: (a) original image [mri.tif], (b) Fourier power spectrum, (c) reconstructed result with the 10% of the terms with largest amplitude, (d) reconstructed with 5% of the terms, (e) reconstructed with 2% of the terms, (f) reconstructed with 1% of the terms (100:1 compression factor).

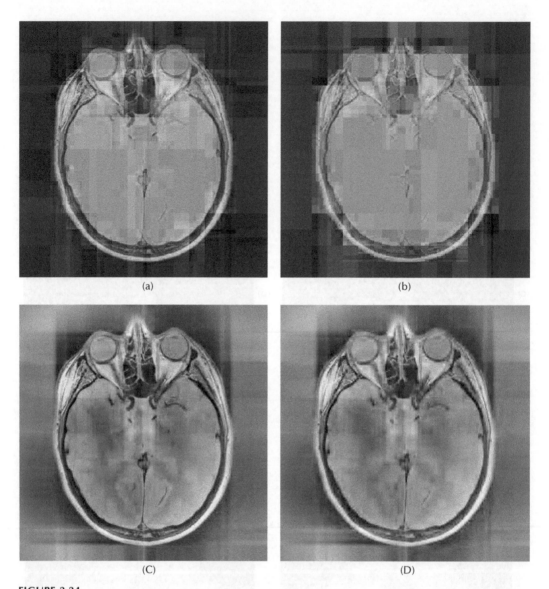

(a)     (b)

(C)     (D)

**FIGURE 3.34**
Wavelet compression: (a) reconstructed result with 10% of the terms with largest amplitude in a Haar wavelet transform, (b) reconstructed result with 5% of the terms from a Haar wavelet transform, (c) reconstructed result with 10% of the terms from a Daubechies-4 wavelet transform, (d) reconstructed result with 5% of the terms from a Daubechies-4 wavelet transform.

### 3.6.4   Problems

3.6.4.1.   Implement a delta- and variable-length code method for compression of gray-scale images, and compare the size of the result to the original and to the theoretical limit defined by the image entropy.

3.6.4.2.   Use the built-in JPEG compression routine in the host program (e.g., Photoshop) to save copies of an image with various levels of compression (usually described as "quality"). Subtract the original image from the compressed version and compare the number of pixels altered, the magnitude of the difference, and the visual fidelity of the result as a function of compression level. For color images, convert the original and compressed versions to a suitable space and compare the intensity and color channels separately.

3.6.4.3.  Implement a program that performs the forward and inverse Fourier transform on an image and produces compression by reducing to zero a fixed percentage of the terms, based on their magnitude. Subtract the original image from the reconstructed result and compare the results when 50, 80, 90, and 95% of the smallest terms are eliminated.

3.6.4.4.  Modify the program from Problem 3.5.2.1 to perform the forward Haar wavelet transform on an image and produce compression by reducing to zero a fixed percentage of the terms, based on their magnitude. Use the inverse Haar transform (Problem 3.5.2.2) to reconstruct the image and subtract the reconstructed result from the original. Compare the results when 50, 80, 90, and 95% of the smallest terms are eliminated.

# 4

## Binary Images

A binary image distinguishes between the features present in an image and the background surrounding them. The term *feature* is used to denote a contiguous group of pixels, which in some image processing texts is called a blob. These features correspond to objects or structures in the scene that the image represents, and will become the subjects for measurements in the next chapter. The ***background*** is simply those parts of the image that are not currently of interest for measurement. In simple cases the background may consist of a more-or-less uniform and neutral surface on which the objects reside, but in more complex situations the background may be at least as varied as the objects themselves (consider picking faces out of a crowd).

Separating the various objects in a scene from one another or from the background is also called segmentation. In some cases this may be accomplished by finding boundary lines (the edges discussed in Chapter 2). However, the most widely used methods for accomplishing this task rely on some unique property of color or brightness that identifies the pixels in the objects as distinct from those in the surrounding background. When it is appropriate, processing using methods introduced in previous chapters may be used to transform characteristics such as local texture into a brightness difference.

In its simplest form, the ***binary image*** represents the features and background by labeling all of the pixels in the background as 0 and all of those in the features as 1. This would require only a single bit per pixel, producing very compact arrays, but such arrays cannot be directly displayed. The display conventions of computers use values of 255 for red, green, and blue to produce white, and so the examples that follow will produce binary images using 0 and 255. For processing, and subsequently for measuring these images, any value less than 0.5 (out of 255.0) will be treated as background, and anything greater than that will be understood as feature. That convention will avoid problems that can arise when attempting to test whether a floating point value exactly equals zero. It will also allow the images to contain values other than 0 and 255, for example, to label features with unique identifying numbers in Chapter 5.

## 4.1 Thresholding

Producing a binary image from an original color or grayscale image is typically accomplished by thresholding. In its very simplest form, ***thresholding*** can be accomplished as shown in Code Fragment 4.1. The brightness (the average of the red, green, and blue values in the example) for each pixel is compared to a threshold value and pixels that are darker are set to black, while those that are brighter are set to white.

```
// Code Fragment 4.1 - Threshold image
RGBPixel WhitePix = {255, 255, 255},
         BlackPix = {0, 0, 0};

for (y = 0; y < height; y++)
{
   ReadOriginalLine(y, line);
   for (x = 0; x < width; x++)
   {
      float brightness = (Line[x].red + Line[x].green + Line[x].blue)/3;
      if (brightness <= Threshold) Line[x] = BlackPix;
      else                         Line[x] = WhitePix;
   } // for x
   WriteResultLine(y, line);
} // for y
```

In some more general cases that will be shown below, there may be two limits, upper and lower, used to select a range of values that distinguish features from background. Also, instead of the average of red, green, and blue, the test may be made using luminance (a weighted combination of red, green, and blue), or any of the other color coordinate representations introduced in Chapter 1.

The processing operations described in the previous two chapters may be applied to the original image to allow the thresholding process to succeed. Typically, steps such as noise reduction, leveling of nonuniform illumination, sharpening of edges, etc., are used to make the pixel values more consistent within the background and feature regions of the image, and more distinct from each other. This is usually accompanied by changes in the histogram that make the peaks narrower and better separated, as shown in Figure 4.1, which facilitates thresholding. Figure 2.25 (Chapter 2) showed another example of the consequences of processing (noise reduction) on the histogram. Even so, thresholding often does not produce a perfect representation of the features within an image, and further processing is required afterward as described later in this chapter.

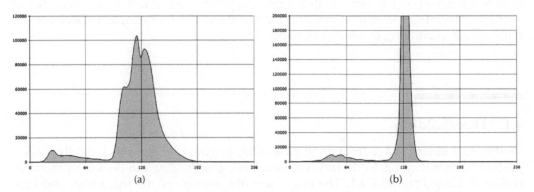

(a)                                                    (b)

**FIGURE 4.1**
Improvement in histogram peak shape and separation by processing: (a) brightness histogram of original [coins.tif] image (shown in Figure 1.25a), (b) brightness histogram after subtraction of nonuniform background illumination (Figure 1.25c).

### 4.1.1 Basics of Thresholding

The thresholding step in Code Fragment 4.1 does not answer the important question of how the value of the threshold should be determined. In rare instances it may be possible to establish it from knowledge about the imaging conditions and apparatus, the nature of the scene or specimen, etc. In most cases, the threshold must be determined from the image contents themselves. Many programs allow this to be done interactively, for instance by the user moving a slider while observing the image.

While this method takes advantage of the powerful human capability for recognition, it creates several problems. For one thing, the same threshold value can rarely be applied to a series of images for automatic processing, because minor changes in illumination, scene contents, camera settings, etc., often require making individual adjustments. Second, human judgment may rely on information about the scene contents that is incomplete, influenced by desire, or only subconsciously applied, which can lead to errors and inconsistency. Third, even if the human selection is correct, the knowledge upon which it was based is hard to communicate to another person, and also may produce different results on Friday afternoon than on Wednesday morning.

For all of these reasons, many efforts have been made to develop algorithmic procedures for determining a threshold value. There is a rich literature in the field (Haralick and Shapiro 1985; Kanpur, Sahoo, and Wong 1985; Kittler and Illingworth 1985; Lee, Chung, and Park 1990; Otsu 1979; Panda and Rosenfeld 1978; Parker 1991; Sahoo et al. 1988; Sezgin and Sankur 2004; Wezska 1978), and several representative approaches are illustrated below.

The application that many of these efforts have been directed toward is the one illustrated in Figure 4.2. Scanned images of printed pages are used as the first step in extracting the text to produce a document that can be transmitted, edited, and otherwise processed in the computer. Optical character recognition (OCR) also requires performing measurements on the individual characters to identify them, as well as lexical and syntactical processing to select the most likely words. The example shown in the figure would not be a good candidate for this procedure because of the equations and graphs, but does illustrate the need for obtaining a good-quality binary image as the initial step.

### 4.1.2 Histogram-Based Thresholding

The image histogram, introduced in Chapter 1, is the starting point for the majority of automatic procedures for determining a threshold value. In the examples that follow, it is assumed that the image has been scanned and the histogram array constructed from the brightness values. The examples are based on a histogram with 256 bins, which is the most common case. Other array sizes can be used but, in general, it is wise to be sure that the number of counts per bin is not too small so that the shape of the histogram and the various statistical values shown below are well defined.

As shown in Figure 4.3, an idealized image histogram suitable for thresholding would consist of two peaks. The peaks may not be completely separated, resulting in an overlap region as shown. When the peaks do overlap, any threshold setting will produce a binary image in which some pixels are misclassified. The tails of each peak that extend beyond the threshold setting correspond to the errors, which it is a goal to minimize. The overlapping tails in the histograms of the feature and background pixels that represent the misclassified pixels are evident in the figure.

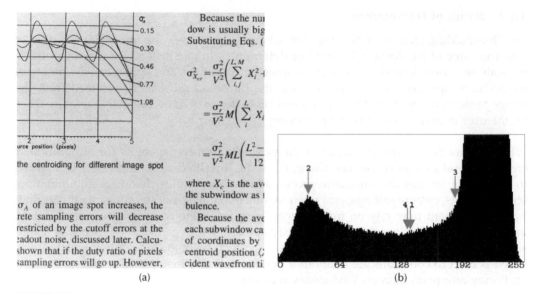

**FIGURE 4.2**
Thresholding printed text: (a) original image [scan.tif]; (b) histogram with several threshold values marked as described in the text (the numbers 1, 2, 3, and 4 correspond to methods and examples shown in the next section).

The initial human response to this type of histogram is typically to set the threshold at the lowest point in the valley between the peaks. This is a flawed strategy, however, because the low point will shift with the relative sizes of the peaks. Making no change in the actual brightness of the pixels associated with the features and background, but simply increasing the fraction of the image area that is background, will increase the height of the dark peak. That results in shifting the low point upward, and vice versa. In addition, the valley has fewer counts in the histogram bins, and the lowest point often wanders because of random noise and brightness fluctuations even when repeated images of the same scene are acquired.

Another approach is to set the threshold midway between the peak locations. This also involves reproducibility problems, and in addition makes an unspoken assumption about the linearity of the camera and image acquisition, as well as any gamma adjustments that may be applied. There is no fundamental reason why the threshold should be at the halfway point.

Any threshold value separates the pixels into two groups, which are identified in this context as features and background. Various statistical tests can be applied to choose a threshold based on the brightness values of the pixels in those two groups. In terms of the histogram, these are the pixels that are darker or lighter than the threshold values. One of the most popular methods for selecting a threshold value automatically finds the threshold value that lies halfway between the mean values for the two groups. This method is used because it is easily programmed, and often produces a result that is consistent, even if (as noted above for the midpoint between the peaks) the midpoint between the means has no particular scientific justification. In many image measurement applications, particularly for machine vision applications in quality control, reproducibility is more important than accuracy.

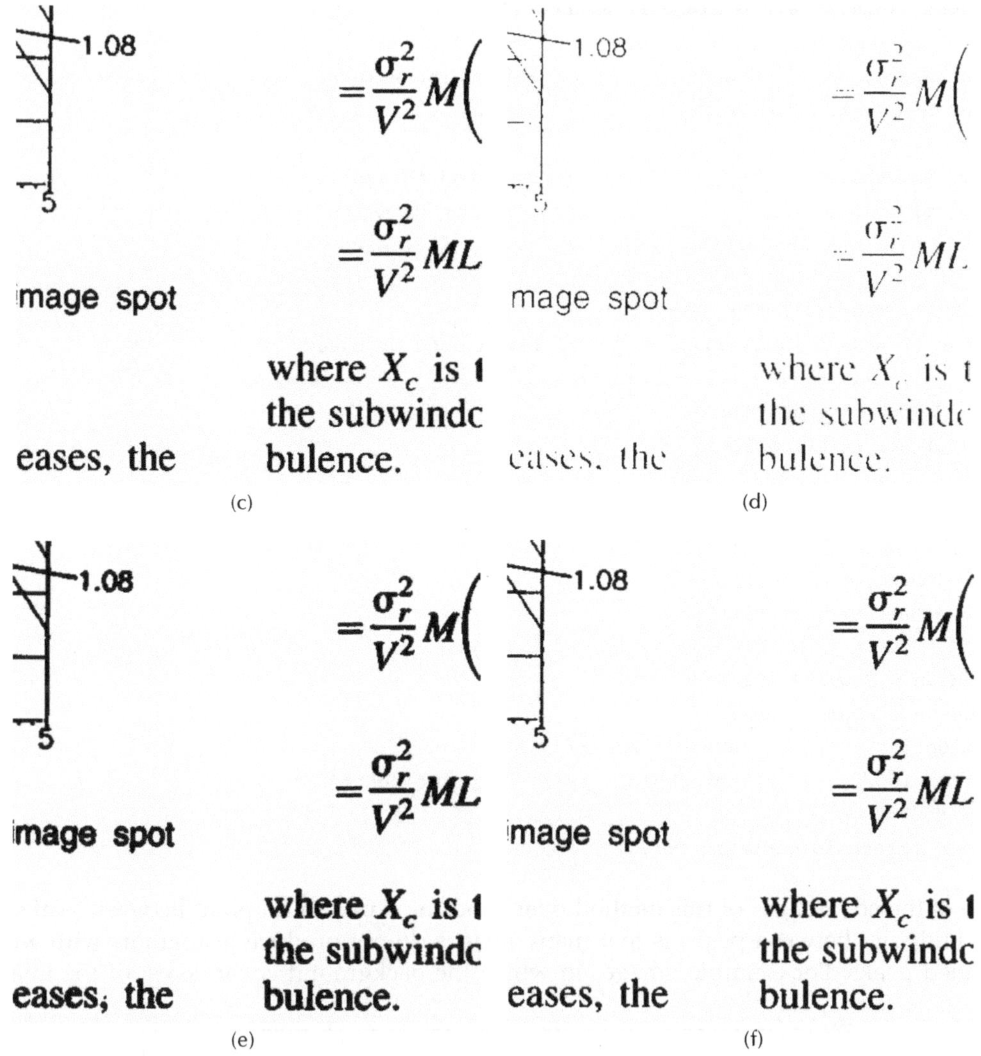

1.08

5

'mage spot

eases, the

$$= \frac{\sigma_r^2}{V^2} M \Big($$

$$= \frac{\sigma_r^2}{V^2} ML$$

where $X_c$ is 1
the subwindc
bulence.

(c)

1.08

'mage spot

eases, the

$$= \frac{\sigma_r^2}{V^2} M \Big($$

$$= \frac{\sigma_r^2}{V^2} ML$$

where $X_c$ is t
the subwindc
bulence.

(d)

1.08

5

'mage spot

eases; the

$$= \frac{\sigma_r^2}{V^2} M \Big($$

$$= \frac{\sigma_r^2}{V^2} ML$$

where $X_c$ is 1
the subwindc
bulence.

(e)

1.08

5

'mage spot

eases, the

$$= \frac{\sigma_r^2}{V^2} M \Big($$

$$= \frac{\sigma_r^2}{V^2} ML$$

where $X_c$ is 1
the subwindc
bulence.

(f)

**FIGURE 4.2 (continued)**
(c,d,e,f) Enlarged detail of the binary images produced by the marked thresholds 1, 2, 3, and 4, respectively

Because each possible candidate threshold value separates the pixels into two groups differently, the mean values of the dark and light groups of pixels will change as the threshold changes. Finding a threshold value that lies halfway between the means is thus an iterative procedure. It can be performed by making an initial guess, for example, 128 = medium gray, calculating the means of the two groups, using the average of those values as the second guess, and so on. This procedure converges rapidly and is often used.

Code Fragment 4.2 uses a different method, the testing of all 255 possible threshold values to find the one that is equal to the average of the means. This brute-force approach is also sufficiently fast, and is used here because it is clear, and the code forms the basis for the additional procedures that follow. The threshold value selected by this algorithm is shown (marked **1**) on the histograms in Figures 4.3 and 4.4. This method is also represented by the results shown (also labeled **1**) in Figures 4.2 and 4.5.

```
// Code Fragment 4.2 - Midpoint method (1)
// ... assumes histo[256] already filled with data
float Wsum1, WSum2, WMean;         // whitepoint values
float BSum1, BSum2, BMean;         // blackpoint values
long  i, t;
long  threshold;                   // computed threshold
long  midpoint;                    // midpoint of means
for (t = 1; t < 255; t++)          // 1..254 inclusive
{
    WSum1= WSum2= BSum1= BSum2= 0;
    for (i = 0; i <= t; i++)        // background or black
    {
        BSum1 += histo[i];
        BSum2 += histo[i]*i;
    }
    for (i = t+1; i <= 255; i++)     // features or white
    {
        WSum1 += histo[i];
        WSum2 += histo[i]*i;
    }
    BMean = BSum2/BSum1;
    WMean = WSum2/WSum1;
    midpoint = (long)((BMean + WMean) / 2);
    if (midpoint == t) threshold = t; // midpoint method (1)
} // for t
// ... threshold contains result
```

One of the advantages of this method over choosing a minimum point between peaks or the midpoint between peaks is that many images do not produce histograms with well-defined peaks. For example, images in which the background covers most of the image

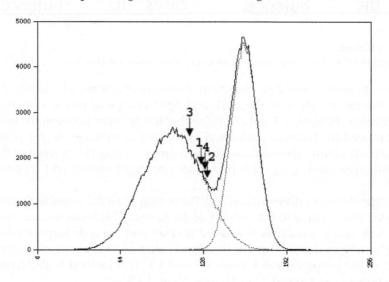

**FIGURE 4.3**
Example of a histogram composed of two overlapped Gaussian distributions. The numbered arrows mark threshold levels selected by algorithms described in the text.

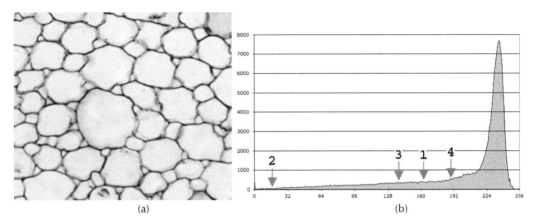

**FIGURE 4.4**
Image with a histogram containing a single peak: (a) original image [corn.tif], (b) histogram of (a) showing the location of threshold points selected by the algorithms described in the text.

area, as is the case with printed text, often produce histograms with a large major peak corresponding to the background and a nearly flat region corresponding to the small number of feature pixels. Figure 4.4 shows another example of this situation. However, finding the midpoint between the means of the two populations of pixels does not require a histogram with two peaks.

The limitation of the approach in Code Fragment 4.2 is that it is only correct (meaning that for the case of two overlapped distributions it produces the threshold setting that misclassifies the smallest number of pixels) for the particular case in which the two groups in the histogram have Gaussian distributions with the same standard deviation and the same number of pixels. Printed text, such as the example shown in Figure 4.2, rarely has more than 10% of the area covered by ink, and the shapes of the light and dark area peaks (if indeed they are both peaks) are quite different. If the two groups do not correspond to the expectations, the errors (the pixels in the two regions with values that lie in the tails of the distributions that extend beyond the threshold) will be larger, and biased toward one group or the other.

There are many other comparisons that can be made between two populations to produce meaningful decisions. Instead of just using the mean values of the groups, the variance values and the size of the groups can also be used to perform a classical statistical test known as the Student's $t$-test. As shown in Equation 4.1, the $t$-statistic depends on the mean values ($\mu$), variances (the squares of the standard deviations $\sigma$), and the number of pixels ($n$) in each group. The larger the value of the $t$-statistic, the greater the probability that the two groups are truly distinct. Consequently, selecting the value of the threshold that maximizes the value of the $t$-statistic is another strategy that can be applied to calculate a threshold value. Code Fragment 4.3 performs this calculation, whose result is shown marked with a **2** on the histograms in Figures 4.3 and 4.4, and for the results shown in Figures 4.2 and 4.5.

$$t = \frac{\left|\mu_1 - \mu_2\right|}{\sqrt{\dfrac{\sigma_1^2}{n_1} - \dfrac{\sigma_2^2}{n_2}}} \tag{4.1}$$

```
// Code Fragment 4.3 — Student's t-test (2)
// ... declare variables
// ... assumes histo[256] already filled with data
// ... test1 is the square of the Student's t-test limit
threshold = 0;                        // resulting threshold value
testval = 0;
cutoff = 1000;                        // to skip empty histo[] areas
for (t = 1; t < 255; t++)
{
   WSum1= WSum2= WSum3= BSum1= BSum2= BSum3= 0;
   for (i = 0; i <= t; i++)
   {
      BSum1 += histo[i];
      BSum2 += histo[i] * i;
      BSum3 += histo[i] * i * i;
   }
   for (i = t + 1; i <= 255; i++)     // 1..254 inclusive
   {
      WSum1 += histo[i];
      WSum2 += histo[i] * i;
      WSum3 += histo[i] * i * i;
   }
   if ((BSum1> cutoff) && (WSum1> cutoff))
      // skip empty portions of histogram
   {
      BMean = BSum2 / BSum1;
      WMean = WSum2 / WSum1;
      BVar = BSum3 / BSum1 - BMean * BMean;
      WVar = WSum3 / WSum1 - WMean * WMean;
      numer = (WMean-BMean);
      numer = numer * numer;
      denom = (WVar/WSum1) + (BVar/BSum1);
      tstat2 = numer / denom;// square of the t-statistic
      if (tstat2 > test1)
      {
          threshold = t;
          testval = tstat; // t-statistic method (2)
      }
   }
} // for t
// ... threshold contains result
```

There are other versions of this type of test, for example, using analysis of variance (ANOVA) rather than the Student's *t*-test. The problem with all of them is that they make the assumption that the brightness values for pixels in the feature and background regions of the image have a Gaussian distribution. In practice, such distributions for the pixel brightness values are not commonly observed. Figures 4.2. 4.4, and 4.5 show images in which the histograms do not have such simple shapes. In such cases the method shown in Code Fragment 4.3 often performs poorly. These cases require a different, nonparametric comparison of the two groups. Substituting a nonparametric statistical test such as a Kolmogorov–Smirnov procedure would be one possible solution.

Entropy is a way of characterizing the amount of information present in an image or a portion of an image (such as the features or the background) (Shannon 1948). Several approaches to finding an optimum threshold try to maximize the entropy that results from various threshold settings. First, it is important to decide just how the entropy value is to be computed. Equation 4.2 shows two slightly different definitions of the entropy (the following examples use version *a*; version *b* [Shannon's definition] is less commonly used) In both cases, the summation covers all of the brightness levels represented by the bins in the histogram (skipping those which are empty), and the *f* value is the fraction of the total number of pixels that are counted in each bin.

$$f_i = \frac{histogram(i)}{Total}$$

$$a)\ Entropy = \sum_i -f_i \cdot \log(f_i) \tag{4.2}$$

$$b)\ Entropy = \sum_i -f_i \cdot \log(f_i) + \sum_i -(1-f_i) \cdot \log(1-f_i)$$

One way to apply this principle to the threshold setting is to maximize the sum of the entropy values calculated for the two groups, i.e., the features and background pixels, as a function of the threshold settings. This should properly be done as shown in Code Fragment 4.4, in which the *f* values for each group are based on the total number of pixels in that group, not (as is sometimes done) on the total number of pixels in both groups (the entire image). Figure 4.2, 4.3, 4,4, and 4.5 also show the threshold values calculated by this procedure, marked **3**.

```
// Code Fragment 4.4 — Maximum entropy (3)
// ... declare variables
// ... assumes histo[256] already filled with data
Total = 0;                          // compute image area (pixels)
cutoff = 1000;                      // skip bins with few counts
for (i = 0; i < 256; i++)
    Total += histo[i];              // count number of pixels in the image
threshold = 0;                      // resulting threshold
testval = 0;                        // max entropy
for (t = 1; t < 255; t++)
{
    WEntr = BEntr= 0;               // group entropy sums
    BCumul= 0;
```

```
for (i = 0; i <= t; i++)
   BCumul += histo[i];
WCumul = Total - BCumul;
for (i = 0; i <= t; i++)
   if (histo[i] > 0)              // avoid zero counts
   {
       ratio = (float)histo[i] / BCumul;
       BEntr = BEntr - ratio * log(ratio);
   }
for (i = t+1; i <= 255; i++)
   if (histo[i] > 0)
   {
       ratio = (float)histo[i]/WCumul;
       WEntr = WEntr - ratio * log(ratio);
   }
if ((BSum1> cutoff) && (WSum1> cutoff)) // skip empty histo[] bins
{
   if ((BEntr + WEntr)>testval)
   {
       testval = (BEntr + WEntr);
       threshold = t;              // max entropy method (3)
   }
}
} // for t
// ... threshold contains result
```

Still another useful way to think about the thresholding problem is to realize that the two groups of pixels have values that are not perfectly separated in the histogram. Because there is almost always some overlap, it makes sense to use a fuzzy logic approach in which each pixel brightness value is assigned a probability of belonging to the group of background or feature pixels. It is possible to use the same entropy approach to sum the probabilities for the two groups, and to determine a minimum point that has the least "fuzziness" or uncertainty in the division of the two portions of the image by the threshold value.

The first requirement is to define a function that expresses the probability that a brightness value belongs to one group or the other. There are several possible candidates for this function, two of which are shown in Equation 4.3 (version *a* is used in the code fragment that follows). The values vary from a maximum of 1.0 when the value matches the mean of the group $\mu$, down to a small value if the difference from the mean is large (at a somewhat greater computational cost, the difference from the median may also be used). Summing the entropy of the probabilities for all gray values selected by a candidate threshold setting, for each of the two groups, produces a test value. Finding the threshold that minimizes this, as shown in Code Fragment 4.5, produces results that are marked as **4** on the histograms in Figures 4.2, 4.3, 4.4, and 4.5.

$$\text{a) } probability_i\left(thresh\right)= \begin{cases} i <= thresh: & 1-\dfrac{\left|i-\mu_1\right|}{255} \\[2em] i > thresh: & 1-\dfrac{\left|i-\mu_2\right|}{255} \end{cases}$$

$$\text{b) } probability_i\left(thresh\right)= \begin{cases} i <= thresh: & \dfrac{1}{1+\dfrac{\left|i-\mu_1\right|}{255}} \\[3em] i > thresh: & \dfrac{1}{1+\dfrac{\left|i-\mu_2\right|}{255}} \end{cases} \qquad (4.3)$$

```
// Code Fragment 4.5 — Minimum Fuzziness method (4)
// ... declare variables
// ... assumes histo[256] already filled with data
Total = 0;                          // compute image area (pixels)
cutoff = 1000;                      // skip bins with few counts
for (i = 0; i < 256; i++)
   Total += histo[i];              // count number of pixels in the image
threshold = 0;                      // resulting threshold
testval = 1e20;                     // fuzzy probability to be minimized
for (t = 1; t < 255; t++)
{
   WSum1 = WSum2 = BSum1 = BSum2= 0;
   for (i = 0; i <= t; i++)
   {
      BSum1 += histo[i];
      BSum2 += histo[i]*i;
   }
   for (i = t+1; i <= 255; i++)
   {
      WSum1 += histo[i];
      WSum2 += histo[i]*i;
   }
   if ((BSum1> cutoff) && (WSum1> cutoff)) // skip empty histo[] bins
   {
      BMean = BSum2 / BSum1;
      WMean = WSum2 / WSum1;
      FSum = 0;
      for (i = 0; i < 256; i++)
      {
         float fuz;                 // fuzzy probability of set membership
         if (i <= t)
            fuz = 1 - (fabs((float)i - BMean) / 255);
```

```
        else
            fuz = 1 - (fabs((float)i - WMean) / 255);
        FSum -= fuz * log(fuz) * (float)histo[i] / Total;
    }
    if (FSum < testval)
    {
        testval = FSum;
        threshold = t;              // fuzziness threshold (4)
    }
  }
} // for t
// ... threshold contains result
```

It is worth pointing out that the threshold values selected by these different criteria do not agree, as shown in the various examples. In some cases, several of the methods produce quite useable delineations of the image into features and background, whereas in others, one or several of the methods do not. Also, there is no single method that is always best, or for that matter always worst. ***Different types of images lend themselves to one or another***

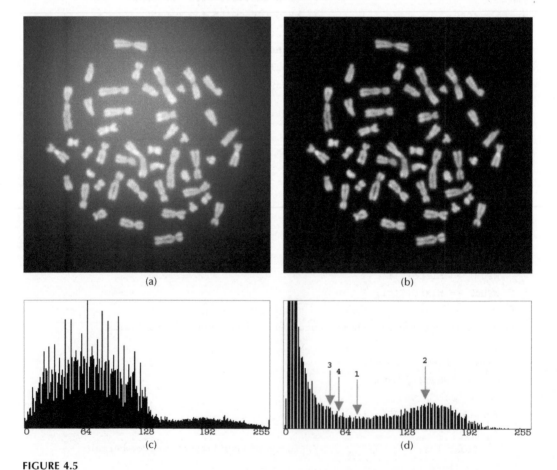

(a)                                                    (b)

(c)                                                    (d)

**FIGURE 4.5**
Thresholding after processing: (a) original image [chromosomes.tif]; (b) image after leveling nonuniform brightness as shown in Chapter 2; (c) histogram of the original image; (d) histogram after leveling, with numbered arrows indicating threshold levels corresponding to algorithms described in the text;

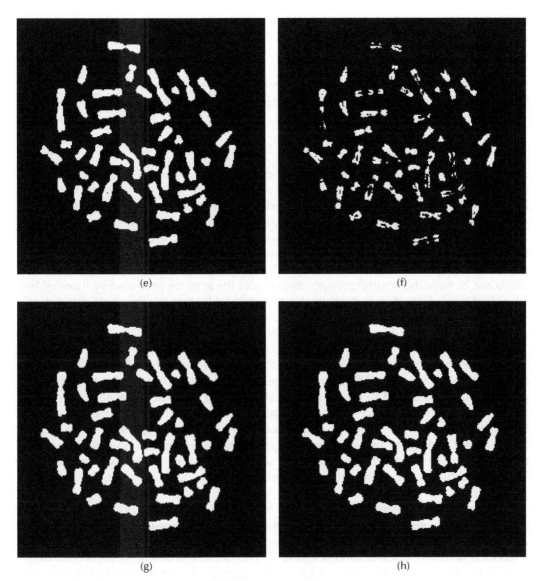

**FIGURE 4.5 (continued)**
(e, f, g, h) Binary images resulting from threshold levels 1, 2, 3, and 4, respectively.

*method for automatically determining a threshold setting.* Once a suitable technique has been found for a given application, it often works well for an entire batch of images, and tracks the changes in threshold value that occur with different imaging conditions, but first it is necessary to compare the performance of the various algorithms and select the one to be used.

It should also be noted that the foregoing methods make the explicit assumption that all of the features are characterized by one brightness value and all of the background is represented by another. This is by no means always the case. Figure 4.11 below shows one example in which the background is uniform but the features vary.

Methods that allow the threshold value to vary across the image function by constructing the histogram of a local neighborhood and applying any of the methods shown to select a threshold value. This is then used to classify only the central pixel in the image as feature or background, after which the neighborhood is moved to the next pixel and the process repeated. In principle, this approach can accommodate images with nonuniform illumination or other spatial variations. However, it is usually much faster, much more straightforward, and just as effective to use the processing methods shown in Chapters 1 and 2 to level the image brightness and contrast before applying a global thresholding method.

### 4.1.3 Other Criteria

All of the code fragments in the preceding section used a common approach (and much common code) to scan through all possible settings using different criteria to select the best value. In the examples shown, the criteria were all based on various statistical comparisons between the groups using the histogram as the basis for calculations. This is very efficient, but ignores some additional information that is present in the images. The individual pixel brightness values are recorded in the histogram, but the histogram contains no information about where the pixels lie, or what the values of the neighboring pixels are, or what the resulting shapes and sizes of the features produced by thresholding are. All of this information is available to the human who sets the threshold interactively while examining the resulting binary image, and is certainly used in making the decision about where to manually position the threshold.

The methods shown in Code Fragments 4.2 to 4.5 can easily be adapted to use other criteria to select the "best" threshold value. For example, locating the boundaries at positions that correspond to the greatest local change in brightness can be accomplished by maximizing the difference between brightness values for the pixels that are adjacent in the image but fall into different groups (i.e., features and background). To implement this, it is not enough to use just the histogram. The selection process must generate a temporary image in which the pixels are labeled as feature or background, and the brightness differences between adjacent pixels with opposite numbers must be tallied.

A similar logical approach to defining the location of the boundaries uses the second derivative (implemented as the Laplacian or difference of Gaussians function, described in Chapter 2). If the maximum gradient of brightness is the proper position for the boundaries between features and background, then the zero-crossings of the second derivative mark the boundaries, and can be used to create a binary image. In this case, the original brightness values of the pixels are not uniquely assigned to feature or background regions. Two pixels with the same brightness value may fall into different groups, depending on the values of their neighbors.

Processing of the original image prior to thresholding, or processing of the binary image after thresholding as discussed later in this chapter, along with the use of simpler histogram-based thresholding algorithms, can often produce equivalent results. For example, Figure 4.6 shows the effect of applying the maximum likelihood filter to the same image from Figure 4.5b. This spatial domain filter from Chapter 2 compares each pixel to its neighbors and replaces the values with the mean of whichever subneighborhood the pixel is most like (in the specific sense that the variance of the brightness values in the subneighborhood is minimum). The effect is to sharpen the steps in the image and improve the delineation of feature boundaries. It also changes the histogram to eliminate counts in the gap between the peaks that correspond to features and background, so that thresholding of the features becomes easier.

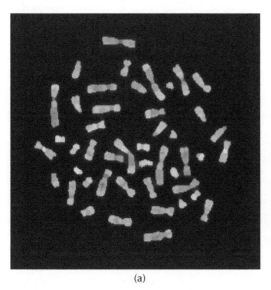
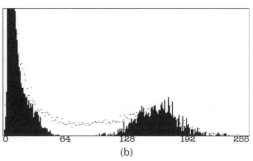

(a)                                              (b)

**FIGURE 4.6**
Processing can simplify thresholding: (a) image from Figure 4.5b after applying the maximum likelihood technique from Chapter 2 to sharpen feature edges; (b) the resulting histogram, showing a broad gap between the feature and background groups of pixels (dots show the histogram of the original for comparison).

The creation of a temporary binary image to evaluate each possible threshold value also allows another approach to optimizing the threshold selection, based on measurements. Measurements are discussed in the next chapter, and fall into two major categories. One class of measurements is global in nature, and the values are determined on the entire image or scene. Examples are the area fraction of features, the total length or smoothness of the boundaries between features and background, the number of features, and so on.

The second class of measurements is performed on individual features, and the values comprise measurements of brightness or color (and derived properties such as density), position, size, and shape. These feature-specific measurements may be summarized for the entire image by descriptive statistical values such as the mean and standard deviation.

Any of these measurements can be used as a test criterion to select an optimum threshold setting, provided there is some known property of the scene that makes them appropriate. For example, if the area fraction or sizes of holes in a part being examined in a machine vision application is known, that can be used to select the threshold. Subsequently, the binary image could be used to measure the positions of the holes for quality control. Similarly, for an image of bubbles, the threshold value that produces the smoothest and roundest features would be appropriate, because the physical force of surface tension produces smooth and round feature boundaries.

Many such scenarios can be devised, but all require a considerable amount of *a priori* knowledge about the scene and how the image was acquired. They also require time, because the various measurement procedures described in Chapter 5 must be performed (many times) in the process of establishing the threshold settings.

Another thresholding approach uses the local brightness gradient to select only those pixels in the image that are not at or near boundaries to produce a histogram of the original pixel brightness values. Because it is the pixels along the boundaries that frequently have

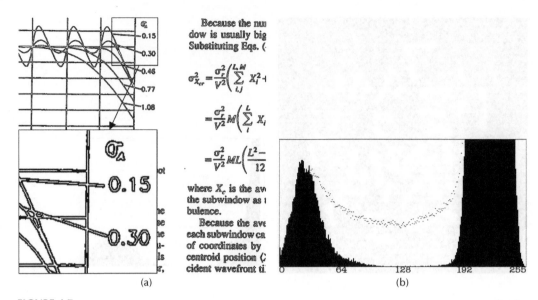

(a)                                    (b)

**FIGURE 4.7**
Using processing to select pixels with a low brightness gradient (i.e., ignoring edges): (a) processed result on the image from Figure 4.2a with enlarged detail showing selected pixels (white areas); (b) histogram of the pixels in the original image that have a small value for the gradient, showing a broad gap between the feature and background regions (dots show the histogram of the original for comparison).

brightness values lying between the characteristic feature and background levels, eliminating them from the histogram reduces, or even eliminates, the counts from the region between the peaks. That makes determining a threshold value using any of the techniques above much easier and the result more robust.

Figure 4.7 shows an example of this method. The brightness gradient vector was determined (using the Sobel operator described in Chapter 2) at each point in the image. Those pixels with a value at least half of the maximum were eliminated from the brightness histogram. Comparing the resulting histogram to that of the full image (Figure 4.2b) shows that the peaks corresponding to the features and background are now distinctly separated.

It must also be pointed out that the examples shown so far have included one (Figure 4.5) in which features are bright and background is dark, so that a brightness threshold applied as in Code Fragment 4.1 can produce a suitable binary image, and another one (Figure 4.2) in which the features are dark and the background is bright. When the features are dark on a light background, the logic can be reversed to select the dark pixels as features. However, in many cases it is easier to threshold in the usual way so that pixels darker than the threshold are set to black and those brighter than the threshold are set to white, and then to reverse the image contrast afterward. This is often described as "inverting" the image, and is accomplished by replacing each pixel value **v** by (**255-v**) as shown in the Introduction.

Figure 4.8 shows a case in which the background is a medium gray and the features contain pixels that are either brighter or darker. In this case, the thresholding requires two values, which are used as upper and lower limits as shown in Code Fragment 4.6. As shown, this procedure would produce an image in which the coins and keys are set to black and the background to white, so that the image must be reversed afterward to produce the correct result.

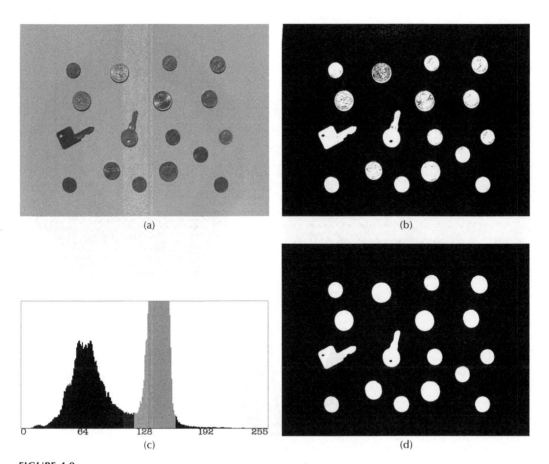

(a)

(b)

(c)

(d)

**FIGURE 4.8**
Thresholding regions with intermediate brightness: (a) original image [coins.tif] after subtracting a background image to level brightness, as shown in Chapter 2; (b) binary image resulting from selection of pixels brighter or darker than background; (c) histogram of image (a) showing in gray the brightness levels of the background pixels; (d) image (b) after morphological processing (closing) described later in this chapter.

```
// Code Fragment 4.6 - Two-level thresholding
// ... for each pixel
float Brightness = (line[x].red + line[x].green + line[x].blue) / 3.0;
if ((Brightness >= LowLimit) && (Brightness < HiLimit))
   line[x] = WhitePix;              // set to {255, 255, 255}
else // not between limits
   line[x] = BlackPix;             // set to {0, 0, 0}
```

Also note in this example that the thresholded result (Figure 4.8b) is imperfect. There are some pixels in the coins that have brightness values that lie within the same range as the background. This can be corrected using the morphological procedures described later in this chapter, for example, by taking advantage of the fact that the mislabeled regions within the coins are small.

### 4.1.4 Color Images

Images of colored scenes can sometimes be satisfactorily thresholded based on the brightness (average of the red, green, and blue pixel values) or luminance (a combination of the

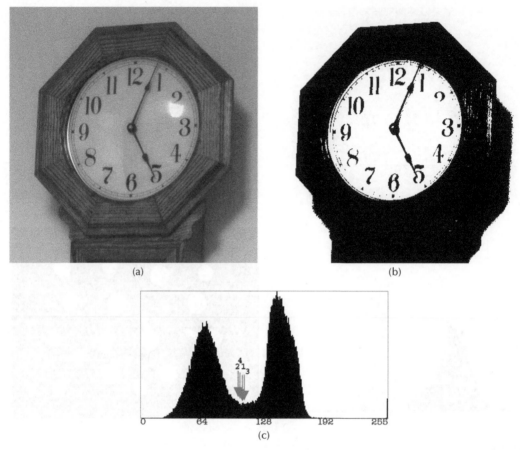

(a)                           (b)

(c)

**FIGURE 4.9**

Thresholding a color image based on the luminance value: (a) grayscale image with the luminance values calculated from the original red, green, and blue [clock1.tif]; (b) thresholded binary image using the value marked 4 on the histogram; (c) histogram of the luminance values showing the threshold values selected by the methods described in the text.

RGB values that models human visual response by weighting green more and blue less, as described in Chapter 1). Figure 4.9 shows an example in which the luminance is used. Note that the histogram consists of large, reasonably well-separated peaks and that all of the automatic procedures for selecting a threshold value produce results that are in quite close agreement.

However, in many cases the distinction between features and background recognized by a human observer is based on color differences rather than brightness or luminance. The RGB color space coordinates do not represent color differences as well as other spaces such as hue–saturation–intensity or LAB. In the example of Figure 4.10, the luminance values do not distinguish between the flower and the background, and the histogram of the luminance values does not show useful peaks or valleys. Successful thresholding of the image to isolate the flower is not possible based on luminance.

Subtracting the green values from the red values produces the image shown in Figure 4.10c. The A axis in LAB space is the red–green axis. A histogram of these values does show two well-separated peaks, and thresholding of the pixels based on these values

**FIGURE 4.10**

(See **Color insert following page 172**.) Thresholding a color image: (a) luminance calculated from the original red, green, and blue [rose.tif]; (b) histogram of image (a); (c) result of subtracting green from red (the A axis in LAB coordinates); (d) histogram of image (c), with arrows indicating the threshold values selected by the algorithms described in the text; (e) thresholded result in which background pixels are set to zero but feature pixels are left unchanged.

can be performed. As shown in Figure 4.10e, in this case the thresholding procedure was modified from that shown in Code Fragment 4.1 so that pixels with values less than the threshold were set to 0 (black) as usual, but ones with values greater than the threshold were left unchanged. The effect is to isolate the features by erasing the background.

It is not always possible to find a single value that can be used to successfully threshold a color image, whether it is the brightness or luminance, or some other color coordinate in the spaces discussed in Chapter 1 such as LAB or hue–saturation–intensity. When multiple channels of color (or other) values are available, combining them to achieve successful thresholding can be an effective solution (Russ 1993). Figure 4.11 provides an illustration using the red, green, and blue values for each pixel, but it should be remembered that the same principles can be applied to any color space coordinates, or to multiple channels of other types of information such as satellite imagery that may include ultraviolet and infrared channels (Russ 1991).

In Figure 4.11, the individual color channels can each be thresholded to select pixels that are bright. The resulting binary images show the regions that have high levels of red, green, and blue, respectively. These do not correspond uniquely to the various object colors, however.

By combining the binary images corresponding to the individual color channels, each of the balloon colors can be isolated. These Boolean combinations are discussed in full detail in Section 4.4 below. Referring back to the color space diagrams shown in Chapter 1, the color yellow can be understood as a combination of green and red. In other words, a pixel may be considered to be yellow if it is selected in both the thresholded red and green binary images. In Boolean terms, this is Yellow = Red AND Green. Similarly, magenta is a combination of red and blue, so keeping the pixels selected in both the thresholded red and blue binary images produces a selection of the magenta pixels, or Magenta = Red AND Blue.

Having isolated the yellow and magenta regions, it is necessary to remove them from the thresholded red binary image to obtain just the red balloons. As shown in the example, this requires combining the thresholded red binary image with the derived yellow and magenta binaries, or Red AND NOT (Yellow OR Magenta). Similar operations can isolate just the blue or green balloons.

In this example, the starting point was the red, green, and blue channels of the original color image but, of course, similar procedures can be based on other color coordinates.

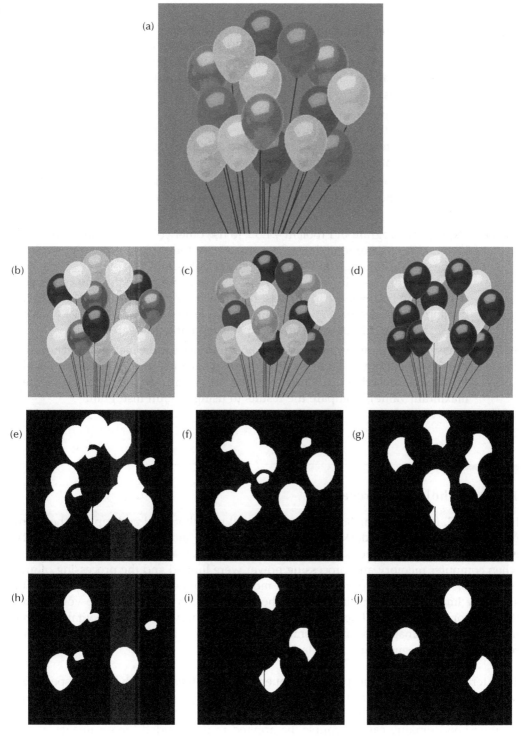

**FIGURE 4.11**

(See **Color insert following page 172.**) Multichannel thresholding: (a) original image [balloons.tif]; (b,c,d) red, green, and blue channels, respectively; (e,f,g) thresholding regions that are bright in the individual color channels (b–d), respectively; (h) pixels that are selected in both the thresholded red (e) and green (f) channels (yellow = red AND green); (i) pixels that are selected in both the thresholded red (e) and blue (g) channels (magenta = red AND blue); (j) pixels in the red balloons only, selected as those in image (e) that are not selected in either image (h) or (i).

### 4.1.5    Problems

4.1.5.1.    Implement a program that thresholds an image using a threshold setting read from a text file and applied to the brightness values (average of red, green, and blue). Modify the program to allow production of either a binary image or one in which the background is erased to zero but the features retain their original values.

4.1.5.2#.   Implement a program that thresholds an image by calculating a threshold setting using the histogram of brightness values, and each of the methods shown in the text (Code Fragments 4.2 to 4.5). Use a control value (1, 2, 3, or 4) to select the method. Compare the results.

4.1.5.3.    Implement a program that uses some of the other optimization routines and calculations shown and discussed in the text. Compare the results for various images.

4.1.5.4.    Modify the program in Problem 4.1.5.2 to use other color coordinates, such as LAB and hue–saturation–intensity axes. Apply the program to various color images by examining the histograms (as calculated in Chapter 1) select the axes that work best in each case. Note that in some cases the binary image will have to be reversed in contrast (black and white interchanged) to select the appropriate features and background. Note also that hue is a tricky scale to use, because it is an angle and wraps around at zero degrees.

4.1.5.5.    Modify the program from Problem 4.1.5.2 to construct the histogram, ignoring pixels with a high value of the brightness gradient (use a Laplacian or Sobel operator as described in Chapter 2 and set a limit as a percentage of the maximum value). Compare the results to using the full histogram.

---

## 4.2    Morphological Processing

In Chapter 2, functions that rank pixels in a small neighborhood and replace the central pixel by the brightest or darkest neighbor were called *morphological operations*. The original development of these morphological techniques was to binary images, at a time when computer memory and processing power were limited and the processing of gray-scale or color images was not common. For most early applications, images were acquired as binary images by performing the thresholding function directly on the voltage from the detector. Morphological operations applied to the binary images were vital to preparing the resulting images for measurement, for example to accomplish purposes such as removing noise or smoothing boundaries.

The literature describing morphological operations is extensive, with its own specialized terminology and mathematical notation. Fortunately, it is possible to understand and apply the basic tools using simpler explanations. The routines shown in Chapter 2 based on neighborhood ranking can, of course, be applied to thresholded binary images as well (and, in fact, in some commercial programs, the same routines are used for this purpose). However, both for reasons of greater clarity and greater flexibility, this section introduces procedures explicitly dealing with binary images.

## 4.2.1 Classic Opening and Closing

The basic tools of morphological operations are ***erosion*** and ***dilation***. As illustrated in Figure 4.12, erosion removes pixels from the boundaries of features and dilation adds pixels along those boundaries. In classic erosion, any feature pixel (displayed as white) that touches a background pixel (displayed as black) is changed to black. In classic dilation, this logic is reversed, and any background pixel that touches a feature pixel is changed to white.

The most common definition of "touching" is that a pixel has eight adjacent, touching neighbors, four sharing sides and four sharing only corners. This is usually called an

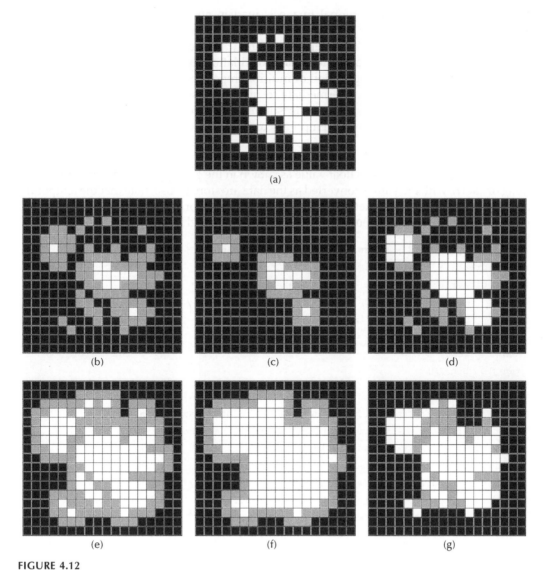

**FIGURE 4.12**
Erosion and dilation (in each step, gray pixels are those that are changed): (a) original image; (b) erosion applied to (a)—gray pixels are removed; (c) dilation applied to (b)—gray pixels are added; (d) opening applied to (a)—gray pixels are removed; (e) dilation applied to (a)—gray pixels are added; (f) erosion applied to (e)—gray pixels are removed; (g) closing applied to (a)—gray pixels are added.

*8-neighbor* relationship. In some cases, a *4-neighbor* relationship is used, meaning that only the four side-sharing neighbors are considered to touch. It will be important for some processing and measurement purposes to distinguish between these different rules and their consequences, but for the present example, logic based on the 8-neighbor relationship is used.

Code Fragment 4.7 outlines the procedure for performing erosion or dilation on an image. In the case of binary images, it is not necessary to rank the values of the pixels, as was necessary for the operations used for processing continuous-tone images. Instead, a simple test to see if any of the neighbors has the opposite state (black or white) is sufficient.

Notice in the code that the original image is read in and stored in a temporary array consisting of a single number for each pixel (the brightness calculated as the average of red, green, and blue values). This step of creating a temporary array is not required for the comparisons involved in a single erosion or dilation, but for opening and closing as described below, or for the application of multiple repetitions of erosion and dilation, the use of a local array to hold the intermediate results is necessary. The temporary array will also be used for other processing and measurement on the binary image in this chapter and the next one. The creation of a temporary array is assumed in all of the examples that follow.

The temporary array may use a float value as shown in the example, or it may use an integer (long or short), with the values converted as the data are stored. Also, note that the temporary array is padded to be one pixel larger on all sides than the original. This is a simple means of providing the necessary boundary protection for the neighbor comparisons.

```
// Code Fragment 4.7 — Classic erosion and dilation
// ... create a temporary image and read from the host
// ... either Erosion==true or Dilation==true
long      Err, x, y, height, width;
float     *LineAbove, *ThisLine, *LineBelow, *Result;
RGBPixel *OrigLine;
float     Test;
Image_ID  Bin_ID;

GetOriginalDimensions(&width, &height);
Err = MakeTemporaryImage(width+2, height+2, sizeof(float), &Bin_ID)
// note: this padding simplifies binary boundary neighborhoods
// ... handle any error (usually indicates insufficient memory)
OrigLine  = CreateAPointer(width,   sizeof(RGBPixel));// for original image
ThisLine  = CreateAPointer(width+2, sizeof(float));   // temp binary image
LineAbove = CreateAPointer(width+2, sizeof(float));
LineBelow = CreateAPointer(width+2, sizeof(float));
Result    = CreateAPointer(width+2, sizeof(float));
for (x = 0; x < width+2; x++)
   ThisLine[x] = 0;     // clear out ThisLine
WriteTempImageLine(Bin_ID, 0, ThisLine);
WriteTempImageLine(Bin_ID, height+1, ThisLine);     // init boundaries
// read in the original image
for (y = 0; y < height; y++)
```

```
{
   ReadOriginalLine(y, OrigLine);
   for (x = 0; x < width; x++)
   {
      float bright = (OrigLine[x].red+OrigLine[x].green+OrigLine[x].blue)/3;
      if (bright < 0.5)   ThisLine[x+1] = 0;  // has +1 shift in x
      else                ThisLine[x+1] = 1;  // to deal with border
   } // for x
   WriteTempImageLine(Bin_ID, y+1, ThisLine);// has +1 shift in y
} // for y
// classic erosion/dilation
for (y = 0; y < height; y++)
{
   ReadTempImageLine(Bin_ID, y  , LineAbove);          // 3 line neighborhood
   ReadTempImageLine(Bin_ID, y+1, ThisLine);
   ReadTempImageLine(Bin_ID, y+2, LineBelow);
   for (x = 0; x < width; x++)
   { // add up number of "ON" pixels in neighborhood
      float NeighborCount = 0;
      NeighborCount += LineAbove[x] + LineAbove[x+1] + LineAbove[x+2];
      NeighborCount += ThisLine[x]                   + ThisLine[x+2];
      NeighborCount += LineBelow[x] + LineBelow[x+1] + LineBelow[x+2];
      Result[x+1] = ThisLine[x+1];                    // keep current value
      // override value depending upon neighbors
      if ((Erosion) && (ThisLine[x+1] == 1))
         if (NeighborCount < 0.5)  Result[x+1] = 0;
      if ((Dilation) && (ThisLine[x+1] == 0))
         if (NeighborCount >= 0.5) Result[x+1] = 1;
   }
   WriteTempImageLine(Bin_ID, y, Result);
} // for y
// write the binary image back to the host, account for (x+1,y+1) positions
for (y = 0; y < height; y++)
{
   ReadTempImageLine(Bin_ID, y+1, ThisLine);
   for (x = 0; x < width; x++)
   {
      if (ThisLine[x+1] < 0.5)                         // has -1 shift in x
         OrigLine[x].red = OrigLine[x].green = OrigLine[x].blue = 0;
      else
         OrigLine[x].red = OrigLine[x].green = OrigLine[x].blue = 255;
   }
   WriteResultLine(y, OrigLine);                       // has -1 shift in y
} // for y
```

Erosion and dilation change the size of features, shifting boundaries from the positions that are located by thresholding. Usually, this is not desirable. If an erosion is followed by a dilation, the boundaries are returned to approximately their original location. However, in the process, the boundaries are smoothed, and fine lines and isolated white noise pixels are removed. The sequence of erosion followed by dilation is called an ___opening___. Figure 4.12d illustrates this process and shows the result in comparison to the original pixels.

When the same two operations are applied in the opposite sequence, a dilation followed by an erosion, the process is called a ___closing___. The result is to fill in isolated black noise pixels and gaps, and to smooth irregular boundaries. This is also illustrated in Figure 4.12g.

Comparing the results in Figures 4.12d and 4.12g, which show many differences, it is natural to ask "which is correct?" The problem is that the original figure is ambiguous. By themselves, the original pixels do not contain enough information to decide whether a closing or an opening is the appropriate operation to apply. Usually, the nature of the original image and the purpose of obtaining the binary image (e.g., for measurement) will indicate whether an erosion or dilation is the proper choice. Indeed, sometimes it is necessary to use both operations.

When applied to a more complex image, the effects of opening and closing are evident. Prior illustrations in Figure 4.8d and the thresholded binary images in Figure 4.11 used these morphological procedures to clean up the images produced by the initial thresholding. In Figure 4.13, an opening can be used to fill in the gaps in the bamboo stems and connect all of the leaves to the branches, while filling in the white holes in the solid black areas. Conversely, a closing removes the branch lines so that the leaves are separated, and removes the black spots in the background. Choosing which result is "correct" depends on the uses for which the image is intended.

The same logic used for erosion to remove the pixels along the boundaries of features can also be employed to remove all pixels except those that are immediately adjacent to the background. This produces a result with just the outlines of features, which will be useful later in this chapter and in Chapter 5 for locating or measuring feature boundaries.

If the outline pixels are defined as feature pixels that touch background pixels by sharing a side (i.e., a 4-neighbor rule), the result is an outline that is 8-connected, meaning that the outline is continuous if pixels are considered to touch any of their 8 neighbors (Figure 4.14a). Conversely, if the outline pixels are defined as feature pixels that touch background pixels if they share either a side or corner (i.e., an 8-neighbor rule), then the resulting outline is 4-connected, meaning that its pixels are continuous when sharing a side (Figure 4.14b). The duality and distinction between 4- and 8-connectivity is important. A 4-connected line will separate 8-connected regions, for example. This subject arises several times throughout this chapter.

Classic erosion and dilation, which flip the status of a pixel that touches any adjacent pixel of the opposite state, can be very effective but suffer from several limitations. Classic morphology works well for smoothing the boundaries of large regions, but can produce too drastic a change when applied to fine lines or narrow features. A modification of the rules offers more control over the results. Changing the procedure to a conditional one that counts the number of touching neighbors, and only alters the state of a pixel if more than $N$ of the neighbors are of the opposite color, can be accomplished simply by testing the sum of touching neighbors in Code Fragment 4.7.

(a)

(b)

(c)

**FIGURE 4.13**
Morphological processing: (a) original image [panda.tif], (b) opening, (c) closing.

The value of $N$ can be set from 0 (classic erosion and dilation) to 7. The latter setting changes a pixel's value only if all eight of the neighbors are of the opposite color, and will only affect isolated points. This offers a way to clean up single-pixel noise without altering anything else in the image. The single white and black pixels in the image in Figure 4.15a can be selectively removed with an opening or closing using $N = 7$.

Other values for $N$ produce different results. Figure 4.16 shows that conditional erosion with a value of $N = 3$ removes lines that are one pixel wide, but not ones that are two pixels wide, although it does shorten them slightly and erase the corners of the 3-pixel-wide squares. This ability to control the conditional erosion can be useful for removing lines of pixels that result from thresholding. This situation arises, for instance, when pixels along boundaries have brightness values that are the same as those in other selected features, as shown in Figure 4.17. Figure 4.15b shows the result of applying a closing with $N = 3$. There are subtle differences from the result of a classic closing shown in Figure 4.13c, including the filling in of white noise pixels.

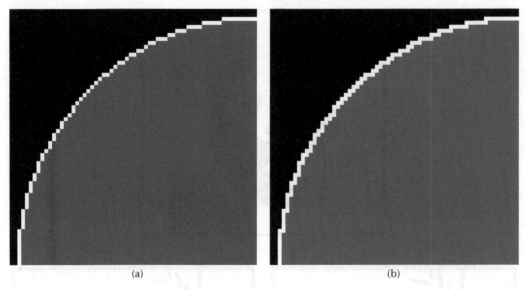

**FIGURE 4.14**
Outline of a feature [circle.tif] enlarged to show individual pixels: (a) 8-connected line of pixels that share a side with a background pixel, (b) 4-connected line of pixels that share either a side or corner with a background pixel.

**FIGURE 4.15**
Conditional morphology applied to the same image as Figure 4.13a [panda.tif]: (a) opening, $N = 7$; (b) closing, $N = 3$.

### 4.2.2 The Euclidean Distance Map

A potentially serious problem with classical and conditional morphological operations is their anisotropy. Treating all of the pixels in the 8-neighbor region equally ignores the fact that the four corner-sharing pixels are farther away from the central pixel than the four side-sharing neighbors. For a single erosion or dilation, this has little consequence. However, sometimes it is desirable to repeat the process several times, for example, to remove protrusions or fill in gaps of greater dimensions. As will be seen, such multiple erosions or dilations may also be used as part of the process by which the measurement of the distances of features from a boundary can be performed.

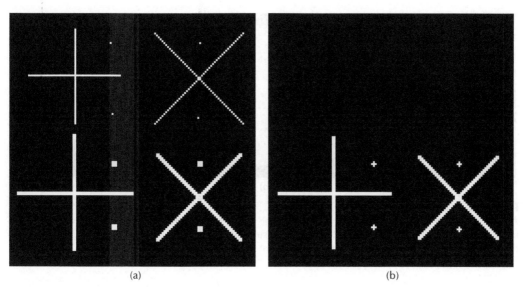

(a)                                    (b)

**FIGURE 4.16**
Conditional erosion: (a) original image [lines.tif], (b) removing white pixels that touch more than three background pixels.

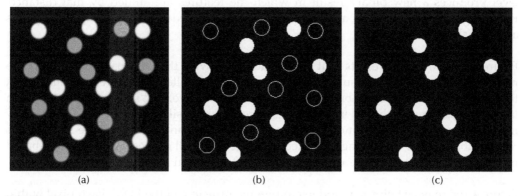

(a)                     (b)                     (c)

**FIGURE 4.17**
Conditional erosion: (a) original image [graycircles.tif] containing slightly blurred circles with two different brightness levels, (b) thresholding the gray circles also selects lines of gray pixels around the white ones, (c) an opening applied to (b) with $N = 3$ removes the lines.

Ideally, repeating an erosion or dilation should remove or add pixels *__isotropically__* (a uniform distance in all directions), preserving shape. Figure 4.18 shows the effect of multiple erosions and dilations on a circle. Instead of producing a larger or smaller circle, other shapes appear. With the value $N = 0$, repeated classical erosion produces a diamond shape because the pixels lying in the 45° directions have a dimension that is $\sqrt{2}$ larger than those in the 90° directions. Conversely, repeated dilation produces a shape that evolves toward a square. Using other values for $N$ in conditional erosion or dilation produces other shapes, but none of these results in an isotropic circle. Note also that the distances covered by the erosion and dilation procedures vary for different values of $N$ as well as varying in different directions, and would be difficult to use for measurement purposes.

Much more consistent and isotropic results can be obtained by using a different approach to erosion and dilation. The *__Euclidean distance map__* (EDM) is an image with grayscale

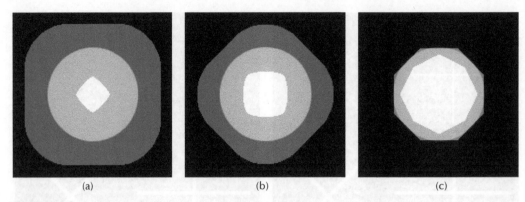

**FIGURE 4.18**
Anisotropic results from morphological processing. The original circle [circle.tif] is shown in gray with super-imposed erosion (white) and dilation (darker gray) results after 25 repetitions using values of $N$ (the number that the number of neighbors of the opposite state must exceed) equal to: (a) zero, (b) one , (c) three.

values that is generated from a thresholded binary, such that each nonbackground pixel is assigned a value that measures its distance from the nearest background pixel, in any direction (Danielsson 1980). Thresholding this grayscale image at any desired brightness level produces a new binary result in which the features are eroded by that distance. Reversing the image contrast and applying the same procedure to the background pixels accomplishes dilation.

Figure 4.19 shows the results that can be achieved with this approach. In addition to producing more isotropic results in which larger and smaller circles are generated from the original figure, the EDM-based approach is not iterative, and so is much faster when large distances are involved. Finally, the distances can be easily controlled, and are directly measured in pixels. With some implementations, the distances can even be specified as real numbers rather than being restricted to integers. For all of these reasons, EDM-based morphological operations are usually preferable to classical methods.

Figure 4.19d may be helpful in understanding the principle of the EDM. Each pixel within the original feature has a value that measures the distance from that location to the nearest point in the background. When these values are plotted as elevations so that the EDM can be visualized as a surface, the result for a circle is a cone whose surface has a slope of 1 (it rises by one pixel for each pixel of radius). Slicing this cone at any elevation, which is accomplished by thresholding, produces a smaller circle with the corresponding amount of erosion.

Generating the EDM from the binary image can be accomplished in several ways. The oldest and simplest method requires just two passes through the image, once downward in the usual scanning order and once upward in the reverse order. On the first pass each pixel within the feature is compared to four of the eight neighbors, three in the line above and one to its immediate left, as shown in Figure 4.20. The pixel value is set to one more than the smallest of its neighbor values, and so counts upward with distance from the top and left sides. On the second pass, which proceeds in the opposite direction, the pixel value is compared to its remaining four neighbors, below and to the right, and is set to one greater than the smallest of those.

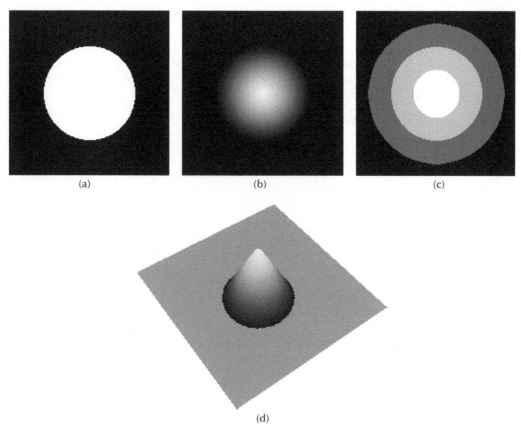

**FIGURE 4.19**
Using the EDM for erosion and dilation: (a) original circle; (b) EDM values assigned to pixels within the circle; (c) result of erosion (white) and dilation (dark gray) by 25 pixels superimposed on the original circle (compare to Figure 4.18); (d) the EDM values in image (b) visualized as a surface, in which the pixel brightness values are represented as elevation.

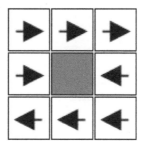

**FIGURE 4.20**
In generating the EDM, the central pixel (gray) is compared to the four neighbor pixels above and to the left on the initial pass which proceeds downward from the upper-left corner. Comparison is made with the four remaining neighbors on the second pass, which proceeds upward from the lower right corner of the array.

Note that in this procedure, unlike the processing performed in Chapter 2 and the erosion/ dilation procedures shown above, the resulting values are NOT stored in a new array. Instead, the results are written back into the same line from which they came, so that the results are propagated across and down (or back and up) through the image as the scan proceeds.

Code Fragment 4.8 implements this procedure. Notice that the temporary image holds the initial binary values of 0 and 1, and is padded by one pixel on all sides to protect the image boundaries.

```
// Code Fragment 4.8 - Generate EDM
// ... declare variables
// ... create pointers
//          (float*)PrevLine, (float*)ThisLine, (RGBPixel*)OrigLine
// ... create temp image 'Bin_ID' in Code Fragment 4.7
//     padded on all sides and filled with binary 0/1 values
// 1. Downwards sweep,
for (y = 1; y < height + 1; y++)                    // has +1 offset in y
{
   ReadTempImageLine(Bin_ID, y-1, PrevLine);
   ReadTempImageLine(Bin_ID, y , ThisLine);
   for (x = 1; x < width+1; x++)                    // has +1 offset in x
      if (ThisLine[x] > 0)
      {  // val = lowest neighbor + 1
         float nbr1, nbr2, nbr3, nbr4;
         nbr1 = PrevLine[x-1] + 1;                    val = nbr1;
         nbr2 = PrevLine[x  ] + 1; if (val > nbr2) val = nbr2;
         nbr3 = PrevLine[x+1] + 1; if (val > nbr3) val = nbr3;
         nbr4 = ThisLine[x-1] + 1; if (val > nbr4) val = nbr4;
         ThisLine[x] = val;
      }
      else
         ThisLine[x] = 0;
   WriteTempImageLine(Bin_ID, y, ThisLine);
} // for y
// 2. Upwards sweep
for (y = height+1; y > 0; y--)                       // reverse loop on y
{
   ReadTempImageLine(Bin_ID, y+1, PrevLine);
   ReadTempImageLine(Bin_ID, y  , ThisLine);
   for (x = width + 1; x > 1; x--)                   // reverse loop on x
   {
      float val = ThisLine[x];
      if (val > 0)
      {  // val = lowest neighbor + 1
         float nbr1, nbr2, nbr3, nbr4;
         nbr1 = PrevLine[x+1] + 1; if (val > nbr1) val = nbr1;
         nbr2 = PrevLine[x  ] + 1; if (val > nbr2) val = nbr2;
         nbr3 = PrevLine[x-1] + 1; if (val > nbr3) val = nbr3;
         nbr4 = ThisLine[x+1] + 1; if (val > nbr4) val = nbr4;
```

```
         ThisLine[x] = val;
    }
    else
         ThisLine[x] = 0;
  } // for x, reverse
  WriteTempImageLine(Bin_ID, y, ThisLine);
}// for y, reverse
// 3. write back result to host
for (y = 0; y < height; y++)
{
  ReadTempImageLine(Bin_ID, y+1, ThisLine);          // has +1 offset on y
  for (x = 0; x < width; x++)
  {
    val = ThisLine[x+1];                             // has +1 offset on x
    if (val>255) val = 255;                          // could autoscale...
    OrigLine[x].red = OrigLine[x].green = OrigLine[x].blue = val;
  } // for x
  WriteResultLine(y, OrigLine);
} // for y
```

The routine shown in Code Fragment 4.8 does not produce a very good EDM. As shown in Figure 4.21, the distances are not measured equally in all directions. Superimposed contour lines drawn at brightness levels of 25, 50, and 75 are not ideal circles. Just as with the iterative classical erosion procedures, the reason is that the corner-touching neighbor pixels are farther away from the central pixel than those that share a side. Changing the comparisons in the code to those shown in Code Fragment 4.9 produces a better result, by using values for the two side-sharing positions of 1.0 and for the two corner-sharing

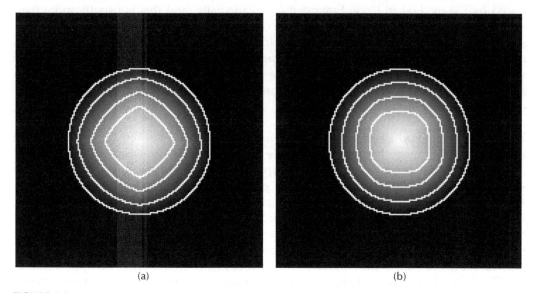

(a)                                                                (b)

**FIGURE 4.21**
Anisotropy of the EDM generated by code fragments 4.8 (a) and 4.9 (b) is shown by superimposing contour lines drawn at distances of 25, 50, and 75 pixels from the boundary.

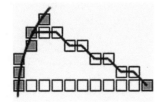

$$\sqrt{9^2 + 4^2} = 9.85$$
$$5 \cdot 1 + 4 \cdot \sqrt{2} = 10.66$$

**FIGURE 4.22**

Cumulative distance error at a nominal radius of 10 pixels. The Pythagorean distance to the marked pixel is slightly less than 10 pixels, whereas the cumulative EDM value is greater.

positions of 1.414 (the square root of 2), but at greater distances errors still accumulate. For example, at a radius of 10 pixels and an angle of about 30°, Figure 4.22 shows the cumulative EDM value of 10.66 pixels, greater than the theoretical result of 9.85 pixels.

```
// Code Fragment 4.9 — Improved isotropy in EDM
if (ThisLine[x] > 0)
{   // compute new val
    float nbr1, nbr2, nbr3, nbr4;
    nbr1 = PrevLine[x-1] + 1.414;                     val = nbr1;
    nbr2 = PrevLine[x  ] + 1.000; if (val > nbr2) val = nbr2;
    nbr3 = PrevLine[x+1] + 1.414; if (val > nbr3) val = nbr3;
    nbr4 = ThisLine[x-1] + 1.000; if (val > nbr4) val = nbr4;
    ThisLine[x] = val;
}
// ... have to do upwards sweep case, too ... ThisLine[x+1]
```

For erosion and dilation by distances up to about 30 pixels, thresholding the EDM produced by the method in Code Fragment 4.9 is adequate, and this is usually more than enough for morphological processing. Some of the other uses of the EDM are particularly sensitive to the accuracy and isotropy with which it is generated. A more accurate method can be implemented by using two temporary arrays, one of which keeps track of the vertical distances from background and the other the horizontal distances. In the final step, these are combined using the Pythagorean theorem to calculate the exact distance of each pixel from the background.

When constructing an EDM with the code examples shown, it is important to consider how to treat the image boundary. The code fragments shown use the same boundary protection method introduced for the morphological operations, namely, copying the image into a temporary array that is one pixel larger on all sides than the original. However, it is necessary to decide what value to write into those bordering pixel locations. If a zero is used for the border, it forces the EDM to measure distances from the image boundary as well as from feature boundaries, as shown in Figure 4.23b. Assigning a large positive value causes the EDM result to measure distances only from the feature boundaries as shown in Figure 4.23c. Which is desired depends on the intended application of the EDM and *a priori* knowledge about the meaning of the image boundaries. When dilation is accomplished by thresholding the EDM of the background, the method shown in Figure 4.23c is appropriate.

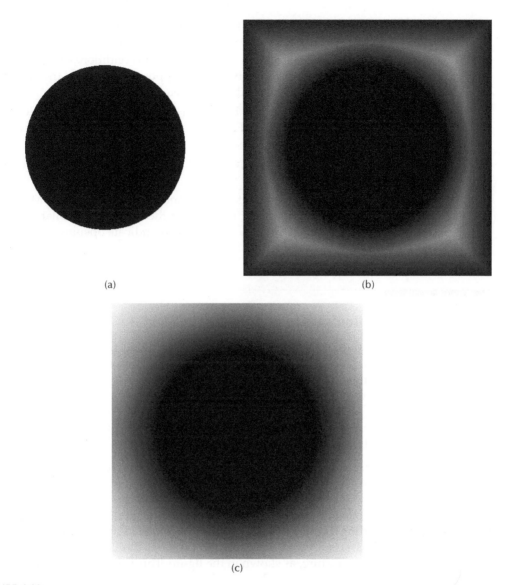

(a)

(b)

(c)

**FIGURE 4.23**
Treating the boundary pixels in EDM construction: (a) original image (the [circle.tif] image with contrast reversed), (b) EDM constructed with the boundary set to zero, (c) EDM constructed with the boundary set to a large positive value.

### 4.2.3 Problems

4.2.3.1. Implement a program that performs classical erosion and dilation on a binary image (treat any pixel value with brightness less than 0.5 out of 255 as background, and ones greater than 0.5 as feature).

4.2.3.2. Implement a program that combines classical erosion and dilation to perform an opening or closing. Note that this will require storing a temporary image to hold the intermediate result.

4.2.3.3#. Modify the program from Problem 4.2.3.2 to allow multiple iterations. Note that a closing of $R$ repetitions is performed by first performing $R$ dilations and then *R* erosions, and vice versa. Read the number of iterations and a flag value to indicate whether opening or closing is intended from a text file.

4.2.3.4.    Modify the program from Problem 4.2.3.3 to perform conditional morphology based on the number of touching pixels of the opposite state. Read the number from the text file along with the other parameters.

4.2.3.5#.  Implement a program to generate a Euclidean distance map using the neighbor comparison scans shown in Code Fragments 4.8 and 4.9. Autoscale the result to display the pixel values.

4.2.4.6.    Modify the program from Problem 4.2.3.5 to generate the EDM of the background rather than the features.

4.2.4.7.    Implement a program that performs opening and closing using the EDM. Read a text file to obtain the distance in pixels for the erosion and dilation (the threshold applied to the EDM) and a flag value to indicate opening or closing.

---

## 4.3   Other Morphological Operations

Conditional erosion and dilation procedures, either based on neighborhood comparisons or on the Euclidean distance map, can be used for additional morphological operations. These include measurement or separation of touching features, and formation of the skeleton, which often represents the topological shape of features.

### 4.3.1   Ultimate Points and Watersheds

When the EDM is calculated for a feature that is not circular, the result can be visualized as shown in Figure 4.24. The slope of the sides of the irregular mountain is still a constant. The highest point in the mountain, corresponding to the maximum value in the EDM, is termed the ***ultimate eroded point***, or UEP. As the name indicates, if erosion is applied to the feature by thresholding the EDM, the last point to be removed would be this maximum. The numerical value assigned by the EDM to the pixels at the UEP is the radius of an inscribed circle within the feature, as shown in the figure. This provides a useful measurement tool in some situations.

When applied to a feature that is not convex, the EDM may generate a set of values that has more than one peak. Most important, when applied to an image that contains several convex but touching or slightly overlapping features, the EDM makes it possible to separate them as individual features for counting or measurement.

The UEPs can be counted directly to determine the number of such features, and for some purposes the values of the UEP pixels (the radii of inscribed circles as noted above) can be used as measures of their sizes. Note, however, that the UEP may not be a single pixel. Depending on the dimension and shape of the original feature, it may be a ridge or a plateau two pixels wide.

For many applications, it is desired to actually draw in lines of separation, as shown in Figure 4.25. This technique is called ***watershed segmentation*** (Beucher and Meyer 1992;

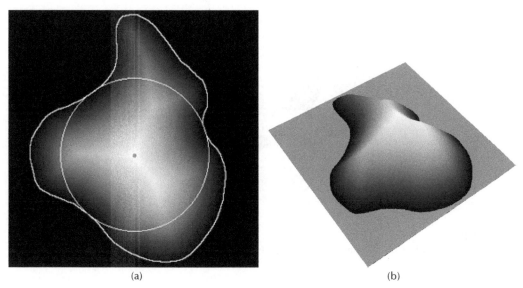

**FIGURE 4.24**
EDM of an irregular feature, shown by the outline in (a). The maximum value, shown as a point in (a) and the peak in the visualization in (b), is the UEP, whose EDM value is the radius of the inscribed circle shown in (a).

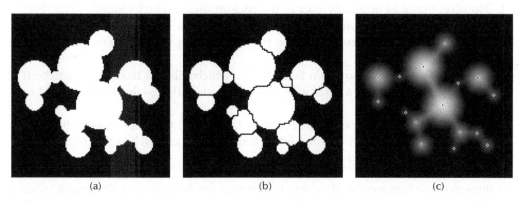

**FIGURE 4.25**
Watershed segmentation: (a) original [overlap.tif], (b) watershed result, (c) EDM of the features in image (a) with superimposed UEPs.

Russ and Russ 1988), and is best explained by using the visualization in Figure 4.26 that represents the EDM of the multiple touching features as a surface in which the elevation is the EDM value. If this surface is considered to represent a mountain range, rising from a background that is considered to represent an ocean, then imagine the process of rain falling on the surface. Raindrops land on a mountain and run down the sides to the ocean. If raindrops from two different directions meet (originating as drops that fell on different mountain peaks), then the line where they meet is the line that separates the mountains into different watersheds. Erasing the lines where this meeting occurs separates the original convex, but overlapping, features. (Note that some texts use an equivalent explanation in which the EDM values represent depressions rather than mountains and the water fills them from the bottom.)

**FIGURE 4.26**
Visualization of the EDM from Figure 4.25 as a surface. Each conical peak represents a convex feature to be separated.

A practical algorithm for implementing a watershed proceeds by dilating from the UEPs. The dilation can be carried out iteratively using classical morphological procedures, but must follow special rules:

1. No pixel that was originally part of the background can be included.
2. Each pixel that the dilation reaches must be given a unique identification that corresponds to the initiating UEP, to prevent dilating features from touching or merging.
3. Only pixels that have not been labeled (i.e., which dilation has not yet reached) and are not adjacent to a pixel with a different identifying label can be added to a dilating feature.
4. Dilation does not begin from the multiple UEPs all at the same time. Rather, the highest values start first, and the lower values begin only when the dilation has proceeded down to their level.

The steps in Code Fragment 4.10 outline this procedure. Beginning at the largest value in the EDM, the pixels are filled in with conditional dilation by assigning them a unique value for each peak, and must not touch a pixel that is already labeled. There are multiple steps involved but most of them such as creating an EDM and performing dilation have been illustrated above, and one step (assigning unique identification numbers to features) is shown in detail in the next chapter. Several optional steps are included that produce improved results in many practical cases.

```
// Code Fragment 4.10 — Watershed separation of convex features
1. Create a temporary image and generate the EDM
2. Optionally, smooth the EDM. This can reduce or eliminate problems with pixels
   along ridges that have EDM values slightly greater than their neighbors and would
   be found as UEPs. Alternatively, require that a pixel exceed its neighbors by a
   defined difference value to be considered a UEP in the next step.
3. Scan through the EDM to find UEPs and the maximum and minimum UEP values. Note
   that each UEP may cover more than a single pixel.
4. Optionally, dilate the UEPs slightly to merge points that are close together and
   presumably represent points along a feature ridge. The amount of dilation can be
```

determined as a fraction (e.g., 1/3 or 1/4) of the minimum UEP value to prevent dilating outside the original feature.

5. Scan through the binary image of the dilated UEPs and assign each one a unique ID number. The procedure for numbering arbitrarily shaped features is shown in detail in Chapter 5.

6. Starting at the maximum UEP value, iteratively decrement a threshold value and scan through the EDM to test values that are at or above the threshold. For each one that is found, test the eight neighbors of that pixel in the binary image. If there is a single ID value present, assign it to the current pixel. If there is more than one ID value present, the pixel lies on the boundary between dilating regions and must be set to black.

7. Finally, rewrite the image with all nonzero pixels set to white.

There are other methods for performing the watershed segmentation that do not require the step (5 in the preceding outline) of assigning unique values to the individual UEPs. These rely on a series of "partial" or "directional" dilations to accomplish an equivalent result. As the iterative decrementing of the threshold level is carried out (step 6) and the dilating UEP image is scanned, the rule for dilation is changed. The ON-or-OFF status of the eight neighboring pixels is used as shown in Figure 4.30 to construct an number from 0 to 255. The resulting number is used as an index into a table (usually called a fate table) that determines whether the central pixel is to be turned ON, or not.

In a typical series of dilations, there are four such tables, which correspond to dilating the feature edges in each of four directions, and only if the consequence would not cause any of the neighboring pixels that are ON to touch if they were not already adjacent. The partial dilation steps in each direction are applied one after another at each iteration of the dilation sequence. This method has the advantage of not requiring the feature identification step, but because the equivalent logic is applied at each iteration by using the multiple partial dilations, it is somewhat more complicated. Both methods produce similar results.

Watershed segmentation is often described as being applicable to convex features, but actually the required condition is that the EDM have only a single UEP for each feature. Figure 4.27 shows images of touching, curved features (a scanned image of cashew nuts, thresholded to produce a binary image) that can be separated using this technique. On the other hand, an image of a line of nominally constant width but with steps along the sides arising from the pixel grid may be broken up by the watershed because the single-pixel variations in width produce multiple peaks in the EDM. Such problems may be dealt with by performing a few repetitions of dilation of the UEPs without the constraint against touching, before continuing with the regular procedure, as shown in step 4 of the outline in Code Fragment 4.10.

In general, the quality of results from a watershed procedure depends strongly on the quality of the EDM that is used, the size of the neighborhood used to define a UEP, and the procedure used to dilate the UEPs without allowing them to merge.

### 4.3.2   Skeletons

The *skeleton* of a feature is a set of connected or branching lines comprising the pixels that are left after a repeated erosion that removes all background-touching pixels until the only remaining ones are those that touch background on both sides. It has been likened

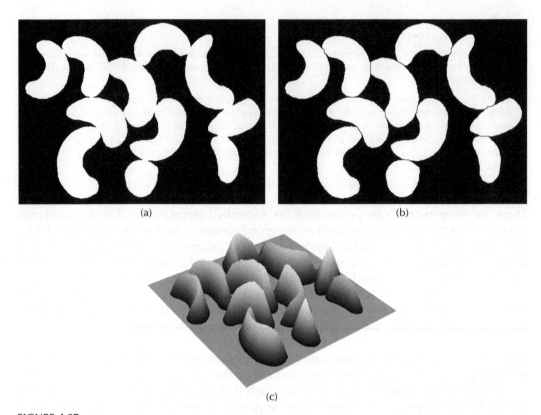

**FIGURE 4.27**
Watershed segmentation of nonconvex features: (a) original binary image [cashews.tif]; (b) watershed result;
(c) visualization of the EDM as a surface, showing the ridgelike shapes with a single peak for each feature.

to the final line of ash that remains after a grass fire that burns into the feature from all sides. In most implementations the skeleton is an 8-connected line of pixels (i.e., each pixel except for those at ends and branch points touches exactly two others, which may be either side-sharing or corner-sharing).

As shown in Figure 4.28, the skeleton of a feature may not correspond to a physical structure within the object. The major veins in the leaf form a set of midlines in the sections of the leaf, for obvious physical reasons, but the skeleton of the shape does not lie on the veins. However, the skeleton does capture the basic topological shape of a feature. As shown in Figure 4.29, the skeleton branches correspond to the teeth, and the number of end points in the skeleton (which will be counted in examples that follow) gives the number of teeth in the gear.

The procedure for generating the skeleton from a binary image is iterative. Every pixel that touches the background is tested. If it does not form part of the skeleton, it is removed. A skeleton pixel is defined as one that either has only a single neighbor that is not background (i.e., the pixel is an end point) or has two or more feature neighbors that are not adjacent to each other. If the neighboring feature pixels touch each other, either by sharing a side or corner, then the pixel being tested can be removed without breaking the feature into separate pieces, and is therefore not an essential part of the skeleton. In addition, the skeleton should be as well centered (equidistant from the original boundaries of the feature) as possible.

(a)            (b)

**FIGURE 4.28**
Skeletonization: (a) original image [leaf.tif]; (b) the skeleton lines (dilated for visibility), which represent the topology of the leaf, but do not coincide with the major veins.

The preceding description of what constitutes a skeleton does not specify a method for creating one. The procedure for removing all of the pixels that are not part of the skeleton can be accomplished in a variety of ways and there is a rich literature on the subject (Davidson 1991; Lam, Lee, and Suen 1992; Levialdi 1972; Lu and Wang 1986; Pavlidis 1980; Russ 1984). The most widely used technique (Zhang and Suen 1984) removes pixels from the feature according to the following set of rules that examine each pixel and its eight immediate neighbors (numbered according to the diagram in Figure 4.30a).

Pixel removal is accomplished in two alternating steps, using slightly different rules that are designed to keep the resulting skeleton centrally located in the feature. The same thing could be done by using a single set of rules but alternately scanning the image from the top-left corner downward, and from the bottom right corner upward, as was done in forming the EDM in the preceding section. The rules for removing a pixel (changing its value to zero and its color to black) are as follows:

1. The original (central) pixel must be nonzero (pixels that are already background can be ignored).

2. At least two and not more than six of the eight neighboring pixels must be nonzero.

3. The number of transitions from feature (nonzero) to background (zero) proceeding around the neighborhood must be exactly one.

4. For the odd-numbered alternating steps, at least one of the pixels at positions 1, 3, and 5 must be zero.

5. For the even-numbered alternating steps, at least one of the pixels at positions 3, 5, and 7 must be zero.

Rule 3 requires some additional explanation. Starting at any position in the neighborhood, for instance, at pixel position 1 in the figure, proceed around the neighbors in order, for instance, clockwise. In traversing the pixels in order 1, 2, 3, 4, 5, 6, 7, 8, and 1, count the number of times that a nonzero (feature) pixel is followed by a zero (background, or black)

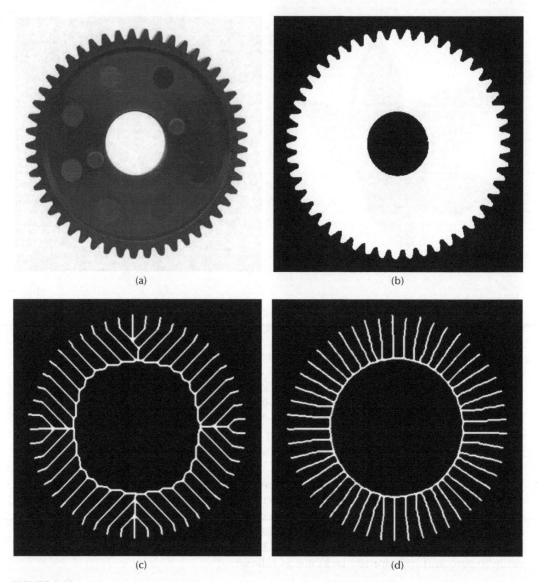

**FIGURE 4.29**
Skeletonization: (a) original image [gear.tif], (b) binary image after thresholding and performing a closing to eliminate noise pixels and smooth boundaries, (c) skeleton produced by the procedure in Code Fragment 4.11, (d) skeleton produced by using four erosion tables rather than two. The skeletons have been dilated for visibility.

pixel. This is sometimes called the transition number. If this value is exactly one, it indicates that the adjacent pixels all touch each other and, consequently, the central pixel is not an essential part of the skeleton and may be removed.

Rules 4 and 5 are applied on alternating sweeps through the image, to prevent bias in the position of the final skeleton. Also note that as the sweep proceeds through the image, any change in a pixel is not used in tests of the subsequent pixels to which it is a neighbor. In other words, the calculation is applied simultaneously to all of the pixels in the image, and it is necessary to keep a copy of at least the three lines of pixels used in the neighbor

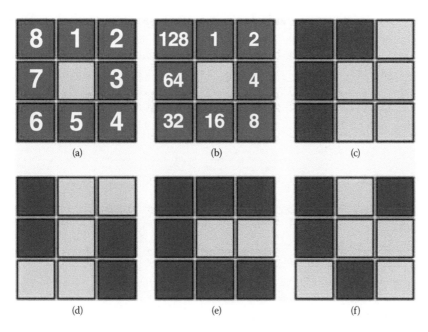

**FIGURE 4.30**
Local pixel neighborhoods used in skeletonization: (a) numbering of the local pixels for identification; (b) index values used in Code Fragment 4.11 (the values for those neighbor pixels that are not background are added together to create the index); (c) a pattern of pixels (index number 30) that does allow removal of the central pixel because the neighbor pixels in positions 2, 3, 4, and 5 are all continuously adjacent; (d,e,f) patterns of pixels that do not allow removal of the central pixel (index numbers 51, 4, and 45, respectively). In (d) the pixels in positions 1 and 2 are not adjacent to those in positions 5 and 6. In (e) there is a single neighbor in position 4. In (f) the pixels in positions 1, 3, and 4 are adjacent to each other, but not to the pixel in position 6.

tests, rather than modifying them as they are used. This is different from the procedure used to generate the EDM above, in which changes are propagated across each line.

It is quite possible to implement these rules directly, but the resulting code tends to be rather specialized and hard to follow. A commonly used trick to simplify the code, which can also be applied to the other erosion and dilation procedures shown above, is to create a lookup table or "fate" table with 256 entries in it, each one corresponding to one of the 256 possible arrangements of ON (white) and OFF (black) pixels in the 8-pixel neighborhood around the central pixel.

To access this table, an index is constructed by adding together the values shown for each neighbor pixel in Figure 4.30b if that pixel is not background. For example, the pixel patterns shown in Figures 4.30c, d, e, and f are 30, 51, 4, and 45, respectively. The table contains either a zero or one value for each index to indicate that the central pixel should be removed or retained. Once constructed, these tables are applied as shown in Code Fragment 4.11. For the skeletonization rules shown above, two tables are needed, which are applied alternately as repetitive passes are made through the image, until no further changes take place. The rules shown tend to produce lines that are oriented in 90° and 45° directions within the features. Alternative sets of rules that require more than two sequential passes through the image for each iteration, and hence more than two fate tables, produce better results, as shown in Figure 4.29.

```
// Code Fragment 4.11 — Apply neighborhood lookup table for skeletonization.
// build tables: these indices are patterns that turn off pixels
// odd_erase = {3,6,7,12,14,15,24,28,30,48,56,60,62,96,112,120,129,131,135,
//         143,192,193,195,199,207,224,225,227,231,240,241,243,248,249};
// even_erase = {3,6,7,12,14,15,24,28,30,31,48,56,60,62,63,96,112,120,124,
//         126,129,131,135,143,159,192,193,195,224,225,227,240,248,252};
// cleanup = {5,13,20,21,22,23,29,52,54,80,84,88,92,116};
// do
// {
// changes = FALSE
// for y // odd loop
//    read three lines from temp image 1
//    for x
//       if pixel is on
//          access neighbors, build index
//          read odd_erase[] table, decide on/off fate, write to result line
//          if pixel was erased, set changes = TRUE
//    write result line to temp image 2
// for y // even loop
//    read three lines from temp image 2
//    for x
//       if pixel is on
//          access neighbors, build index
//          read even_erase[] table, decide on/off fate, write to result line
//          if pixel was erased, set changes = TRUE
//    write result line to temp image 1
// } while (changes == TRUE);
// for y // cleanup pass
//    read three lines from temp image 1
//    for x
//       if pixel is on
//          access neighbors, build index
//          read cleanup[] table, decide on/off fate, write to result line
//    write result line to temp image 2
// write result image from temp image 2
```

The skeleton that these rules produce preserves topology and consists of lines that are centered within the features, but in some cases the lines are not entirely thinned to the ideal 8-connected minimal line. After the procedure is complete, it is desirable to make an additional pass through the image to remove any 4-connected pixels that remain. These are pixels that share sides with two or more adjacent neighbors, such as shown in Figure 4.31. This can also be accomplished with a lookup table using the same indexing procedure as already shown.

The 8-connected skeleton can be used in several ways to perform measurements of feature shape and size. Some of these will be encountered in Chapter 5. As indicated in Figure 4.30, the number of end points in the skeleton can be used to count the number of teeth

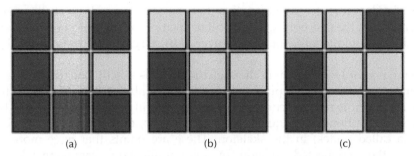

**FIGURE 4.31**
The central pixel in each of these cases can be removed to leave a minimal 8-connected skeleton.

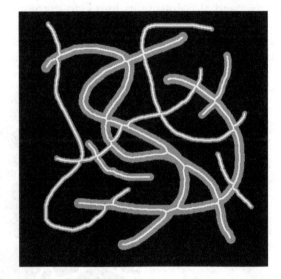

**FIGURE 4.32**
Skeletonizing the features in this image [fibers.tif] and counting the ends (18) provide a convenient way to determine that there are nine fibers present. The skeleton is shown dilated for visibility and superimposed on the original fiber image.

in a gear. Likewise, a modest tangle of fibers as shown in Figure 4.32 can be difficult to count except by skeletonizing, counting the number of end points, and dividing by two. Note that in the case of an image in which some of the fibers extend outside the image boundaries, half the number of end points is still the correct estimate for determining the number of fibers per unit area, because the other end of the fiber appears in another field of view. Chapter 5 discusses the effects of image boundaries on counting and measurement in more detail.

Another example of the use of end points is reading the time from an image of a clock face. Skeletonizing the hands and isolating the end points provides a set of coordinates for each. The distance of each from the center of the face identifies which is the minute and which is the hour hand, and the angles provide a direct way to determine the time. Reading of traditional pre-digital gauges using video cameras and computer image analysis was often accomplished in this way.

It is desirable, therefore, to have a routine that can scan through an skeletonized image to locate and count the end points, and it will be seen below that selecting them and either assigning them a unique value or erasing everything else can be useful as well.

An end point is simply any point in the skeleton that has exactly one neighbor. (Note that it is important to NOT count a pixel with a single neighbor if it is adjacent to the image boundary since that is not an actual end to the skeleton.) Similarly, it is useful to be able to locate, count, optionally uniquely assign a value to, erase, or keep only the branch points (also called nodes) in the skeleton. These are points that have more than two neighbors in the 8-connected skeleton. In some applications, removing the nodes can separate the individual branches for measurement. Note that depending on the arrangement of pixels around the node, removal of the node pixel itself may not separate the branches (which may still touch by sharing the corners of two pixels). This can be solved by removing the node pixel and also erasing its four side-sharing neighbors.

Figure 4.33 shows an example in which the skeletonization is useful for delineating the cell structure, but because of the width and irregularities in the original thresholded image

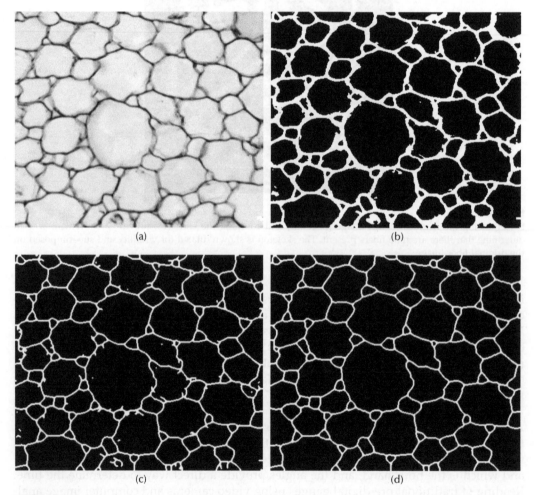

(a)    (b)

(c)    (d)

**FIGURE 4.33**
Pruning a skeleton: (a) original image [corn.tif], (b) thresholded, (c) skeletonized, (d) pruned. The skeletons have been dilated for visibility.

there are some branches that are not a part of the continuous tessellation that correctly represents the structure. This type of problem can be corrected by ***pruning*** the skeleton, or removing those branches that have an end point, as shown.

Pruning takes place after the skeleton has been constructed, and there are two different approaches that can be taken to implement it. The first one scans through the image to locate an end point (a white point having exactly one white neighbor). This point is removed, and then the neighbor point is examined to see if it is an end point. If so, it is also removed and the process repeated. This trims that one branch back to the node from which it comes. Then the scan resumes at the location where the initial end point was found.

Although it is logical and clear this method is often not the preferred approach because of the way that pixels are accessed in memory. To examine a pixel (to see if it is white) and to test the eight neighbors (to see if one of them is white) requires addressing pixels in three adjacent lines of the images. A second approach loads three lines at a time, and scans through the entire image to find and remove end points (if a white pixel is found, its neighbors are counted). If any end points are found, the entire scan is repeated. The process ends when none are found. This method is potentially slower, because many pixels that are not white, or are not end points, must be examined on each scan, but the overall programming may be simpler.

In a few situations, a 4-connected skeleton is needed rather than an 8-connected one. Typically, this circumstance arises when the background is skeletonized, and then the image contrast is reversed so that the skeletonized lines can be used to segment the image into regions or features (for example, the cells in Figure 4.33d). The distinction between 4- and 8-connectedness is also important in counting features and filling holes, as described in the Chapter 5. As shown in Figure 4.34, if pixels are considered to be touching in an 8-connected sense, then reversing the contrast for an image with an 8-connected skeleton does not separate the features on either side, which are also touching in the same 8-connected sense. It is possible to convert an 8-connected skeleton to a 4-connected one. Another lookup table is used, one that fills in black pixels having white neighbors that are 8-connected.

The conflict between 4- and 8-connectedness is inherent in the use of a square pixel grid. Historically, a few systems have tried to use hexagonal pixels in which no such confusion can arise because all of the six neighbors are equivalent and share a side, but because images are captured with square pixels these systems have not proved practical.

### 4.3.3   Outlines and Holes

The skeleton marks the midlines of features, and will be used for several marking and measuring operations in following sections. Similarly, the outline marks the boundary of features. It is defined as just those pixels within a feature that touch the background, as shown above. If the touching rules include background pixels that share a corner, the resulting boundary line will be 4-connected. If the touching rules require that the background pixel share a side, the resulting boundary line will be 8-connected. The procedure is carried out just as with erosion, except that the sense of which pixels are removed and which are retained is reversed. Figure 4.35 shows an example of 8-connected outlines produced by keeping any pixel in a feature that shares a side with a background pixel.

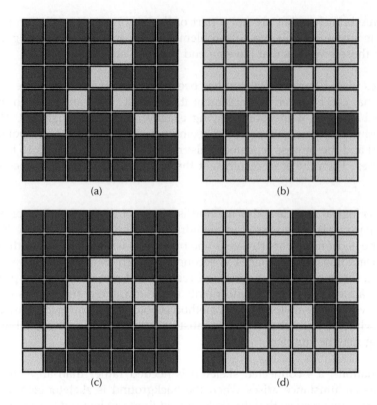

**FIGURE 4.34**
If the pixels in the lines in (a) are treated as connected based on 8-connected logic, then when the image contrast is reversed (b) the lines do not separate the features, which also touch in an 8-connected sense. Converting the line to 4-connected (c) and reversing the image contrast (d) does separate the features.

**FIGURE 4.35**
The 8-connected outlines of the feature area (including outlines of single pixel noise) in the image from Figure 4.13a [panda.tif]. The image has been reversed in contrast to show outline pixels in black for better printing.

A final morphological operation that is often useful on images is to fill any internal holes. This could be used, for example, to fill in the interior of feature outlines that might be obtained by thresholding an image processed to highlight feature edges, as shown in Chapter 2. A hole is defined as any contiguous set of background pixels that do not have a connected path to the boundary of the image. The procedure for performing the hole-filling operation will be deferred until Chapter 5 takes up the task of identifying and labeling features, and determining whether they touch the edges of the image. The process consists of reversing the image contrast so that the holes become features, identifying the features that do not touch the image boundaries, and then adding those pixels back to the original image to fill the holes.

### 4.3.4  Problems

4.3.4.1#.  Implement a program to perform watershed segmentation as described in the text.

4.3.4.2#.  Create lookup tables that implement the rules shown for skeletonization and implement a program to apply them.

4.3.4.3.  Implement a program to process a skeletonized image and apply unique brightness or color values to end points and branch points for display, and to count the end points.

4.3.4.4.  Implement a program to prune a skeletonized image.

4.3.4.5.  Create a lookup table and implement a program to convert an 8-connected skeleton line to a 4-connected one. Hint: To facilitate visualizing the various pixel neighbor patterns and create the lookup table, try arranging markers on a checkerboard.

4.3.4.6#.  Implement a program to produce the outlines of regions by removing pixels that are not adjacent to the background (i.e., have no background neighbors). Compare the results when adjacency is defined as side sharing only, or based on either side or corner sharing.

## 4.4  Boolean Operations

The math routines for combining a previously selected reference image with the current image, shown in Chapter 1, did not include ***Boolean*** operations. These work the same way as, for example, adding the two images together, but are specifically intended for use with binary images. Similar to the mathematical operations shown in Chapter 1, these functions will combine a previously stored reference image with the current image. Because these images still consist of values in the 0 to 255 range, for consistency and to avoid arithmetic precision problems, the procedures shown here treat any value less than 0.5 as black, or background, and any value greater than that as white, or foreground. For purposes of describing Boolean operations, background pixels may be described as "OFF," and foreground pixels described as "ON."

Boolean operations are typically used to combine images that represent selection of regions within the image based on more than one criterion. They are also used to apply various grids of lines or points to images, to facilitate measurement.

## 4.4.1 Multiple Criteria for Selection

Figure 4.11 showed an example of applying multiple selection criteria to isolate a specific structure in a binary image. Beginning with the three binary images produced by thresholding pixels with high values in the red, green, and blue channels, it is possible to isolate the yellow, magenta, red, green, and blue balloons by combining the binaries in various combinations. For example, yellow can be defined as pixels that are selected in both the red and green channels. This is written in Boolean notation as Yellow = Red AND Green, the AND operator meaning that the resulting pixel is ON (white) only if both of the original pixels are ON.

The principal Boolean operators are AND, OR, and Exclusive-OR (abbreviated as XOR). Their meanings are summarized in Table 4.1 and illustrated in Figure 4.36. The OR function selects pixels that are ON in either of the original images, and XOR selects pixels that are ON in either of them, but not both. In other words, the XOR function reveals the differences between images. It is common to include an additional term in the Boolean lexicon, NOT. This simply reverses the contrast in an image, changing all black (background) pixels to white (feature) and vice versa.

In the balloons example of Figure 4.11, the selection of the red balloons was shown to be the Boolean combination of the original red binary image with the derived yellow and magenta images, according to the relationship RED AND NOT (YELLOW OR MAGENTA). Substituting the relationships that defined yellow and magenta would make this even more complicated:

RED AND NOT ((RED AND GREEN) OR (RED AND BLUE))

In general, creating complex expressions with Boolean logic should make liberal use of parentheses to make very explicit the scope and order of the various operations. Also,

**TABLE 4.1**

Boolean Operations

| Original | | Result | | |
|---|---|---|---|---|
| **A** | **B** | **AND** | **OR** | **XOR** |
| ON | ON | ON | ON | OFF |
| ON | OFF | OFF | ON | ON |
| OFF | ON | OFF | ON | ON |
| OFF | OFF | OFF | OFF | OFF |

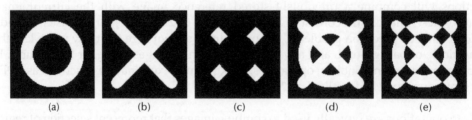

(a)        (b)        (c)        (d)        (e)

**FIGURE 4.36**

Boolean combinations: (a,b) original pixel selections [boolX.tif and boolO.tif], (c) a AND b, (d) a OR b, (e) a XOR b.

note that the four expressions (AND, OR, XOR, and NOT) are redundant, and it is possible to express the same result in more than one way, for example, A XOR B = (A OR B) AND NOT (A AND B).

Code Fragment 4.12 shows the combination of a previously defined binary reference image (which is not altered in the process) with the current image, easily extended to accommodate any of the Boolean combinations. Note that the C-language Boolean and bitwise operators should not be used with floating point variables. Floating point values must be converted to either integer or Boolean values before performing the various logical or bitwise operations.

```
// Code Fragment 4.12 — Boolean combination
// ... declare variables
// ... get both image dimensions and verify that they are the same
// ... allocate (RGBPixel*)OLine, (RGBPixel*)RFile
// ... BooleanMode is (0 = OR, 1 = AND, 2 = XOR, 3 = AND NOT, 4 = NOR, etc.)
for (y = 0; y < height; y++)
{
   ReadOriginalLine(y, OLine);
   ReadRefImageLine(y, RLine);
   for (x = 0; x < width; x++)
   {
      Boolean  orig = (OLine[x].red+OLine[x].green+OLine[x].blue) > 127.5*3);
      Boolean  ref  = (RLine[x].red+RLine[x].green+RLine[x].blue) > 127.5*3);
      Boolean  result;
      switch (BooleanMode)
      {
         case 0: result = (orig || ref);      break;    // OR
         case 1: result = (orig && ref);      break;    // AND
         case 2: result = (orig ^ ref);       break;    // XOR
         case 3: result = (orig && (!ref));   break;    // AND NOT
         case 4: result = !(orig || ref);     break;    // NOR
         // ... etc
      } //switch
      if (result)
         OLine[x] = WhitePix;       //{255, 255, 255} = TRUE
      else
         OLine[x] = BlackPix;/      /{0, 0, 0} = FALSE
   } // for x
   WriteResultLine(y, Oline);
} // for y
```

Of course, any combination of selection criteria, not just colors, may be combined with Boolean operations. The various texture operators shown in Chapter 2 are often candidates for this (e.g., to isolate as regions those pixels that have brightness or color in a certain range AND a high texture value, or directionality in a specific range). Each criterion is represented by a corresponding thresholded binary image, and the Boolean operations are applied sequentially.

### 4.4.2    Grids for Measurements

One of the important uses of the Boolean AND function is combining binary images with various types of grids to facilitate measurements. As a simple example, Figure 4.37 shows a cross section through several layers of paint. The original image is in color, and the different layers can be easily thresholded; in the example, the red layer has been isolated. It is not perfectly uniform, and measurement of its thickness and the variation in that thickness can be accomplished very nicely by creating a grid of vertical lines, spaced at whatever separation is convenient for the purpose. Combining this grid with the binary image of the layer using a Boolean AND produces a set of line segments, which (as discussed in Chapter 5) are easily measured. The mean value, minimum and maximum, and the standard deviation of the line lengths provide the desired information about the layer thickness.

One requirement for using grids to obtain measurements is the ability to generate the appropriate type of grid. Some grids consist of points, either in a regular array or randomly distributed. Other grids consist of lines. These may be straight lines arranged vertically, horizontally, radially, randomly, or at some selected angle. In other applications, circles are useful (either an array of circles, or concentric circles). Other curves are also used in particular situations. The second requirement for using grids is obtaining (and processing) the image in a controlled way so that the desired information is accessible.

It is not the purpose here to present a comprehensive list of the various types of grids nor to enumerate all of their uses, but a few examples may indicate typical cases. The layers in

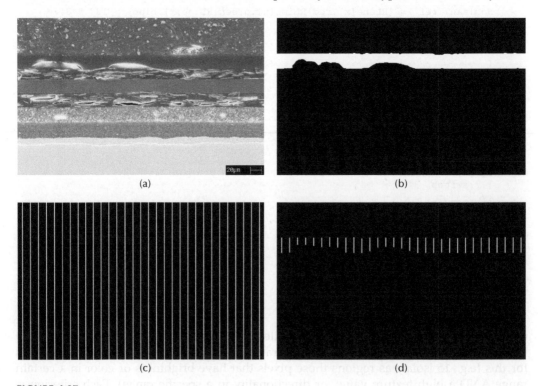

(a)                                                                                        (b)

(c)                                                                                        (d)

**FIGURE 4.37**
(See **Color insert following page 172**.) Measurement with a grid: (a) section through layers of paint [paint.tif], (b) thresholded binary image of a single layer, (c) grid of vertical lines, (d) intersection of the lines with the layer.

the image of Figure 4.37 are horizontal, so a grid of vertical lines is an appropriate measurement tool. If the image was instead a cross section through a cylinder, such as a pipe whose wall thickness was to be measured, the appropriate grid would be a set of radial lines.

In order to sample a beef roast to determine the amount of fat and bone, a set of points that can be ANDed with the binary image of the meat in the roast will allow quick counting to measure the fraction of the image area that is meat (which is the same as the volume fraction of the roast that is meat). Note that this latter example is superior to using the histogram of the original image, because the binary image can be subjected to morphological processing after thresholding to produce a corrected representation of the structure. Also, there are statistical advantages to counting a small number of separated points as compared to counting all of the pixels. Chapter 5 illustrates the use of point grids to measure areas.

In many microscopy applications, the structure of metals and ceramics is characterized by superimposing circle grids and counting the number of intersections the grid makes with the boundaries of the individual grains or crystals. These methods have been used as industry standards for more than a century. In some modern biological applications, such as determining the surface area of structures in the liver or the lungs, grids that have other specific shapes such as a cycloid are used for similar purposes.

Given the desirability of being able to combine a grid with a binary image, it is useful to consider how such grids can be produced. In most cases, a grid image the same size as the binary image with which it is to be combined can be generated in a temporary array, and either used directly or saved on disk for use as required. The work flow for using it in the latter case is to save the grid image as the designated reference image, where it will remain unaltered until it is replaced, and then AND it with a series of binary images to isolate the details of interest and perform the measurement operations.

The grid of straight vertical lines in Figure 4.37c can be produced by a program such as that shown in Code Fragment 4.13 to create a grid with the specified line spacing. A similarly straightforward method can be used to produce horizontal lines.

```
// Code Fragment 4.13 — Vertical line grid
// ... allocate Line of type RGBPixel (width)
long spacing = 20; // horizontal spacing of the lines (can read from a file)
long xoffset = (width - spacing * (long)(width / spacing)) / 2;
// xoffset centers the line grid in the image
for (y = 0; y < height; y++)
   for (x = 0; x < width; x++)
     if (x == (xoffest + spacing * (long)(x / spacing))
        Line[x] = WhitePix;    // {255, 255, 255} = lines
     else
        Line[x] = BlackPix;    // {0, 0, 0} = voids
   WriteResultLine(y, Line)
} // for y
```

Lines at arbitrary angles are typically drawn using code such as the example shown in Code Fragment 4.14. Two points **(x1,y1)** and **(x2,y2)** are specified, and the line is drawn between them. This method can also be adapted to drawing lines through a single point at a specified angle, or to create a set of parallel lines with fixed spacing at an angle, and so on. Circles and circular arcs can also be drawn using the method shown in Code Fragment 4.15. For point grids, either regular spacings or random locations may be desired depending on the application and the nature of the image.

```
// Code Fragment 4.14 — Line drawing
// ... read (x1,y1) (x2,y2) coordinates for end points
// ... create a temporary image (Temp_Image), filled with black (zero)
// ... allocate Line of type unsigned char (width)
dx = x2 - x1;                          // horizontal component
dy = y2 - y1;                          // vertical component
if (dx > dy)
{
   float delta = dy/dx;               // line slope
   float yval = y1;
   for (x = x1; x <= x2; x++)
   { //plot a pixel
     ReadTempImageLine(Temp_Image, (long)yval, line);
     Line[x] = 255;
     WriteTempImageLine(Temp_Image, (long)yval, line);
     yval += delta;                  // advance to next pixel
   } // for x
} // < 45° slope
else
{
   float delta = dx/dy;              // line slope
   float xval = x1;
   for (y = y1; y <= y2; y++)
   { //plot a pixel
     ReadTempImageLine(Temp_Image, y, line);
     Line[(long)xval] = 255;
     WriteTempImageLine(Temp_Image, y, line);
     xval += delta;                  // advance to next pixel
   } // for y
} // > 45° slope

// Code Fragment 4.15 — Circle drawing
// ... read xcenter, ycenter, radius values
// ... get image dimensions: width, height
// ... create a temporary image (Temp_Image), filled with black (zero)
// ... allocate Line of type unsigned char (width)
for (y = ycenter - radius; y <= ycenter + radius; y++)
   if ((y >= 0) && (y < height)) // protect against running off boundaries
   {
     ReadTempImageLine(Temp_Image, y, line);
     dy = y - ycenter;
     for (x = 0; x < width; x++)
     {
        dx = x - xcenter;
```

```
        r2 = dx*dx + dy*dy;
        if (abs(r2 - radius*radius) < radius)
            Line[x] = 255;
    }
    WriteTempImageLine(Temp_Image, (long)yval, line);
} // if within bounds
```

### 4.4.3   Other Combinations

Classic morphological dilation as described above changes any background pixel that is adjacent to a foreground pixel to white. If this procedure is combined with a Boolean operation, a more restrictive type of dilation can be created that is sometimes described as constrained or conditional dilation. Two binary images are involved, one containing the initial points or "seeds" from which dilation will be performed, and a second one that represents the features that will limit the dilation. Figure 4.38 shows a simple example. Some of the features contain one or more seed points, and others do not. Dilation of the seed points is limited to those pixels that are white in the second image, and continues until no further changes occur. The result is an image that selects only the regions that contained the original seed points.

The procedure for performing this conditional dilation uses a designated reference image (corresponding to Figure 4.38a) and processes the seed image (corresponding to Figure 4.38b). Code Fragment 4.16 outlines the steps involved.

```
// Code Fragment 4.16 — Conditional dilation
// ... assumes binary image is stored in the reference image
// ... and temporary image contains the seed points
changes = FALSE;
do
{
    // ... read three lines from temporary image array
    // ... read line from reference array
    // ... for each pixel perform dilation logic from code fragment 4.7
    // ... perform Boolean AND of result with pixel from reference image
    // ... if pixel has changed set changes flag to TRUE
    // ... write line to temporary image array
} while (changes == TRUE);
```

Conditional dilation provides a means of implementing an approach to thresholding called "hysteresis thresholding," which is particularly useful for obtaining images of edges highlighted by processing as shown in Chapter 2. The usual description of the method is that it uses two threshold values. Pixels above the upper threshold are always selected, and pixels below the lower one are always rejected. Pixels between the two thresholds are selected if they are connected to other pixels that lie above the upper threshold or to other pixels that are selected. The success of this method depends on the choice of the thresholds, which are generally adjusted manually.

Figure 4.39 shows how this procedure is carried out using conditional dilation. Processing the image of the coins to highlight edges also highlights some other pixels, some corresponding to detail inside the coins and some at isolated noise in the background.

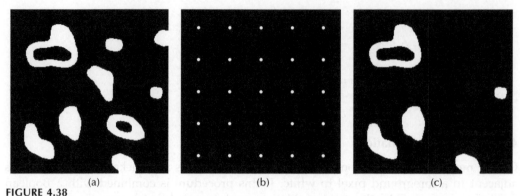

(a)                                          (b)                                          (c)

**FIGURE 4.38**
Conditional dilation: (a) binary image containing various features, (b) grid of seed points, (c) result of performing conditional dilation of the seed points from (b) limited by the features in (a).

Thresholding the image (Figure 4.39c) to select all of these points (the lower threshold in the preceding description) produces a result that does not cleanly delineate the edges of the coins. A second threshold with a higher threshold (Figure 4.39d) isolates only points along the coin edges, but does not completely outline those edges. Dilating the regions in Figure 4.39d, but with the condition that restricts the dilation to those pixels selected in Figure 4.39c produces the desired result (Figure 4.39e).

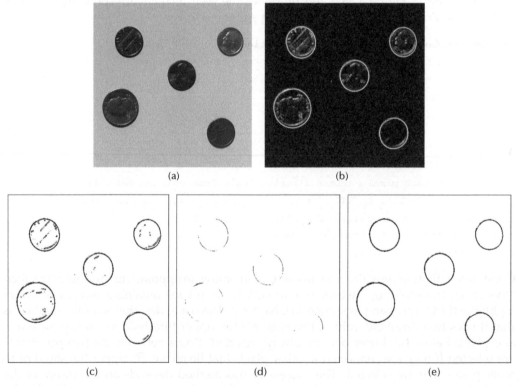

**FIGURE 4.39**
Conditional dilation used to implement hysteresis thresholding: (a) fragment of the [coins.tif] image after subtracting the background, (b) edges highlighted by a Sobel derivative operator, (c) thresholded result from (b) showing all high gradient points, (d) thresholded result from (b) showing only a few highest gradient points, (e) conditional dilation of image (d) limited by image (c). Images (c), (d), and (e) have been reversed in contrast to show the pixels in black for better printing.

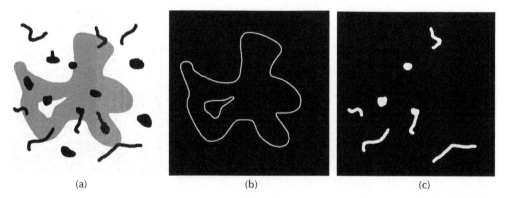

**FIGURE 4.40**
Feature selection as described in the text: (a) composite image showing black features and a gray region, (b) outline of the gray region, (c) selected features that touch or cross the boundary.

Conditional dilation is sometimes called "marker logic," because it allows marker pixels in one image to select features in a second image. Sometimes, these markers may be manually created, for example, by the user clicking on a single point within features that are of interest for measurement. In many cases, the marker is either a set of points or lines obtained from the original image by previous processing. For example, Figure 4.40 shows the selection of dark features that touch or overlap the boundary of the gray region. The outline of the gray region is obtained by morphological processing and used as the marker image. Conditional dilation using the image of the dark features as the reference image to limit dilation produces the desired result.

By using dilation along with conditional dilation, it is also possible to select features within a specified distance of a line or boundary. Figure 4.41 shows schematically a map that might represent major highways (black) and shopping malls (gray) around a typical city. By thresholding the roads and dilating the image by a set distance, a suitable marker image can be created. Performing conditional dilation using the image of the malls as a limiting template, an image of just those malls within the set distance is obtained. If a simple Boolean AND of the dilated highways and malls is used instead, the result might be only a portion of a mall, whereas the conditional dilation recovers the entire feature. Note that for this purpose, the initial dilation of the highway lines using the EDM method

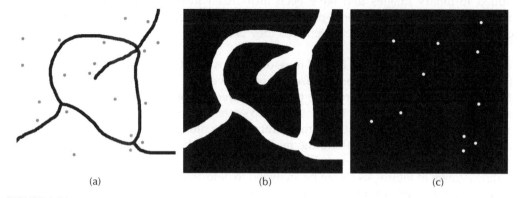

**FIGURE 4.41**
Selection of features within a set distance: (a) original image [map.tif], (b) binary image of roads dilated by a set distance, (c) features selected by conditional dilation using (b) as the marker and the gray features in (a) as the limiting template.

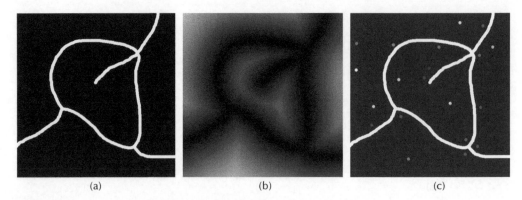

**FIGURE 4.42**
(See **Color insert following page 172**.) Distance measurement using the EDM: (a) binary image of the roads from Figure 4.41, (b) EDM of the pixels outside the roads, (c) result of combining the EDM values with a binary image of the malls in Figure 4.41. The brightness values are shown with false colors that vary from magenta to red, and the roads from image (a) have been superimposed, to assist in visual interpretation of the distances.

is preferred to generate the most isotropic result and to enable proper specification of the distance.

Selection of features by conditional dilation of markers is a powerful tool. However, for the situation shown in Figure 4.41 there is an even more flexible technique that can be used. With the Euclidean distance map, measurement of the distance of every mall from the nearest highway is possible. Figure 4.42 shows the procedure. The binary image of the highways is processed to generate the EDM of the background, which assigns a value to every pixel that represents its distance from the road. Combining this image with the binary image of the malls, keeping whichever pixel value is darker (keeping the brighter or darker value for each pixel is one of the mathematical combinations of two images introduced in Chapter 1, Figure 1.32), produces an image of features in which each one has a brightness value that represents its distance from a highway. Measurement of the brightness values, as discussed in Chapter 5, provides distance information for all of the features. In Figure 4.42c, the grayscale values of the features are shown with a false color rainbow to make it easier to see the variations.

Many of the measurements discussed in Chapter 5 make use of the processing operations applied to binary images that have been introduced in this chapter. It is worthwhile, therefore, to illustrate here some of these procedures in terms of the binary image processing steps used and the measurements they enable.

The EDM is useful for another type of measurement when it is combined with the skeleton of a feature. Figure 4.37 showed the use of a grid of vertical lines to measure the thickness of a horizontal layer. If the structure to be measured is not a simple shape, a grid is not useful. The example shown in Figure 4.43 shows a feature of variable width and irregular shape. The EDM of the feature has maximum values along the feature midline, and the value assigned to each pixel along the midline is the distance to the boundary on either side, which is the radius of an inscribed circle at that location. The pixel values along that midline (shown in Figure 4.43f) can be analyzed to determine the mean value of the width, the minimum and maximum, the standard deviation, and so on.

Combining the skeleton of the feature with the EDM, keeping whichever pixel is darker, will erase to black all pixels other than those along the midline and assign the EDM values

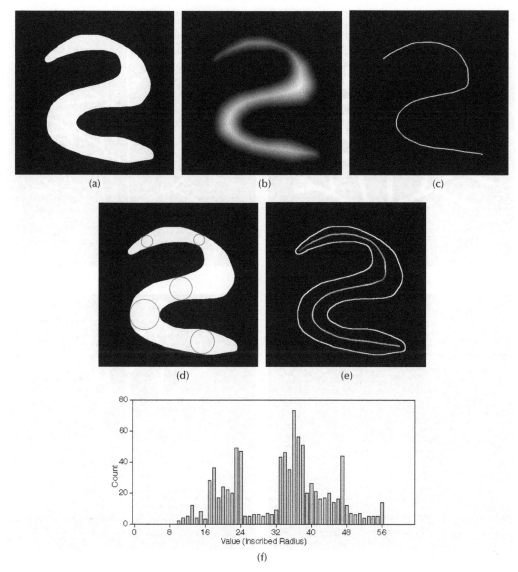

**FIGURE 4.43**
(See **Color insert following page 172**.) Measuring variable feature width: (a) original image of an irregular feature [irregular.tif], (b) EDM of the feature, (c) skeleton of the feature, (d) superposition of a few inscribed circles on the feature to illustrate the measurement made by the technique shown, (e) the EDM values assigned to the skeleton, (f) histogram of values. The skeleton has been dilated for visibility and the pixel grayscale values are shown in false color with the original feature outline superimposed to show the relationship between the feature width and the resulting values.

to the pixels in the skeleton. The histogram of the resulting image contains the values of interest. In Figure 4.43e, the skeleton is shown dilated to greater width for visibility, and the grayscale values have been replaced with a false color rainbow to make it easier to see the variation of values along the midline of the feature, whose outline shape is superimposed.

As a final example of combinations of the various tools shown previously in this chapter, Figure 4.44 shows a branching structure. As mentioned above, the skeleton of this feature can be separated into the individual branches by removing the node points. Restricting

**FIGURE 4.44**
Selecting terminal branches: (a) original image [branches.tif], (b) skeleton, (c) end points, (d) separated branches selected by end points as described in the text, shown superimposed on the original image. The end points and skeletons have been dilated for visibility.

the measurement of the branch length (using techniques shown in Chapter 5) to only the terminal branches is possible by the procedure shown in the figure. Terminal branches are those that terminate at an end point in the original skeleton, as opposed to internal branches that extend from node to node. A separate image of the end points in the original skeleton can be used as markers for the conditional dilation technique shown above to select only those separated branches that have an end point.

These examples represent only a few of the interesting possibilities that arise when morphological erosion and dilation, outlines, the skeleton, and the Euclidean distance map are combined in various ways using Boolean or arithmetic combinations. They enable the selection and measurement of structures of interest in a broad variety of images and situations.

### 4.4.4  Problems

4.4.4.1#.  Implement a program to perform Boolean operations combining the current image and a reference image. Read values from a text file that specify the operation (AND, OR, XOR) and whether to apply a NOT operator to reverse the values in either the current image, reference image, or both.

4.4.4.2.  Implement a program that generates a grid of either vertical or horizontal lines in a temporary array. Read a text file to determine the line orientation and the spacing. Combine the lines with the current image using a Boolean AND. Note: To see the entire line grid, perform the AND of the lines with an image that is filled with white.

4.4.4.3.  Implement a program that generates a grid of arbitrary lines in a temporary array. Read a text file to determine the number of lines and the coordinates of the end points for each. Combine the lines with the current image using a Boolean AND.

4.4.4.4.  Implement a program that generates a grid of radial lines in a temporary array. Read a text file to determine the $x,y$ coordinates of the center point and the number of lines, which should be drawn with uniform angular spacing from the center point to the borders of the image. Combine the lines with the current image using a Boolean AND.

4.4.4.5.  Implement a program that generates a grid of concentric circles in a temporary array. Read a text file to determine the $x,y$ coordinates of the center point and the spacing of the circles. Draw the circles with radii that increment by the spacing value. Be sure to prevent the routine from attempting to draw beyond the image boundaries. Combine the grid with the current image using a Boolean AND.

4.4.4.6.  Implement a program that creates a regular grid of points in a temporary array. Read the vertical and horizontal spacing of the points from a text file, and center the grid in the image area. Combine the grid with the current image using a Boolean AND.

4.4.4.7.  Produce the outlines of regions by performing an erosion using the EDM and then Exclusive-ORing the result with the original image. Vary the width of the erosion, and compare the results to those from Problem 4.3.4.6.

# 5

## *Measurements*

Measurements are usually performed on images after appropriate thresholding to define and remove the background, isolating the structure(s) of interest. There are two principal types of measurements that can be performed. One, which is discussed in Section 5.2, obtains numerical measures of position, size, shape, or color for each separate feature present. As introduced in Chapter 4, a ***feature*** is a contiguous region of nonzero (non-background) pixels, which is presumed for the purposes of this chapter to represent some object in the image.

The other type of measurement is concerned with the entire field of view, and reports ***global*** parameters such as the number, total area, and total perimeter of all features. These measurements are introduced first.

## 5.1   Global Measurements

For measurements such as total area, perimeter length, and number of separate features, the starting point is a thresholded image that defines the background (consisting of black pixels) and the features. These pixels may be white (as produced by the thresholding operations in Chapter 4) but, as will be seen, can be better defined as any pixels that are "not black." This convention will allow for situations in which the pixels have either been labeled with a feature identification or other information, or may have preserved brightness or color values from the original image.

In the examples that follow, the assumed starting point will be the binary image, a temporary array of values created by the thresholding routines in Chapter 4 and possibly operated on by the morphological processing routines shown there. That array contains values of 0 (black = background) and usually either 1 or 255 (feature), and may be larger by one pixel on all sides than the original image, to deal with boundary addressing.

As in Chapter 4, for programming purposes any value greater than 0.5 (out of 255) will be interpreted as feature rather than background. That avoids numerical precision problems that may arise with tests such as **if (value != 0)**, instead of using the test **if (value > 0.5)**.

These global measurements are especially useful when images represent two-dimensional sections through three-dimensional structures, such as are commonly encountered in those applications that utilize microscopy (but may also be appropriately applied in geology, forensics, food science, medical imaging, astronomy, many industrial quality control situations, and other disciplines). The mathematical relationships between the areas, lengths, and numbers measured on the two-dimensional images to the volumes, surfaces, and lengths that they represent in the three-dimensional structure are described by stereological rules (Baddeley and Vedel Jensen 2006; Howard and Reed 2005; Russ and Dehoff 2002; Weibel 1979).

The stereological interpretations of image measurements are not detailed in this text, but generally can be applied with little more than simple arithmetic to convert the measured values to the desired structural parameters. The derivations of many of these rules are not so simple, involving geometric probability. A few representative relationships from *stereology* that apply to images of randomly oriented sections through three-dimensional structures are the following:

- The expected value of the volume fraction can be measured by the area fraction, which is the ratio of the total area of all of the features to the total area of the image.
- The expected value of the surface area per unit volume can be calculated as $(4/\pi)$ times (total perimeter length/image area); the total perimeter length is the length of the line formed by the intersection of the image plane with the surface present in the volume.
- The expected value of the length of a linear structure per unit volume can be calculated as twice the number of times the structure passes through the image plane divided by the image area.
- The number of objects per unit volume can be calculated as the number of features seen in the image divided by the product of the area of the image times the mean object diameter.

Most modern stereological emphasis is on appropriate sampling of the structure to obtain unbiased results.

### 5.1.1   Area and Perimeter

The total area of non-black pixels can be determined by scanning through the image and counting pixels. The resulting value may be reported as the total area (either as pixels, or converted to some real dimension if a suitable calibration factor is available), or as the *area fraction*. This is simply the ratio of the total of the non-black pixels to the total number of pixels in the image. The total area may either be determined by counting all pixels as part of the same routine or simply calculated as the product of the image height and width.

The measurements of size introduced here (and those of position, dealt with below) are all carried out in terms of pixels. Calibration of an image in other units, whether kilometers, inches, or micrometers, is typically performed by including a scale in the image, or by acquiring an image (under the same conditions as those to be measured) of some known standard object such as a ruler, and then measuring the distance between fiducial marks. The resulting calibration (pixels per inch, etc.) can be applied to any subsequent measurement as required.

Many scanners come with software that automatically calibrates images in pixels per inch or per centimeter. Some file formats (e.g., tiff images) may optionally include spatial calibration values, but these are usually intended for printing applications, to specify the size of the printed image, and may not be meaningful for measurement purposes. Measures of shape are formally dimensionless so as to be independent of size, and calibration of brightness or color values is considered separately in Section 5.2.4.

It is possible in principle to determine the area or area fraction without performing the step of creating a binary image, by summing the area of the histogram that is selected by the threshold values. In practice, the need to apply morphological processing to remove isolated noise pixels, smooth boundaries, or otherwise clean up the binary image as shown in Chapter 4, makes these values less reliable than performing the count on the binary image after thresholding and subsequent processing.

It is also worth noting that human judgment of area fraction is often wrong, and may be strongly influenced by the size, shape, and clustering of the nonwhite regions. As shown in Figure 5.1, an image with a single large feature does not present the same visual effect as one with the same total area scattered as many small features. Fine lines may be overlooked and long ones interpreted as being larger than their actual area. If the brightness or color of features or their surroundings vary, efforts to visually judge their size can be further complicated.

The total perimeter or boundary length of features can also be measured by performing a single scan through the image. There are several different measurement techniques, some of which are more accurate than others, or make fewer assumptions about the nature of the image. One of the most common methods approximates the boundaries as a series of links connecting pixels that lie along the feature periphery (i.e., pixels within the feature that share a side with a background pixel). Each link connects a pixel to an adjacent touching pixel in either a 90° or 45° direction, as shown in Figure 5.2. This representation of the boundary is called ***chain code*** (also sometimes called Freeman code) (Cederberg 1979; Freeman 1961).

For the example shown in the figure, which approximates a circle as well as possible on a square pixel grid, the chain code perimeter consists of 24 links in a 90° direction (i.e., along a row or column in the pixel array connecting a pixel to its side-sharing neighbor) and 16 links in a 45° direction (diagonally connecting a pixel to its corner-sharing neighbor). Assigning a length of 1.0 to the 90° links and a length of $\sqrt{2}$ to the 45° links produces

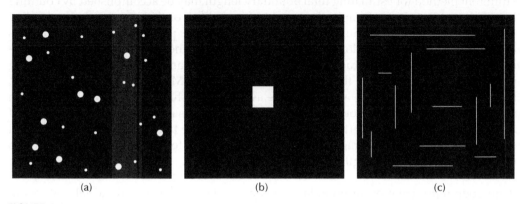

(a)                              (b)                              (c)

**FIGURE 5.1**
Three images with the same total area of white pixels, which are not usually perceived visually as being equal.

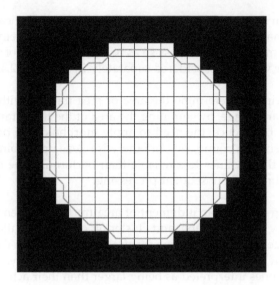

**FIGURE 5.2**
The chain code perimeter consists of segments connecting the centers of the outer pixels. The 90° links have length 1.0, and the 45° links have length $\sqrt{2}$.

a total length for the chain code that is slightly shorter than the calculated diameter of a perfect 15-pixel-diameter circle as shown in Equation 5.1a. However, this should be expected because the chain code links the centers of the pixels and thus lies inside the circle. By this method, a single pixel would have a perimeter of zero! Equation 5.1 compares three methods for calculating the perimeter of the circle in Figure 5.2: (a) the chain code length; (b) the intercept method calculation; and (c) the ideal value.

$$\text{a) } 24*1.0+16*\sqrt{2} = 46.63$$

$$\text{b) } \left(\pi/2\right)*30 = 47.12 \tag{5.1}$$

$$\text{c) } 15*\pi = 47.12$$

A different method for estimating total boundary length may be accomplished by counting the number of transitions from background to foreground, or the reverse, along each line of pixels. Multiplying this number times $(\pi/2)$ gives the perimeter. The result of this method, also shown in Equation 5.1, is exact for the circle. The method makes the assumption that the boundary is isotropic, meaning that it has the same amount of length running in each direction. Obviously, the circle fits this definition. Many images contain features that are reasonably isotropic, at least on the average. A technique that may be used as a partial test of isotropy is to count the number of transitions along each horizontal row of pixels and also along each vertical column. Comparing the two values provides a partial test for isotropy, and the average of the two may be useful for applying the calculation to determine the boundary length.

Chain code representation of boundaries is one of the ways that individual features in binary images are sometimes represented, as discussed later in this chapter. However, it is not necessary to follow the boundary of each feature separately and connect the chain links in order to measure the total perimeter length. Instead, it is only necessary to compare each pixel to its four neighbors above and to the left, as shown in Figure 5.3, to determine the local links and sum their lengths. This requires only a single pass through the image, comparing the contents of two lines (the current line and the line above).

For each pixel, the four neighbor pixels shown in Figure 5.3 are examined to construct an index value. For each neighbor pixel that is ON (i.e., not background), a corresponding weight value is added to form a total. There are 16 possible values of the index, as shown in Table 5.1. Depending on whether the central pixel whose neighbors are examined is itself ON or OFF, the corresponding chain code length is added to the total perimeter length.

**FIGURE 5.3**
Index weight factors for neighboring pixels.

**TABLE 5.1**

Perimeter Values for Pixel Neighbor Patterns

| Pixels | Pattern | Index | ON | OFF |
|---|---|---|---|---|
|  | 0000 | 0 | 0 | 0 |
|  | 0001 | 1 | $\sqrt{2}$ | 0 |
|  | 0010 | 2 | 1 | 0 |
|  | 0011 | 3 | 1 | 0 |
|  | 0100 | 4 | $2\sqrt{2}$ | 0 |
|  | 0101 | 5 | $3\sqrt{2}$ | 0 |
|  | 0110 | 6 | $\sqrt{2}$ | 1 |
|  | 0111 | 7 | $\sqrt{2}$ | 1 |
|  | 1000 | 8 | 1 | 0 |
|  | 1001 | 9 | $1+\sqrt{2}$ | 0 |
|  | 1010 | 10 | 0 | $\sqrt{2}$ |
|  | 1011 | 11 | 0 | 0 |
|  | 1100 | 12 | $\sqrt{2}$ | 1 |
|  | 1101 | 13 | $2\sqrt{2}$ | 0 |
|  | 1110 | 14 | 0 | $\sqrt{2}$ |
|  | 1111 | 15 | 0 | $\sqrt{2}$ |

Code Fragment 5.1 shows the calculation procedure, including the assigning of the table values and the construction of the index value. The code fragment also sums the total feature area and counts the transitions along each line, referred to as intercept counts. As written, the code fragment operates on the temporary images created in Chapter 4, but it can be easily adapted to perform measurements on the current image if it has been thresholded beforehand and appropriate boundary protection is incorporated.

```
// Code Fragment 5.1 — Measure Perimeter
{
    // scan through binary image to get area fraction,
    // and perimeter by chain code & intercept methods
    // uses the binary images, pointers, etc., created using
    // fragments in Chapter 4
    long  x, y, i, j, val, area = 0, icept = 0, index;
    float ONTable[16]  = {0, 1.414, 1,      1, 2.828, 4.243, 1.414, 1.414,
                          1, 2.414, 0,      0, 1.414, 2.828, 0    , 0    };
    float OFFTable[16] = {0, 0,      0,      0, 0,      0,      1    , 1    ,
                          0, 0,      1.414, 0, 1,      0,      1.414, 1.414};
    float perim = 0;
    // ... allocate (float*)PrevLine, (float*)ThisLine (width+2)
    for (y = 1; y < height+1; y++)
    {
        ReadTempImageLine(Bin_ID, y-1, PrevLine);
        // ... assumes the usual boundary protection of one pixel from Chap. 4
        ReadTempImageLine(Bin_ID, y, ThisLine);
        for (x = 1; x < width + 1; x++)         //    4  2  1      PrevLine
        {                                       //    8  x  .      ThisLine
            long   index = 0;                   //    .  .  .
            float  val = ThisLine[x];
            nbr8 = ThisLine[x-1]; if (nbr8 > 0.5) index += 8;
            nbr4 = PrevLine[x-1]; if (nbr4 > 0.5) index += 4;
            nbr2 = PrevLine[x  ]; if (nbr2 > 0.5) index += 2;
            nbr1 = PrevLine[x+1]; if (nbr1 > 0.5) index += 1;
            if (val > 0.5) perim += ONTable[index];
            else           perim += OFFTable[index];
            if (val > 0.5) area ++;
            if (((val > 0.5) && (nbr8 < 0.5)) || ((val < 0.5) && (nbr8 > 0.5)))
                icept++;
        } // for x
    } // for y
    // ... report area, perim, 1.5708*(float)icept
}
```

Applying the measurement procedure in the code fragment to the simple test shapes shown in Figure 5.4 produces measurement results as shown in Table 5.2. For each shape, the ideal geometric area and perimeter values are shown. Because of the limitations of a square pixel grid, the areas of the circle and the diamond (the square rotated by 45°) are not perfect. These small errors are an inherent result of working with images constructed of finite pixels.

**FIGURE 5.4**
Simple shapes for area and perimeter measurement. The circle diameter and the side of each square are 200 pixels.

**TABLE 5.2**

Measurements on Simple Shapes

| Shape | Ideal Area | Measured Area | Ideal Perimeter | Measured Perimeter (Intercept) | Measured Perimeter (Chain Code) |
|---|---|---|---|---|---|
| Circle | 31416 | 31757 | 628.32 | 631.46 | 664.05 |
| Square | 40000 | 40000 | 800 | 628.32 | 796.00 |
| Diamond | 40000 | 40044 | 800 | 889.07 | 796.67 |

Notice in the table that the intercept length method of determining perimeter fails badly for features such as the square and diamond, which have boundaries that are extremely non-isotropic. When applied to typical images, such as those shown in Figure 5.5, the perimeter length results by the two techniques are in somewhat better agreement (Table 5.3). The table shows results obtained by performing the measurement by processing vertical columns of pixels as well as using horizontal rows.

The gear and the leaf image were thresholded, and then a morphological closing applied to clean up noise and smooth boundary irregularities. For the fiber image (Figure 5.5c) the image was skeletonized before measurement. The total length of the skeletons is half the total perimeter length reported (which is measured along each side of each line). Measuring the length of skeletons is a preferred method for determining the length of features that are not straight, such as these irregularly curved fibers.

Chapter 4 introduced the possibility of using image measurements as a criterion for selecting an optimum threshold value. One technique that seems to mimic the human

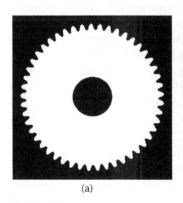
(a)

(b)

(c)

**FIGURE 5.5**
Binary images for perimeter measurements: (a) gear, (b) leaf, (c) fibers (the white skeleton lines were measured; the gray features are shown only for reference).

**TABLE 5.3**

Perimeter Measurements

| Figure | Intercept | Chain Code |
|---|---|---|
| Gear | 2318.5 | 2391.0 |
| Gear (vertical) | 2277.7 | 2391.0 |
| Leaf | 2741.1 | 2857.2 |
| Leaf (vertical) | 2264.6 | 2858.2 |
| Fiber skeletons | 8275.0 | 7957.1 |
| Fiber skeletons (vertical) | 7307.4 | 7972.3 |

visual judgment involved in manual threshold selection uses the measurement of the length of perimeters. As the threshold selection is adjusted, points at which the change of the threshold value produce little change in the image, as judged by the length of the boundaries of the thresholded region, are interpreted as producing smooth boundaries which (at least in some circumstances) correspond to the correct delineation of features.

Figure 5.6 illustrates the results. The original image contains several regions with different brightness values, and the expectation that they should have smooth boundaries is reasonable. A plot of the change in the total length of perimeter of the thresholded region as a function of threshold setting is shown. Setting the threshold level at the value where the graph rises through zero produces a binary image that does a good job of separating the elements in the original image. However, this technique and all others that use image measurements as thresholding criteria depend critically on *a priori* knowledge about the images, what they represent, and how they were obtained.

## 5.1.2   Number of Features

Determining the number of separate features in an image, and labeling them with an identity that will be needed to allow their individual measurements to be obtained in Section 5.2, is a basic procedure in image analysis. The logic involved begins with the comparison of each feature pixel to its four neighbors above and to the left, to determine whether the pixel is connected to others. The same method for constructing an index number is used as shown above in the context of perimeter measurement.

If a pixel has no touching neighbors above and to the left (index number 0 in Table 5.1), it marks the beginning of a feature and is assigned a unique nonzero identifying number. This number will be propagated to subsequent pixels that are connected to this one. If a pixel being examined has one or more neighbors above or to the left that have the same previously assigned identification number, that number is repeated for the current pixel. There are three cases in which the neighbor pixels can have previously assigned identification numbers that are not all the same, as shown in Figure 5.7. In these cases, the current pixel joins together the two previously separate parts of a feature. This situation cannot arise if the feature shape is absolutely convex.

As shown in Code Fragment 5.2, in the case that a pixel joins together two parts of a feature that previously had been given different identifying numbers, it is necessary to renumber the previous pixels so that one of the numbers is replaced by the other. The code appears moderately complex, but the three parts of the routine are simply the construction of the index number, a determination of the proper pixel from which to take the new feature identification number, and the renumbering of preceding pixels.

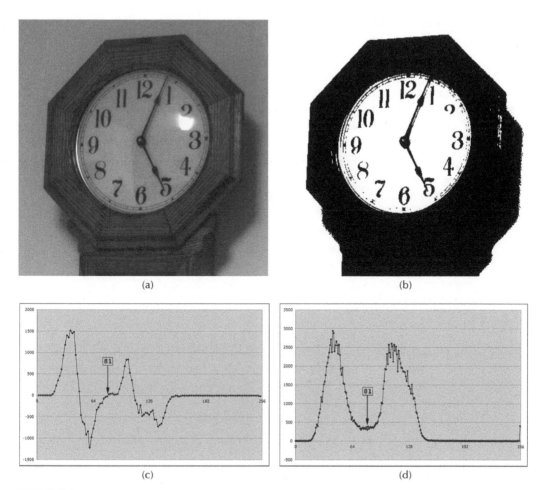

**FIGURE 5.6**
Thresholding based on image measurements: (a) original image (mean brightness of the RGB values in the original color image [clock.tif]), (b) result of thresholding based on change in perimeter length, (c) plot of the change in total perimeter length as a function of threshold setting (the zero crossing is marked), (d) histogram of the original image with the threshold setting marked.

**FIGURE 5.7**
Local pixel configurations in which the current pixel (marked X) can join two parts of a feature with different prior identifications (index numbers 5, 9, and 13).

```
// Code Fragment 5.2 — Count features
// scan through image to count number of features
// uses the binary images created using procedures in Chapter 4
long x, y, i, j, val, index;
long number = 0, featID = 1;
```

```
long newval, oldval, nbr1, nbr2, nbr4, nbr8;
// ... allocate (float*)ThisLine, (float*)PrevLine, (float*)TempLine
for (y = 1; y < height+1; y++)          // traverse padded temp image
{
   ReadTempImageLine(Bin_ID, y-1, PrevLine);
   ReadTempImageLine(Bin_ID, y  , ThisLine);
   for (x = 1; x < width+1; x++)
   {
      long  index = 0; // construct the index number from the neighbors
      float val = ThisLine[x];
      nbr8 = ThisLine[x-1]; if (nbr8 > 0.5) index += 8;
      nbr4 = PrevLine[x-1]; if (nbr4 > 0.5) index += 4;
      nbr2 = PrevLine[x  ]; if (nbr2 > 0.5) index += 2;
      nbr1 = PrevLine[x+1]; if (nbr1 > 0.5) index += 1;
      if (val > 0.5) // feature counting
      {
         switch (index)
         {
            case 0: // new feature, count and assign a unique ID num
               number++;
                  newval = featID;
               featID++;
            break;
            case 1: case 3: case 7: case 11: case 15:
               // otherwise propagate ID number from a previous pixel:
               newval = nbr1;
            break;
            case 2: case 6: case 10: case 14:
               newval = nbr2;
            break;
            case 4: case 12:
               newval = nbr4;
            break;
            case 8:
               newval = nbr8;
            break;
            // more than one separated previous pixel ID is nonzero, must
            // test if they are the same. only possible index values with
            // different neighbor IDs are 5, 9 and 13...
            default:
               newval = nbr1; // choose an existing ID number
               if ((index == 9) || (index == 13)) oldval = nbr8;
               else                                oldval = nbr4;
               if (oldval != newval) // oldval is ID to be replaced
               {  // feature IDs are not the same, so must renumber
                  number--; // decrement feature count because of merge
                  for (i = 1; i <= y; i ++) // update all prior lines
```

```
            {
                ReadTempImageLine(Bin_ID, i, TempLine);
                for (j = 1; j < width+1; j ++)
                    if (TempLine[j] == oldval) TempLine[j] = newval;
                WriteTempImageLine(Bin_ID, i, TempLine);
            }
            for (j = 1; j < x; j ++) // change current line
                if (ThisLine[j] == oldval) ThisLine[j] = newval;
            ReadTempImageLine(Bin_ID, y-1, PrevLine); // reload
        }
      break;
    } // switch
    ThisLine[x] = newval;           //mark this pixel with value
  } // if it is a feature
 } // for x
 WriteTempImageLine(Bin_ID, y, ThisLine);
} // for y
// ... report number. the temp image now has features uniquely labeled
```

An alternative method for handling the cases in which differently numbered portions of features merge is to build a table of the numbers that meet and to defer the renumbering until after the entire image has been scanned. This gives equivalent results.

One of the consequences of the routine shown is the labeling of the pixels within each separate feature with a unique identifying number. This number will be used in subsequent parts of this chapter to perform feature-specific measurements. One brute-force (but straightforward) method of accomplishing such measurements is to scan through the entire image performing, for example, the summation of area and perimeter using the routine shown previously, but only for pixels having one particular identifying number, and then repeating this for each feature number. The other way is to accumulate the sums for each separate feature during a single pass through the image. The two methods differ in time and complexity, but give identical results.

It is also worth noting that the method shown for feature identification works with images in which the full pixel array is present to show the features and background. That is chosen here as the representation that is most obvious (after all, the image started out as an array of pixels), but it is not the only possibility. Two other representations are commonly used, both of them offering certain advantages for the measurement of some feature parameters (but not others), and both of them requiring less storage space for the binary image data.

One method, already mentioned in connection with the measurement of the perimeter, is chain code. The chain of links around the perimeter of each feature forms a closed path in which each link can extend the chain in one of just eight directions (multiples of 45 degrees). For each feature, a record consisting of the $x,y$ coordinates of the first pixel encountered followed by a series of numbers from 0 to 7 as shown in Figure 5.8 can provide an exact representation. From such a record, the perimeter can be quickly determined by counting the number of even and odd digits, multiplying the number of odd digits by the square root of 2, and adding the number of even digits. On the other hand, determining area and position, or accessing the individual pixels to measure feature color, is awkward with the chain code representation. Problems also arise with holes in features,

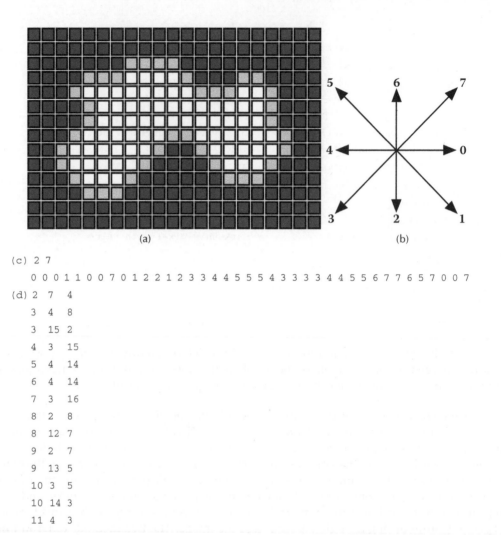

(a)                                                      (b)

(c)  2  7

　　　 0  0  0  1  1  0  0  7  0  1  2  2  1  2  3  3  4  4  5  5  5  4  3  3  3  3  4  4  5  5  6  7  7  6  5  7  0  0  7

(d) 2  7  4

　　　 3   4   8

　　　 3   15  2

　　　 4   3   15

　　　 5   4   14

　　　 6   4   14

　　　 7   3   16

　　　 8   2   8

　　　 8   12  7

　　　 9   2   7

　　　 9   13  5

　　　 10  3   5

　　　 10  14  3

　　　 11  4   3

**FIGURE 5.8**
Representations of a feature in a binary image: (a) pixel array (gray pixels mark the boundary, (b) direction codes used for the chain code in (c), (c) chain code (the first two values identify row 2 and column 7 as the coordinates of the initial pixel, and each subsequent digit represents the direction of the link to the next boundary pixel in the chain), (d) run-length encoding (each triplet gives the row and column of the beginning of a run and its length in pixels).

which produce separate chains that must be combined, and become more complex when one feature resides within a hole in a second feature.

The second popular method is ***run-length encoding***, also shown in Figure 5.8. This lists the line number and row number of the first pixel in the feature, followed by the length of the horizontal run or chord that extends across the feature. This pattern is repeated for each subsequent intercept. Run-length encoding is used in facsimile-transmission equipment as described in Chapter 3. For a color or grayscale image, the run-length table would also include the values for the pixels in each horizontal chord. From the list of run-lengths or chords, the total area of the feature and its position (introduced below) are easily calculated, but determining the perimeter requires additional logic.

### 5.1.3 Counting and Image Boundaries

When counting features in images, two situations can arise as shown in Figure 5.9. In one, the image covers a field of view that includes all of the features, and counting them is straightforward. In the other, the image field of view is a sample of a much larger population, and it is important to know the number of features per unit area so that an extrapolation can be made from the sample that has been counted to the entire population. In the latter case, some features are likely to intersect the boundaries of the image, and a special procedure for obtaining a correct count must be employed.

The area covered by the image within a large population should be positioned at random to avoid bias in the counting results. If every possible image was obtained, they could be arranged like tiles to cover the full area of the population. Because the area covered by a single image is one of the many possible positions, it is proper to consider it as a sample of the entire area. The number of features counted in this image multiplied by the number of possible images (which is the total area covered by the population) estimates the total number of features. This method is used in fields as wide-ranging as astronomy (to estimate the number of galaxies in the universe), crowd control (to estimate the number of people from a few photos), food science (estimating the density of bubbles in whipped cream), and medicine (counting bacteria in a petri dish), as well as many more.

In these instances, some features are likely to cross the boundaries of the image. Counting the features seen in every image field would result in double-counting of the boundary-crossing features. The correct procedure is to count the features that cross two boundaries, for example, the bottom and right borders, and not to count those that cross the other two (the top and left). Those features would potentially be counted in other (adjacent) fields of view in which they cross the bottom or right borders.

To implement this correct counting procedure, any feature that touches the top or left boundary of the image must not be counted. There are two equivalent ways of implementing this requirement. One is to perform an additional scan of the image after the routine in Code Fragment 5.2 is completed, just examining the boundary row and column

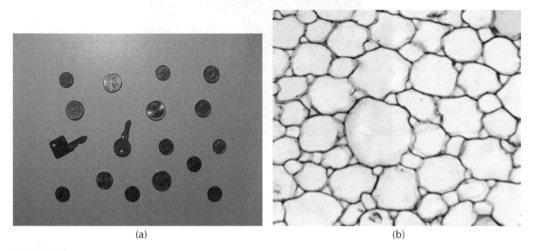

(a)                                                          (b)

**FIGURE 5.9**
Different classes of images for feature measurement and counting: (a) the [coins.tif] image includes all of the objects to be measured arranged within the image area, with no features touching the feature boundaries; (b) the [corn.tif] image is a sample of a larger field, in which features extend beyond the image boundaries.

of pixels along the top and left sides. If any of those pixels are not background, it indicates that a boundary-touching feature has been detected. For each unique ID number found, the count of features should be decremented by one. Optionally, it is also possible when a non-background boundary pixel is found to scan through the image and change all pixel values with that ID number to a reserved value (such as –1) to indicate the boundary-touching condition. When this method is used, it eliminates any possible confusion about whether the feature touches the boundary at more than one pixel.

Alternatively, it is possible is to modify the procedure shown in Code Fragment 5.2 so that when a non-background boundary pixel (along the top and left sides) is encountered, the ID number of the feature is immediately changed to the reserved value, and this ID number is propagated to all of the pixels in the feature, both those that have not yet been encountered and those that have already been assigned some other ID number. The feature count must also be decremented.

For feature measurements, discussed in Section 5.2, a feature that crosses any of the image boundaries (on all four sides, not just the top and left) cannot be measured because the extent of the feature beyond the boundaries is unknown. Not counting or measuring the features that intersect image boundaries would bias the resulting measurements by systematically undercounting large features (which are more likely to intersect the boundaries) as compared to small ones. There are two different solutions to this problem, which produce results that are statistically equivalent.

One method, which was primarily used with manual measurements, places a smaller measuring frame inside the image boundaries as indicated in Figure 5.10. Any feature that crosses the top or left boundary of the image is ignored, as before. Any feature that crosses the bottom or right border of the measuring frame must be counted and measured in its entirety. Any feature outside that frame, even if it is within the full image boundaries and

**FIGURE 5.10**
Using a measuring frame within an image. Features lying in the grayed-out guard region are not counted or measured. Any feature entirely within the active measuring region or that crosses its bottom or right borders is measured and counted. Any feature that intersects any image boundary cannot be measured and is not counted.

could be measured, must be ignored. With this requirement, the width of the ***guard region*** between the border of the counting frame and the bottom or right boundaries of the image must be large enough that no feature can cross both the counting frame border (meaning that it must be measured) and also the image boundary (in which case it cannot be measured). The use of a guard region can reduce the useful area of an image significantly.

The second method does not reduce the image area by using a guard region. It measures every feature in the image that does not intersect any of the image boundaries, but compensates statistically for the fact that randomly placed large features are more likely to intersect a boundary. Instead of counting features in the usual way with integers, features are assigned a calculated ***adjusted count*** that is used as a weighting factor in any subsequent data analysis, such as calculating a mean value or constructing a distribution of feature sizes or shapes. The value of the adjusted count, shown in Figure 5.11, depends on the extent of the feature and the size of the image. For a feature that is very small compared to the image size, the value is close to 1.0, indicating that such features are not likely to intersect a boundary of the image. As the feature size increases, the denominator of the fraction becomes smaller and the adjusted count increases.

Keeping track of the minimum and maximum values of $x$ and $y$ coordinates for pixels in each labeled feature during the same scan that does the labeling, area measurement, etc., is easy to add to the code. These allow the feature extents to be calculated and combined with the image dimensions as shown in the figure, to determine the adjusted count for each feature.

The adjusted count is used as a weight when computing statistical parameters such as the mean value, as shown in Equation 5.2. When distribution histograms are constructed, instead of adding one to the appropriate bin for each feature, the adjusted count is added to the bin total. Calculating statistical summaries or preparing distributions can be

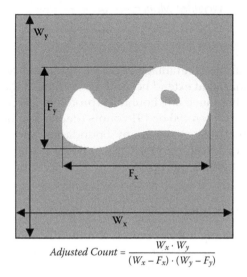

$$\text{Adjusted Count} = \frac{W_x \cdot W_y}{(W_x - F_x) \cdot (W_y - F_y)}$$

**FIGURE 5.11**
The adjusted count depends on the dimensions $F$ of each feature in the $x$- and $y$-directions, and the width $W$ of the image in those directions.

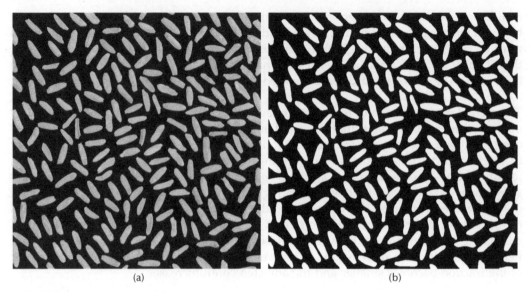

(a)                                                                              (b)

**FIGURE 5.12**
Counting boundary-touching features: (a) original image [rice.tif], (b) after automatic thresholding and application of a closing.

performed in the analysis program, or carried out separately in a spreadsheet or statistics program.

$$Conventional\ Mean = \frac{1}{N} \sum_i Value_i$$

$$Weighted\ Mean = \frac{\sum_i Weight_i \cdot Value_i}{\sum_i Weight_i} \tag{5.2}$$

Figure 5.12 shows an example of an image of objects (grains of rice, imaged by spreading them onto a flatbed scanner) that extend beyond the bounds of the image. Counting every feature, including those that touch any boundary, produces 209 counts. Counting features that intersect two sides produces 190 or 191 counts (depending on which sides are used). There are 174 features that do not touch any boundaries, and summing their adjusted counts produces a total of 196.86 counts. The fractional result is a statistically accurate measure of the number of grains per unit area.

### 5.1.4   Measurements with Grids

The creation of grids containing lines or points, and their Boolean AND combination with a binary image, was introduced in the preceding chapter. Grids have frequently been used for determination of global properties because they can often reduce a procedure from one involving measurement to one requiring a simpler measurement, or just counting. This was particularly important when the procedures were carried out manually, but even with computer processing the grid-based approaches offer advantages in some cases.

The example shown in Chapter 4, Figure 4.37, using a grid of vertical lines to measure the thickness of a horizontal layer, is particularly easy because the length of the individual vertical line segments can be determined simply by counting the pixels in each one or finding the top and bottom of each line. For other types of grids (diagonal lines, circular arcs, etc.), measuring the length may require the procedures shown above for perimeter measurement. However, many measurement procedures involving grids require only counting.

If a grid of points is combined with a binary image, the fraction of the points that fall onto the features provides a quick estimate of their total area. Figure 5.13 shows an example, in which a 20 × 20 point grid has been superimposed on the binary image of the leaf shown above. One hundred and thirty-three of the four hundred points in the grid lie on the leaf. Multiplying the fraction 133/400 by the image area results in an estimate of the feature area 109,550 pixels. Counting all of the pixels in the original binary image gives an area of 110,862 pixels, a difference of slightly over 1%.

With computer processing, the total area can be determined very easily, but if manual measurement had been used, the advantage of using a grid and counting the points is obvious. The typical manual procedure for measuring area would have been to (carefully) cut out the features from a photograph, and weigh them. The ratio of the weight of the features to the total weight of the photograph measures the area fraction, assuming of course that the paper used to print the image is uniform in density.

It is worth noting that the pixel array itself is a grid. In many images the value of the pixel represents an average over a finite area, and the thresholding operation selects a level that determines whether that value is considered to be feature or background. In a grid such as the one shown in the figure, each point similarly represents an area, in this case a larger one equal to the spacing between the points. Depending on whether or not the grid point falls onto a feature pixel, the result is used to represent the entire area. The number of points can be adjusted depending on the measurement precision required, just as the number of pixels in the original may be selected based on the required resolution.

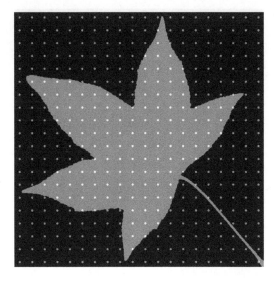

**FIGURE 5.13**
Superimposing a grid of points (dilated for visibility) on a binary image to measure the area.

A grid of lines can be used to measure perimeter length with a counting procedure, as well. A grid of lines of total length $L$ will intersect the edges of features, producing $N$ intersections. The total length of the perimeter can then be estimated using Equation 5.3. This relationship was used above to determine the total perimeter by counting the total number of transitions, treating every horizontal row of pixels as a measurement line. However, by using a generated grid of lines the problem revealed there of assuming that the perimeter is isotropic can be overcome. If the grid of lines is isotropic, such as a set or radial lines or concentric circles, it does not matter whether the feature peripheries or other linear structures that they intersect are isotropic or not.

$$Length = \left( \frac{\pi}{2} \right) \cdot \left( \frac{N}{L} \right) \cdot Image\ Area \qquad (5.3)$$

Figure 5.14 illustrates the procedure. The binary image being measured is the skeletons of the fibers shown in Figure 5.5c. The superimposed grid consists of three concentric circles, which have a total length (calculated from their diameters of 150, 300, and 450 pixels, respectively) of 2827.4 pixels. There are 28 intersection points (labeled with circles), and the image size is $512 \times 512$ pixels. Putting these values into Equation 5.3 produces an estimate for the fiber length of 4,077.8 pixels, which is in good agreement with the value of 3,978.6 pixels obtained above by the chain code method (this is half the value in Table 5.3, because that value was obtained by measuring the perimeter along both sides of each skeleton line).

It is important in this procedure involving grids to understand that it is the number of intersections that must be counted, not the number of pixels that are intersected. In the example shown, many of the intersections consist of a single pixel. However, depending on the angle at which the grid intersects the line, some intersections extend over several pixels. The feature-counting procedure shown above must be used to count the number of intersections (usually called "events" or "hits"), not the number of pixels.

**FIGURE 5.14**
Measuring length with a grid.

### 5.1.5 Problems

In addition to the various example images shown in the text, the accompanying CD includes several images that can be used to test the programs written for this set of problems (which perform measurements on the entire image) and the next set (which perform measurements on each individual feature):

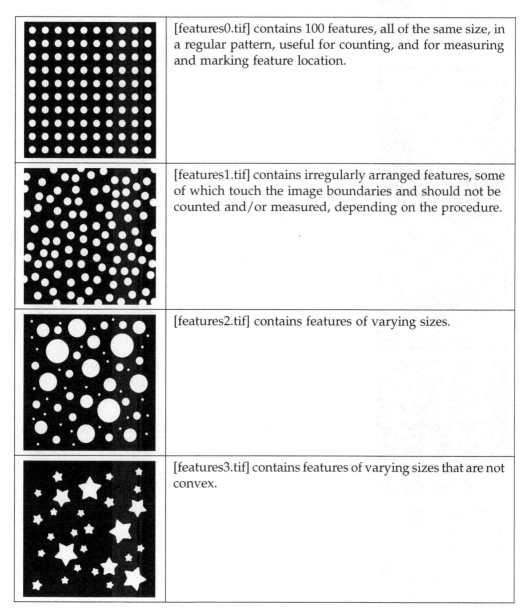

| | |
|---|---|
| | [features0.tif] contains 100 features, all of the same size, in a regular pattern, useful for counting, and for measuring and marking feature location. |
| | [features1.tif] contains irregularly arranged features, some of which touch the image boundaries and should not be counted and/or measured, depending on the procedure. |
| | [features2.tif] contains features of varying sizes. |
| | [features3.tif] contains features of varying sizes that are not convex. |

| | |
|---|---|
| 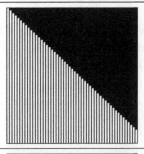 | [features4.tif] contains a series of vertical lines of varying lengths. |
|  | [features5.tif] contains a series of horizontal lines of varying lengths. |
|  | [features6.tif] contains a series of 45° lines of varying lengths. |
| 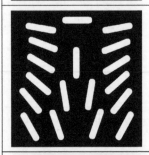 | [features7.tif] contains a series of lines of approximately the same length oriented in steps of approximately 10°. |
| 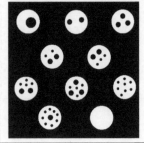 | [features8.tif] contains features with different numbers of internal holes. |

[features9.tif] contains features with varying shapes, many of which are not convex, that can be used for shape measurement and convex bounds measurement, as well as testing counting procedures.

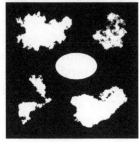

[featuresA.tif] contains five features with varying fractal dimensions, including an ellipse that is Euclidean (not fractal).

5.1.5.1#. Implement a program to measure the area and total perimeter of features as shown in Code Fragment 5.1. Write the results to a text file.

5.1.5.2. Implement a program that measures total perimeter using horizontal row intercepts and vertical column intercepts. Write the results to a text file and compare them for various images.

5.1.5.3#. Implement a program to count features and label the pixels with a unique identifying number as shown in Code Fragment 5.2. Display the feature identification as a brightness value.

5.1.5.4. Modify your program from Problem 5.1.5.3 in one of the ways described in the text, so that it does not count features that touch the top or left image boundaries, or to not count features that touch any boundary.

5.1.5.5. Modify your program from Problem 5.1.5.3 to use 4-connected logic rather than 8-connected logic for features; that is, devise a table of rules for local pixel matching that defines pixels as touching and forming part of the same feature only if they share a side, not a corner.

5.1.5.6. The program from Problem 5.1.5.3 assigns features unique identification values, but because of renumbering where different partial features meet, the ID numbers are not necessarily sequential. Modify the program so that sequential numbers result. This can simplify subsequent procedures to perform measurements on each individual feature.

5.1.5.7. In Chapter 4, the problem of filling internal holes was deferred. The solution is to invert the logic to treat any black (background) pixel as being a pixel of interest and to use the logic from Problem 5.1.5.4 to select all these background regions that do not touch any image boundary. These are the holes. It is important that the holes be identified using 4-connected logic as used in Problem 5.1.5.5 if the features are defined using the usual 8-connected logic. Combine these holes with the original image. If the procedure is executed by making a duplicate of the original image and reversing the values, then the combination is performed with a Boolean OR. Create a program that fills holes in features.

5.1.5.8.   Implement a program to create various point and line grids and combine them with binary images (using a Boolean AND) to measure area and boundary length. Compare the results to those produced by the methods used in the program from Problem 5.1.5.1.

---

## 5.2   Feature Measurements

Four basic classes of measurements can be performed on individual features: position, color, size, and shape. The results of these measurements may be used to produce descriptive statistical summaries or distributions, or to look for relationships (such as shape varying with size, or orientation varying with position, etc.). They may also be used for classification, for example, by setting limits on various measurement results that correspond to groupings or to specifications established beforehand. Within each of the four classes of measurements, there are many different specific parameters that can be chosen. Representative examples are discussed in the following sections.

### 5.2.1   Size

Two examples of size measurements have already been described. The measurement of area and perimeter shown above can be applied to a single feature just as well as to all of the features in an image. As noted previously, this is done after features have been assigned a unique identifying number. The measurements can be performed either by scanning through the image once for each feature, or by accumulating the various sums separately for each feature in a single scan. The same choice can be made for most of the other measurements described below, and the process may be made simpler by assigning a sequential set of ID numbers to the features.

The two methods are equivalent, but each has different advantages. Scanning through the entire image for each feature involves examining a very large number of pixel addresses multiple times, when most of those pixels are not part of the feature of current interest. This is inefficient and time consuming, but may be clearer in implementation and can avoid problems in keeping track of the data for the various features. It also simplifies the logic in case two different features are close enough together that a single background pixel has two different feature pixels as neighbors. In the case of adding a perimeter, this could require adding different partial values to each feature's sum.

A popular measure of size is the equivalent circular diameter. This is simply the diameter of a circle with the same area as that measured for the feature, and requires no additional measurements on the image. It can be calculated from the measured area result as shown in Equation 5.4, and this may be done either in the program itself before the values are reported or summarized, or externally in a spreadsheet.

$$Equiv.\ Diam. = \sqrt{\frac{4\ Area}{\pi}} \tag{5.4}$$

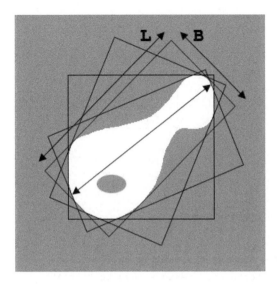

**FIGURE 5.15**
Rotating the orientation of the bounding box in four steps, with the length *L* and breadth *B* shown for the orientation producing the maximum projected dimension. The arrow inside the feature shows the true maximum caliper dimension.

The calculation of the adjusted count, described above, uses the horizontal and vertical projected dimensions (also called the Feret's diameters or caliper dimensions) of the feature. These define a bounding box around the feature. In some cases these dimensions are reported as measurements of size, but they are sensitive to the orientation of the feature in the image. A preferred method of characterizing the **_length_** of a feature can be obtained by rotating the bounding box to fit the longest dimension of the feature as shown in Figure 5.15.

This is easier than it might at first seem. If the projected dimension is measured at a series of fixed rotation angles, the true maximum caliper dimension can be estimated rather accurately, The measured projected dimension is only slightly less than the true maximum; specifically, it is equal to the true value times the cosine of the difference between the angle of the measurement and that of the true dimension. For example, if only four rotational steps are used (each of 22.5°), the maximum difference in angle between the exact orientation of the maximum projected length and the side of the box would be 11.25°. The cosine of that angle is 0.98, and the resulting underestimate of the maximum length is 2% in the worst case. More steps result in smaller angles and even smaller maximum errors.

Determining the projected dimensions for a feature is performed by keeping track of the minimum and maximum *x*- and *y*-coordinates as the scan progresses through the pixels. For each fixed rotation angle, the previously calculated and stored values of the sine and cosine are used to calculate the $(x',y')$ coordinates in the rotated coordinate system as shown in Figure 5.16. This only needs to be done for the first and last pixels (minimum and maximum *x*) in the feature in each line (*y*) in the original image. Keeping track of the maximum and minimum values of $(x',y')$ for each rotation angle produces an array of values for the feature. Code Fragment 5.3 outlines the procedure.

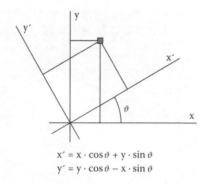

$$x' = x \cdot \cos \vartheta + y \cdot \sin \vartheta$$
$$y' = y \cdot \cos \vartheta - x \cdot \sin \vartheta$$

**FIGURE 5.16**
Calculating pixel coordinates in rotated axes.

```
// Code Fragment 5.3 — Finding the maximum caliper dimension
// after counting the features, create arrays to hold the min and max
// values of projected x and y values for each angle, for each feature
// then scan through the image and,
//    for each line (y) get the x value for the first and last pixel
//       for each angle
//          calculate x', y' as in Figure 5.16
//       keep the maximum and minimum values
// for each feature
//    for each angle
//       calculate length = max - min
//    keep greatest length value for each feature
```

The rotated bounding box is then defined by the projected dimensions with the greatest length, and provides the reported length (and optionally the breadth) dimensions, as shown in Figure 5.15. Note that these dimensions are not used for the adjusted count calculation, which depends on the projected dimensions of the feature parallel to the boundaries of the image.

The breadth of the bounding box shown in Figure 5.15 is the projected dimension of the feature in a direction perpendicular to the direction in which the length was determined. This is one of many possible meanings of the term "breadth," and there is no general consensus on what the best definition is. Another possibility is the minimum caliper dimension, the shortest projected dimension in any direction. There is no reason for the shortest projected dimension to be perpendicular to the longest — consider, for example, a square, for which the shortest dimension is the side of the square (Figure 5.17a) and the longest is the diagonal.

While the length of the bounding box is a very good estimate of the maximum caliper dimension of the feature, the breadth of that box is not generally a useful estimate of the minimum caliper dimension. As a simple example noted previously, a square has a maximum dimension along its diagonal, and the bounding box oriented in that direction would report both a length and breadth equal to the diagonal, rather than the actual minimum dimension, which is the side of the square (Figure 5.17a). Furthermore, a long narrow feature (Figure 5.17b) whose axis was between the stepped angles used for finding

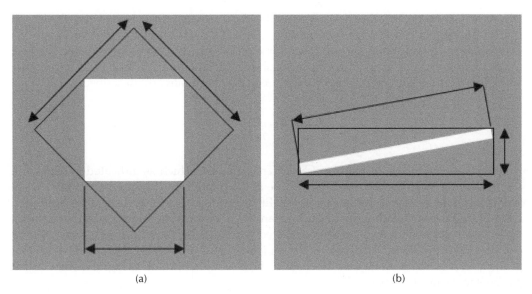

(a)                                                              (b)

**FIGURE 5.17**
Problems with using the oriented bounding box to determine the minimum caliper dimension: (a) for a square feature the breadth of the bounding box with the greatest length is 41% greater than the minimum dimension (the side of the square); (b) for a long, thin feature at a small angle (10° in the example) the length is closely estimated (197 versus 200 pixels) by the bounding box, but the minimum dimension is badly overestimated (45 versus 10 pixels).

the maximum projection would be estimated to have a breadth that is too large, and the error is not a cosine error (as with the length) but a sine error. The magnitude of the error is dependent on the length of the feature and can be quite large.

Figure 5.18 shows a plot of the distribution of length values determined using the maximum caliper dimension for the rice grains in Figure 5.12b. Measurements such as this are used to determine that the rice can be sold as long-grained rather than short-grained rice. The distribution plots the total adjusted count (as defined in the previous section) for the grains in each bin of the histogram based on their lengths.

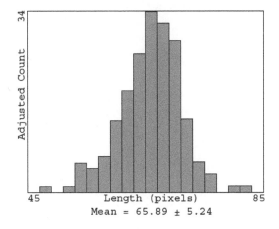

**FIGURE 5.18**
Length distribution measurement for the rice grains in Figure 5.12b.

For many feature types (other than elongated but curved shapes such as fibers), the length (maximum caliper dimension) is commonly used as a measure of size, but various ways to estimate the breadth may be employed. One possibility it to use the length and the area to calculate the minor axis of an ellipse, as shown in Equation 5.5. This does not require any additional measurements beyond the area and length already described.

$$Breadth = \frac{4 \cdot Area}{\pi \cdot Length} \tag{5.5}$$

Another approach uses the radius of an inscribed circle, which as indicated in Chapter 4 can be determined from the maximum value of the Euclidean distance map of the feature (the value at the ultimate eroded point, or UEP). This does require additional computation if the EDM has not already been generated, and also requires combining information from one image representation (the features labeled with identification numbers) and a second one (the EDM or UEP). Furthermore, for some feature shapes the EDM may have several peaks, and it is not obvious that the inscribed circle radius is a meaningful measurement parameter for most shapes.

For a curved or elongated feature such as a fiber, an entirely different approach to length and breadth is needed. As shown previously, the length of a curved feature is conveniently measured as half the perimeter of the skeleton. Chapter 4 described the combined use of the Euclidean distance map and the skeleton to determine the width, which might be reported as twice the mean value of the EDM (the radius of an inscribed circle) at every point along the skeleton.

Another measure of size that is useful in many circumstances is the ***convex hull*** area of a feature (Graham and Yao 1983). This is also known as the rubber-band or taut-string area, and corresponds to the size of the feature with all indentations and interior holes filled in, as illustrated in Figure 5.19. It can be closely approximated by a polygon produced by connecting the maximum and minimum points located by the procedure described above

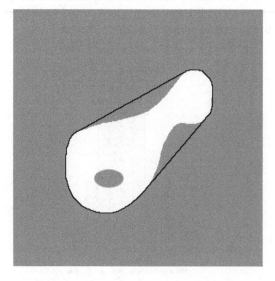

**FIGURE 5.19**
The convex hull polygon using the maximum and minimum points (located as described in the text) as vertices.

for finding the maximum projected dimension with a series of rotations of the coordinate system. The area and perimeter of this polygon are calculated as shown in Equation 5.6. This calculation can be combined with the procedure for finding the maximum caliper dimension. The summations in the equations use the sequentially numbered $(x,y)$ coordinates with the final pair (subscript $n$) repeating the initial point (subscript 0) to complete a closed path.

$$Area = \tfrac{1}{2} \sum_{k=0}^{n-1} \left( x_k \cdot y_{k+1} - x_{k+1} \cdot y_k \right)$$

(5.6)

$$Perimeter = \sum_{k=0}^{n-1} \sqrt{ \left( x_{k+1} - x_k \right)^2 + \left( y_{k+1} - y_k \right)^2 }$$

### 5.2.2 Position

Measuring the location of features is done by calculating the $(x,y)$ coordinates of a point that characterizes the feature. The problem is choosing which point to use. One of the simplest locations to use is the center of the bounding box, either the horizontal box that is used to calculate the adjusted count, or the oriented bounding box that aligns with the longest projected dimension. The $(x,y)$ coordinates of the center are just the averages of the $(x,y)$ coordinates of the four corners.

A more robust measure of location, and probably the most often used, is the ___centroid___ of the feature. This is the point at which the cutout shape will balance, assuming that it is made from uniform, rigid material. The centroid coordinates can be calculated from sums accumulated during a scan through the image, to determine the average $(x,y)$ position of the pixels shown in Equation 5.7. Usually, all of the pixels are considered equal so that the weights $(w_i)$ are all 1, and the denominator is simply the feature area. In some cases a value can be associated with each pixel, for example, a density value calculated from the pixel brightness value. In that case the density value is used as a weight, and the density-weighted centroid is located. The denominator is then the integrated optical density of the feature.

$$C.G._x = \frac{\sum\limits_i w_i \cdot x_i}{\sum\limits_i w_i}$$

(5.7)

$$C.G._y = \frac{\sum\limits_i w_i \cdot y_i}{\sum\limits_i w_i}$$

The location of the feature centroid is calculated with sub-pixel precision. The coordinates as shown are based on an origin at the upper-left corner of the image, and it may be appropriate to add some offset values that represent the location of the image itself (latitude and longitude, position of a scanning specimen stage, etc.). If the feature shape is not convex, there is no guarantee that the centroid actually lies within the feature boundaries. When a point within the feature is needed, the location of the maximum value in the Euclidean distance map may be used. As noted previously, this ultimate eroded point (UEP) is the center of the largest inscribed circle that can be drawn within the feature.

Closely related to the location of the feature is the measurement of an ***orientation*** angle. The angle selected for the oriented bounding box is at best a crude approximation of this orientation. A preferred value is the angle of the line about which the feature has the lowest moment of rotation, again treating it as being cut from uniform density rigid material. This angle also corresponds to the line that would be fit through all of the pixels in the feature by regression. The angle is determined by accumulating sums as shown in Equation 5.8, which are used to calculate the moments about the $x$- and $y$-axes, and then the angle for the line with the lowest moment. The sums include those shown in Equation 5.7, and can be similarly modified if weight factors are assigned to the pixels.

$$S_x = \sum x_i$$

$$S_y = \sum y_i$$

$$S_{xx} = \sum x_i^2$$

$$S_{yy} = \sum y_i^2$$

$$S_{xy} = \sum x_i y_i \tag{5.8}$$

$$M_x = S_{xx} - \frac{S_x^2}{\text{Area}}$$

$$M_y = S_{yy} - \frac{S_y^2}{\text{Area}}$$

$$M_{xy} = S_{xy} - \frac{S_x \cdot S_y}{\text{Area}}$$

$$\vartheta = \tan^{-1}\left\{\frac{M_{xx} - M_{yy} + \sqrt{\left(M_{xx} - M_{yy}\right)^2 + 4 \cdot M_{xy}^2}}{2 \cdot M_{xy}}\right\}$$

Of course, once it has been determined, the orientation angle may be used to recalculate a more accurate projected length measurement (and if it is considered useful for the application, a projected breadth measurement) by repeating the rotated coordinate transformation shown above.

The location coordinates and orientation angle calculated by the methods shown are absolute values relative to the image. In some cases that is useful information, but in others it is the location of one feature with respect to others that is of interest. Chapter 4 (Figure 4.42) showed one example of this, in which the Euclidean distance map was used to determine the distance of features from an arbitrary line. Lines such as feature skeletons or outlines are often useful for this purpose.

It may also be important to characterize the distribution of features relative to one another. In the examples that follow, the neighbor distances are calculated from interior points, such as the feature centroids. It is also possible to determine the minimum distances between the edges of features, but this is a more difficult procedure. One way to perform that calculation is by combining the skeleton and Euclidean distance map of the background to locate minima, which are the points of closest approach between feature edges.

The key to interpreting the distribution of distances between feature centroids lies in understanding a random spatial distribution (Schwarz and Exner 1983). Figure 5.20a shows an example of an image in which 200 features are randomly located within the field of view. This was done by using a random number generator, but it corresponds to many physical processes such as sprinkling sand onto a table. If each feature (or sand grain) is independent of all the others, the distribution is described as a Poisson random process.

The name comes from the fact that a distribution plot of the distances from each feature to its nearest neighbor has the shape of a Poisson distribution. Figure 5.21 shows an example of such a distribution, in this case produced by 1200 randomly generated points in a 900 × 1200 pixel image. The nature of the Poisson distribution is such that the mean value of the nearest neighbor distance is determined only by the number of features and

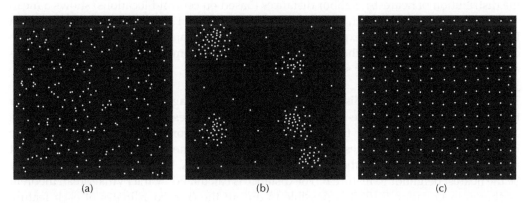

(a)  (b)  (c)

**FIGURE 5.20**
Illustrations of feature distributions: (a) random, (b) clustered, (c) self-avoiding.

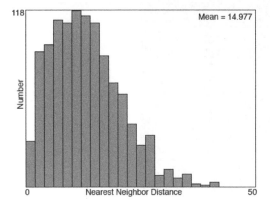

**FIGURE 5.21**
Distribution plot of nearest neighbor distances for 1200 randomly generated points in a 900 × 1200 pixel image.

the image area, as shown in Equation 5.9. The measured mean value in the example is 14.977 pixels, very close to the expected value based on the equation of 15.00.

For a distribution of features that is not random, the mean value of the nearest-neighbor distance is often an efficient way to detect and characterize it. ***Clustering*** and ***self-avoidance*** are illustrated in Figure 5.20. A variety of physical forces can produce nonrandom distributions, including electromagnetic attraction or repulsion, chemical depletion, surface tension, and so forth. For instance, cacti in the desert are typically self-avoiding because each plant collects water and nutrients and prevents others from growing nearby. But, cottonwood trees cluster along streambeds where there is moisture. Students in a classroom are self-avoiding, but they cluster together at parties.

$$Distance = 0.5 \cdot \sqrt{\frac{Area}{Number}} \tag{5.9}$$

The mean nearest-neighbor distance in the case of a clustered distribution (Figure 5.20b) is smaller than that for a random distribution of the same number of features. Conversely, a self-avoiding distribution (Figure 5.20c) has a mean nearest-neighbor distance greater than that for the random case. For example, in the image of coins shown in Figure 5.22 the distribution of nearest-neighbor distances (based on centroid locations) shows a mean value of 168.69 pixels, much greater than the value of 126.03 pixels calculated from Equation 5.9 that would be expected if the features were randomly distributed. The ratio of the measured distance to the value expected for a random distribution is often used as a descriptor for clustering or self-avoidance.

The mean nearest-neighbor distance is determined by averaging the individual values for the nearest-neighbor distance for each feature. Those values in turn are calculated by an exhaustive sorting procedure that calculates the distance from each feature to every other feature and keeps the smallest value. Note that nearest-neighbor relationships are not reciprocal: feature B may be the nearest feature to feature A, but a different feature C may be the nearest neighbor to feature B. The distance is calculated by the Pythagorean theorem as shown in Equation 5.10. It is possible to output the $(x,y)$ coordinates of each feature and offload the distance calculations, sorting, and averaging to a spreadsheet or data

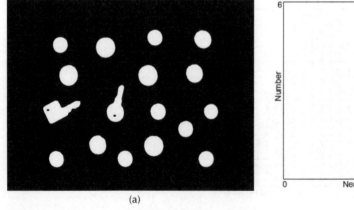
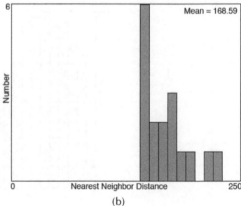

(a)                                              (b)

**FIGURE 5.22**
Self-avoiding distribution: (a) features from the coins image, Figure 4.8d; (b) plot of the nearest-neighbor distribution based on feature centroids.

analysis program, but it is generally more efficient to perform the process within the same program that measures the feature locations.

$$Distance = \sqrt{\left(x_1 - x_2\right)^2 + \left(y_1 - y_2\right)^2} \qquad (5.10)$$

It is also possible to use the nearest-neighbor directions to determine whether a distribution of features has preferred (nonrandom) alignments. For both distance and direction analysis, if a feature's nearest-neighbor distance is greater than the distance from the feature to a boundary of the image, it should not be used in any calculations because the actual nearest neighbor may not be present in the image.

It is useful in many situations to measure the alignment of features. In satellite imagery, the towers along electric transmission lines are visible, although the wires themselves are not. Also, geologic faults are not typically visible as solid continuous lines, but the alignment of visible sections still defines their locations. Similar circumstances arise at smaller scales. One of the first uses of the procedure described below was to determine the angle of the pointers on analog dial gauges seen in video images.

It is practical in some cases to use the location coordinates of features with analytical geometry to fit equations that describe the lines that pass through multiple points. However, there is a more direct way based on image processing that can also be used (Duda and Hart 1982; Hough 1962). The ***Hough transform*** is illustrated here for the simplest case of fitting straight lines. Human vision is very good at linking together a series of points to form continuous boundaries. This is the basis for several visual illusions, such as the one in Figure 5.23. The Hough transform is a method by which computer processing can accomplish some of the same results.

The principle of the transform is that each pixel whose coordinates are used to perform the fitting procedure votes for all of the lines that can pass through it, and the lines that get many votes are the ones that pass through many pixels. Figure 5.24 shows a simple example. The original image (Figure 5.24a) consists of five pixels (labeled A through E). For each of these points, a line can be drawn at any angle through the point. To characterize these lines, it is convenient to use polar coordinates and to specify each line by the angle (0 to 2π) and its perpendicular radius from the origin, which in this example is placed at

**FIGURE 5.23**
Kanisza's triangle is a visual illusion produced by the linking of points to form a perceived contour or boundary defining a central white triangle that may appear to float above and be brighter than its surroundings.

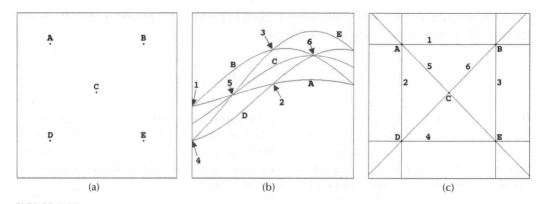

**FIGURE 5.24**
The linear Hough transform: (a) five original points; (b) the Hough transform with sinusoids corresponding to each of the five original points (A–E), and their six intersections marked (1–6); (c) the lines corresponding to each of the locations in Hough space superimposed on the original points in image (a).

the upper-left corner of the image. Using that convention, the set of lines through each point is represented in Hough space by a sinusoid, as shown in Figure 5.24b. Each sinusoid is labeled to correspond to the point through which the corresponding lines pass.

The voting mentioned above is accomplished by adding together the sinusoids in Hough space. Consequently, where two or more sinusoids cross, the value is increased to the number of original points that lie on the corresponding line. Points in the Hough space correspond to lines in the original image, so finding the points where multiple sinusoids cross defines the alignments in the image. In the example, the six crossing points in the Hough space plot (labeled 1 through 6 in Figure 5.24b) define the lines that pass through two or more of the original points. These are shown with the same identifying labels in Figure 5.24c.

The angles and spacing of the lines can be directly measured in the Hough space plot, because the coordinates are the angle and radial distance values that characterize the lines. The array used for Hough space can be as large as needed to produce the desired precision. Fitting circles instead of straight lines requires three parameters (the radius and the $x,y$ coordinates of the center). In that case, the Hough space (also called an accumulator space because it sums up the votes) is three dimensional, and each pixel in the original image generates a cone in the Hough space. The method can be extended to other shapes, including ellipses and arbitrary functions, but the more parameters that are used, the greater the dimensionality of the accumulator space that is used to construct the transform (Ballard 1981; Guil and Zapata 1997).

In the example shown, each pixel has an equal vote, appropriate for a thresholded binary image or a plot of feature centroids. For a grayscale or color image, each pixel may have a vote proportional to its brightness value. In that case, the accumulator space must be able to handle floating point numbers to sum the contributions from all of the pixels in the image. The result in that case is usually called the Radon transform, and corresponds to the way that CT scans are recorded in medical imaging (Cormack 1963; Cormack 1964). By rearranging the data into Cartesian coordinates, filtering the Radon transform (for instance, with the ideal inverse high-pass filter shown in Chapter 3, Figure 3.11a) and performing an inverse Fourier transform, the familiar cross-sectional slice images can be obtained (Herman 1980; Kak and Slaney 2001).

### 5.2.3 Shape

Feature shape is often an important factor in distinguishing between different classes of objects, but it is not always easy to find measurable numeric parameters that correspond to those differences in shape that are so readily perceived by a human (Ferson, Rohlf, and Koehn 1985; Loncaric 1998). Figure 5.25 shows the image of coins (and two keys) that was processed to level the nonuniform illumination in Chapter 2 and thresholded to produce a binary image of features in Chapter 4. The circular coins are very different in shape from the keys. In the image, they have been distinguished by assigning a grayscale brightness to each feature that is proportional to a shape descriptor, the ***formfactor***. This is calculated as shown in Equation 5.11.

The formfactor depends on the area and perimeter of each feature, and Table 5.4 lists the measured area and perimeter and the calculated formfactor for each of the individual features. The coins all have values for the formfactor between 0.90 and 0.92, but the two keys have values close to 0.4. (The coin values are less than the theoretical value of 1.0 for a circle because of slight departures of the features from perfect circles owing to the inclusion of shadows, the finite sizes of the pixels making up the features, and the limited accuracy of the chain code perimeter measurement.)

$$Formfactor = \frac{4\pi \cdot Area}{Perimeter^2}$$

$$Roundness = \frac{4 \cdot Area}{\pi \cdot Length^2}$$

(5.11)

This example shows a case in which the two classes of objects (coins and keys) can be easily separated based on the calculated formfactor. However, it is easy to find or construct examples in which features with identical values of the formfactor have shapes that appear quite different to a human. More shape descriptors are needed. Equation 5.11 shows another, the roundness. Both the formfactor and the roundness calculate values of 1.0 for a perfect circle (and values close to that for circular features digitized as an array of pixels). However, as illustrated in Figure 5.26, they measure characteristics that correspond to the different ways in which shapes can deviate from a circle.

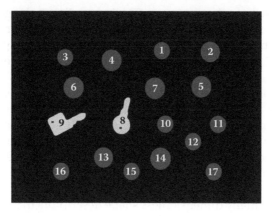

**FIGURE 5.25**
Color-coding of features to represent measured values. The calculated formfactor shape values shown in Table 5.4 were scaled arbitrarily between 100 and 200 (on the 0 to 255 scale) to color the features in shades of gray.

**TABLE 5.4**

Measured Area and Perimeter, the Calculated
Formfactor, and the Assigned Gray Display Value
for the Features in Figure 5.25

| Feature* | Area | Perimeter | Formfactor | Gray Value |
|---|---|---|---|---|
| 1 | 4260 | 241.36 | 0.919 | 100 |
| 2 | 5368 | 273.74 | 0.900 | 104 |
| 3 | 4158 | 239.36 | 0.912 | 101 |
| 4 | 6989 | 309.88 | 0.915 | 101 |
| 5 | 5458 | 275.74 | 0.902 | 103 |
| 6 | 6899 | 308.71 | 0.910 | 102 |
| 7 | 7142 | 313.54 | 0.913 | 101 |
| 8 | 8389 | 497.17 | 0.426 | 196 |
| 9 | 10249 | 565.42 | 0.403 | 200 |
| 10 | 4480 | 249.02 | 0.908 | 102 |
| 11 | 3733 | 226.53 | 0.914 | 101 |
| 12 | 4317 | 244.78 | 0.905 | 103 |
| 13 | 5749 | 282.81 | 0.903 | 103 |
| 14 | 7061 | 312.71 | 0.907 | 102 |
| 15 | 4413 | 246.19 | 0.915 | 101 |
| 16 | 4245 | 241.36 | 0.916 | 101 |
| 17 | 4207 | 240.78 | 0.912 | 101 |

\* The feature numbers correspond to the labels shown.

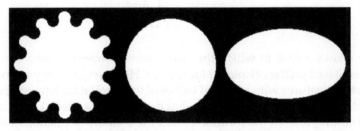

**FIGURE 5.26**
Shapes at left and right that vary from a circle in different ways.

The shapes in the figure have the same area. One is smooth but elongated, with a departure from circularity that can be measured by the roundness parameter in Equation 5.11. The other is equiaxed but has boundary irregularities that increase the perimeter. This type of noncircular shape can be characterized by the formfactor parameter.

The equations show only two of the practically unlimited number of shape descriptors that can be produced by combining various size measurements in such a way that the units of measurement cancel out. For instance, you could define one as the ratio of feature area to the area of the oriented bounding box, or the ratio of feature area to the convex hull area, or the ratio of the inscribed circle radius to the length, etc. The ones shown in Equation 5.11 are probably the most widely used.

Finding one of these formally dimensionless shape descriptors that is useful in a particular situation, for instance, to classify objects as shown in Figure 5.25, is very convenient because of the ease with which they can be calculated, but there are no general guidelines. The additional problem they present is a complete lack of any standardization. Some software packages, for example, define the formfactor as the reciprocal of the expression shown, and others omit the numeric constants.

Nor is the nomenclature standard. Some software packages use the name "roundness" for the expression for formfactor shown in Equation 5.11, for example. None of these terms have a fixed common definition, and human languages do not have many adjectives that describe shape. The word "circular" means "like a circle," but "non-circular" does not specify whether the deviation from circularity is similar to one of the shapes in Figure 5.26, or has some other form (e.g., trilobed, polygonal, etc.).

One of the most important characteristics of shape from the point of view of human recognition is concerned with topological characteristics such as the number of holes and branches. As shown in Figure 5.27, the number of end points in the feature skeleton provides a direct way of determining the number of points in each of the stars, which people recognize instinctively. Indeed, the skeleton contains a great deal of topological information that can be used for shape classification. The example in Chapter 4, Figure 4.29, the skeleton of the image of a gear, showed that counting the number of end points provided a way to determine the number of teeth on the gear.

There are several ways to extract the information from the skeleton for each feature in the image:

1. The code fragment from Chapter 4 that performs skeletonization can be modified for application to features after the pixels in the binary image have been assigned unique identification numbers, so that those numbers are preserved for all of the pixels in each separate skeleton. Then, a scan through the image can count the number of end points, for example, for each separate feature.

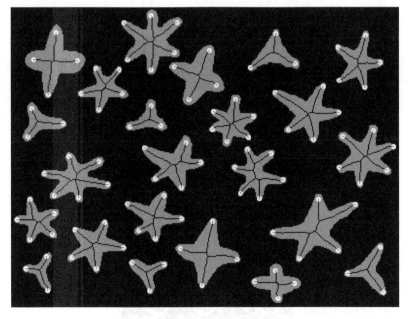

**FIGURE 5.27**
Counting the end points of the feature skeleton characterizes the important topological shape of these features [stars.tif].

2. Two separate temporary image arrays can be created in memory, one with the features to which identification numbers are assigned and the other containing the skeletonized pixels. Scan through the skeleton array to locate end points, for example, and for each one found look at the same location in the second array to get the feature identification number. Count up the end points for each separate feature.

3. Create two separate images, one having the features with their pixels labeled with unique identification numbers and the other containing the skeletons. Combine them pixel by pixel using a math operation that keeps whichever value is darker at each location. That will erase all of the pixels to 0 (background) except those in the skeletons, which will retain the feature identification. Scan though the image to count the number of end points, for example, for each separate feature.

4. Create a temporary image in memory using a defined structure for each pixel that can hold multiple values, such as the feature identification, the skeleton, and potentially other information such as original pixel brightness or color values, the EDM values, etc. The various pieces of information can then be accessed and combined as in method 2.

These methods will give equivalent results, and the choice between them is one of convenience and clarity. Because there will be other situations introduced below, such as determining pixel brightness or color information from individual features, that require combining various pixel values with the feature identification numbers, method 2 is recommended as being the most general and usually the least likely to produce confusion.

A second topological shape property that is very important to human characterization and recognition of features is the presence of holes. The letters shown in Figure 5.28 are in different fonts, some with serifs or scrolls and some without. The number of holes in each letter is a robust characteristic that can be used to distinguish them.

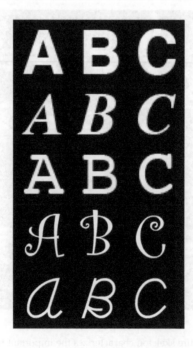

**FIGURE 5.28**
Even in different fonts, these letters can be distinguished because they contain 1, 2, and 0 holes, respectively.

The number of holes in a feature is identical to the number of loops in the feature's skeleton, and this is related to the number of branches and ends by Euler's formula (Equation 5.12). Hence, it is possible to measure it by using one of the methods listed above to obtain the skeleton information for each feature. Some caution in applying this relationship is needed as there are a few situations in which the "apparent" topology of the skeleton can be deceptive. A ring (e.g., the letter "0") skeletonizes to a single circular branch that has one loop, a single branch, no ends, and no apparent node. However, the rules of topology require that there be a "virtual" node someplace on the ring where the two ends of the linear branch are joined. Likewise, a perfectly symmetrical circle skeletonizes to a single point which, having fewer than two neighbors, might be counted as an end. In reality, this point represents a short branch with two ends. Special tests in the analysis code can deal with these cases.

$$\text{Loops} = \text{Branches} - \text{Ends} - \text{Nodes} + 1 \qquad (5.12)$$

It is also possible to count the number of holes using the procedure described above (in Problem 5.1.5.7) for filling them: A copy of the image is made, and the contrast reversed so that the features become background and vice versa. Then, the feature-labeling algorithm is applied, with the requirement that the features have 4-neighbor connectivity (pixels are considered to touch only if they share a side). An additional step of identifying the features that do not touch any image boundary is also necessary. These features in the complemented image are the holes, and combining them with the original image accomplishes hole filling as described previously. Once the image of the features with the holes filled and the image of the holes are available, it is straightforward to scan through them to count the number of holes that were originally in each feature.

Either of these methods can be used, but in practice a simpler way to perform the hole counting is to do it as part of the feature-labeling algorithm in Code Fragment 5.2. In that routine, the neighboring pixels above and to the left of each feature pixel are examined to get the identification number that is propagated to the current pixel. There are exactly three cases in which it is necessary to test whether the previous pixels have different identification numbers assigned, indicating that two partial features are being joined together. The code fragment as written detects this for index numbers 5, 9, and 13 and if the previous numbers are different, it replaces one of them with the other. However, if the previous numbers are the same, it does nothing.

If the two previous identification numbers are the same, the joining of the two parts of the same feature indicates that a hole within the feature has just ended. Counting these events for each feature counts the holes. Because much of the code in Code Fragments 5.1 and 5.2 is already common, combining them and adding this hole-counting algorithm can, by scanning through the image, count and label the features, measure the area and perimeter of each, and count the number of holes in each image.

In addition to topological shape descriptors, there is a particular kind of "roughness" of feature edges that occurs often in nature, is visually recognized (but, of course, not quantified — human vision is comparative rather than metric), and can be measured as a ***fractal dimension***. This is a value greater than 1.0 (the dimension of a smooth Euclidean line) but less than 2.0, which is the dimension of the entire plane in which the feature resides.

Mandelbrot, who introduced the term fractal, points out that "clouds are not spheres, mountains are not cones, and lightning does not travel in a straight line." There are some cases in which boundaries are smooth and can be described (at least over a considerable range of dimension) by Euclidean geometry, such as bubbles whose surface is controlled by surface tension, and man-made objects such as machined parts, buildings, and roads. However, most natural objects reveal more detail and longer perimeters as image resolution is increased.

Consider the following experiment to measure the length of a coastline: Using a chain 1 km long, count the number of straight chain lengths required to walk along the coast as indicated in Figure 5.29. Then, repeat the operation with a shorter chain, 100 m long. Obviously, this will follow the irregularities more closely and produce a longer estimate of the length. Repeating the operation again with a 10-m chain, a 1-m ruler, etc., continues to increase the measured length. For many natural shapes, plotting the total measured length versus the ruler length on log–log axes produces a straight-line graph. The slope of this graph, called a Richardson plot, describes the roughness, which would be greater in this example for the fjords along the coast of Norway than for the beaches along the coast of Florida (Mandelbrot 1967; Mandelbrot 1982).

There is a substantial mathematical literature about fractal dimensions, and a much smaller body of publications about the relationships found between this measure of roughness and various physical processes. None of this interesting background is needed for the purpose of using the fractal dimension as a measure of feature shape that is in many ways complementary to the topological descriptors. The fractal dimension is purely concerned with the local irregularities of the feature edges, which the topology ignores.

There are several different methods that can be used to measure the fractal dimension of feature boundaries (Kaye 1986; Russ and Russ 1989). For an image made up of pixels, the Richardson (stride-length) method, which requires finding a location an exact distance away and lying on the periphery, is not a good choice. By far the most accurate, and also one of the easiest to calculate, method is based on the Euclidean distance map. To illustrate the procedure, Figure 5.30 presents an ideal fractal shape in which an endless regress of self-similarity decorates the feature edges (the method works equally well on natural

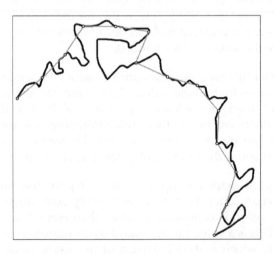

**FIGURE 5.29**
Basis for constructing a Richardson plot. The length of the boundary line or coastline (black) is estimated by "striding" along it with dividers set to a fixed length and counting the number of steps, shown in gray. Reducing the stride length allows the irregularities to be followed more closely and produces a longer estimate for the length.

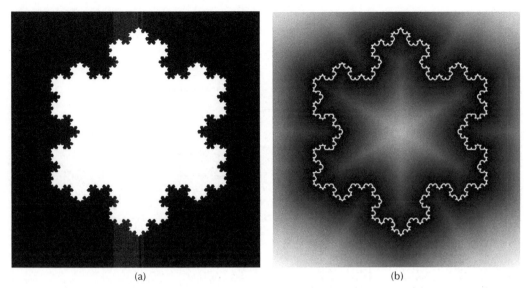

(a)                                    (b)

**FIGURE 5.30**
A self-similar Koch "snowflake" [koch.tif], and the EDM of pixels inside and outside the feature periphery.

features such as coastlines, but for ideal fractals such as this one, the exact dimension is known). The EDM for pixels both inside and outside the feature boundary is shown.

Saving the histogram of the EDM image to a file that can be opened in a spreadsheet presents a tabulation of the number of pixels with each brightness value, which is just their distance from the feature edge. By creating a few additional columns in the spreadsheet, and plotting the results, the fractal dimension of the feature can be obtained. Figure 5.31 shows the steps. First, the numbers of pixels with each distance value are summed to create a cumulative column that shows the total number of pixels lying within each distance from the feature edge (an area). Two additional columns are then created, one with the logarithm of the distance value and one with the logarithm of the area. The values in these columns are then plotted for all values greater than 0 (which are the pixels marking the edge of the feature), as shown in the figure. This results in a straight line whose slope is related to the fractal dimension as shown. For this example, the procedure calculates a dimension for the Koch snowflake of 1.283. The theoretical value is 1.262, and the minor discrepancy results from the limitations of representing the feature with the finite resolution of the pixel array.

This entire procedure can be carried out within a measurement routine by using an EDM image. It is important to use the EDM of pixels both inside and outside the feature, and to generate the EDM for one feature at a time so that distances are measured from the boundary of the feature, and not the distance from some other nearby feature. A good fit to determine the slope of the log-log plot can be obtained with as few as 8 to 10 points, so it is only necessary to carry the calculation out to that distance.

### 5.2.4 Density and Color Measurement

The preceding sections have dealt with the measurement of feature size, position, and shape. In a few cases the measurements have involved pixel values (for instance, using the EDM values for fractal dimension calculation), but the procedures have started with

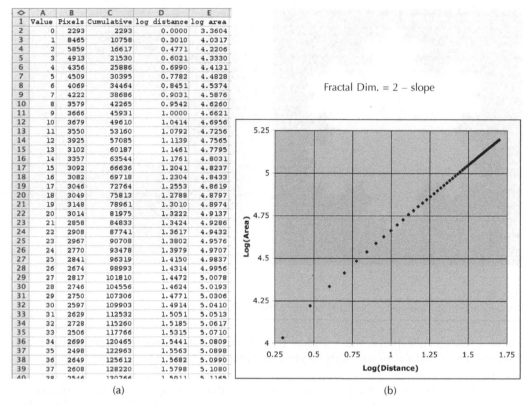

| | A | B | C | D | E |
|---|---|---|---|---|---|
| 1 | Value | Pixels | Cumulative | log distance | log area |
| 2 | 0 | 2293 | 2293 | 0.0000 | 3.3604 |
| 3 | 1 | 8465 | 10758 | 0.3010 | 4.0317 |
| 4 | 2 | 5859 | 16617 | 0.4771 | 4.2206 |
| 5 | 3 | 4913 | 21530 | 0.6021 | 4.3330 |
| 6 | 4 | 4356 | 25886 | 0.6990 | 4.4131 |
| 7 | 5 | 4509 | 30395 | 0.7782 | 4.4828 |
| 8 | 6 | 4069 | 34464 | 0.8451 | 4.5374 |
| 9 | 7 | 4222 | 38686 | 0.9031 | 4.5876 |
| 10 | 8 | 3579 | 42265 | 0.9542 | 4.6260 |
| 11 | 9 | 3666 | 45931 | 1.0000 | 4.6621 |
| 12 | 10 | 3679 | 49610 | 1.0414 | 4.6956 |
| 13 | 11 | 3550 | 53160 | 1.0792 | 4.7256 |
| 14 | 12 | 3925 | 57085 | 1.1139 | 4.7565 |
| 15 | 13 | 3102 | 60187 | 1.1461 | 4.7795 |
| 16 | 14 | 3357 | 63544 | 1.1761 | 4.8031 |
| 17 | 15 | 3092 | 66636 | 1.2041 | 4.8237 |
| 18 | 16 | 3082 | 69718 | 1.2304 | 4.8433 |
| 19 | 17 | 3046 | 72764 | 1.2553 | 4.8619 |
| 20 | 18 | 3049 | 75813 | 1.2788 | 4.8797 |
| 21 | 19 | 3148 | 78961 | 1.3010 | 4.8974 |
| 22 | 20 | 3014 | 81975 | 1.3222 | 4.9137 |
| 23 | 21 | 2858 | 84833 | 1.3424 | 4.9286 |
| 24 | 22 | 2908 | 87741 | 1.3617 | 4.9432 |
| 25 | 23 | 2967 | 90708 | 1.3802 | 4.9576 |
| 26 | 24 | 2770 | 93478 | 1.3979 | 4.9707 |
| 27 | 25 | 2841 | 96319 | 1.4150 | 4.9837 |
| 28 | 26 | 2674 | 98993 | 1.4314 | 4.9956 |
| 29 | 27 | 2817 | 101810 | 1.4472 | 5.0078 |
| 30 | 28 | 2746 | 104556 | 1.4624 | 5.0193 |
| 31 | 29 | 2750 | 107306 | 1.4771 | 5.0306 |
| 32 | 30 | 2597 | 109903 | 1.4914 | 5.0410 |
| 33 | 31 | 2629 | 112532 | 1.5051 | 5.0513 |
| 34 | 32 | 2728 | 115260 | 1.5185 | 5.0617 |
| 35 | 33 | 2506 | 117766 | 1.5315 | 5.0710 |
| 36 | 34 | 2699 | 120465 | 1.5441 | 5.0809 |
| 37 | 35 | 2498 | 122963 | 1.5563 | 5.0898 |
| 38 | 36 | 2649 | 125612 | 1.5682 | 5.0990 |
| 39 | 37 | 2608 | 128220 | 1.5798 | 5.1080 |
| 40 | 38 | 2546 | 130766 | 1.5911 | 5.1165 |

Fractal Dim. = 2 − slope

(a)                                             (b)

**FIGURE 5.31**
Calculation of the fractal dimension as described in the text.

a thresholded binary image delineating the objects to be measured. This section deals instead with measurement of the brightness or color of the pixels from the original image.

It is still necessary to restrict the measurement to individual features, which is accomplished by erasing the background pixels between features to black (pixel value = 0). This can be done by thresholding a copy of the original image to obtain the usual binary image, and if necessary applying morphological operations to process that image so that features are accurately delineated. Combining this image with the original using a mathematical operation that keeps whichever pixel is darker will replace the background pixels with black while leaving the values of pixels within features unchanged.

Figure 5.32 shows an example. The original color image was thresholded as described in Chapter 4, by selecting pixels with color saturation greater than zero (the background and balloon strings are gray and black, respectively). The resulting binary image was eroded by a single pixel to produce a result in which all of the balloons are separated and delineated as shown in Figure 5.32a. An alternate method of producing this image is to threshold each color separately, as was done in Chapter 4, and then use a Boolean OR to combine the individual binary images. Combining the image from Figure 5.32a with the original and keeping whichever pixel is darker produces the result shown in Figure 5.32b, in which each separate feature retains the original color values for the pixels.

Using the binary image as a guide that labels each feature, the pixel values within each feature were read from the original image and the RGB values converted to

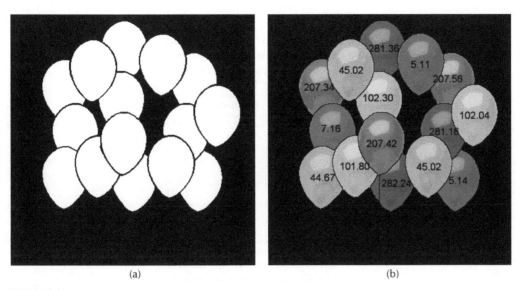

(a)                         (b)

**FIGURE 5.32**
(See **Color insert following page 172**.) Measuring the mean hue in each feature [balloons.tif]: (a) thresholding the individual features as described in the text, (b) combining the original image with the binary image in (a). The numeric values correspond to an angle (0 to 360°) on the color wheel, and are shown superimposed on the original image for convenient reference.

hue–saturation–intensity as shown in Chapter 1. The hue values for the pixels in each feature were averaged to obtain the results that are shown superimposed on the features in the figure. The hue values represent angles on the color wheel (0 degrees = red, 120 degrees = green, 240 degrees = blue).

Chapter 1 pointed out that three values are required to locate a point in color space, and that hue, saturation, and intensity values are usually more meaningful than red, green, and blue. Figure 5.33 shows a cross section through a series of paint layers with lines of separation between them. The lines were obtained by thresholding a copy of the image after processing the image as shown in Chapter 2 to highlight edges (discontinuities in the image marked by abrupt changes in brightness). This produced lines around some of

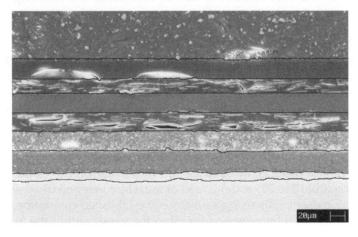

**FIGURE 5.33**
(See **Color insert following page 172**.) Layers in a cross section of paint, measured to produce the data in Table 5.5 [paint.tif].

**TABLE 5.5**

Measuring the Mean Hue and Saturation,
and the Mean and Standard Deviation of
Intensity, in Each Layer of Paint Shown
in Figure 5.33 (from Top to Bottom)

| Hue | Saturation | Intensity | Std. Dev. |
|---|---|---|---|
| 3.5 | 0.03 | 119.7 | 20.72 |
| 0.4 | 42.2 | 117.0 | 36.98 |
| 240.6 | 43.9 | 141.3 | 36.20 |
| 117.9 | 51.4 | 103.8 | 7.21 |
| 299.8 | 42.8 | 155.0 | 39.80 |
| 59.2 | 46.5 | 168.3 | 26.43 |
| 178.8 | 43.0 | 142.7 | 15.38 |
| 11.0 | 100.2 | 232.9 | 6.50 |
| 7.8 | 0.01 | 227.2 | 5.04 |

*Note:* The hue is measured as an angle (0 to
360 degrees), whereas the saturation,
intensity, and standard deviation are
measured on a 0 to 255 scale.

the internal bright and dark features, but only those lines extending across the full width of the image (touching both left and right boundaries) were kept. These lines were used as boundaries to separate the various paint layers.

Measuring the hue, saturation, and intensity values from the pixels within each layer produces the results shown in Table 5.5. For the intensity values, the mean and standard deviation were calculated to characterize the texture present within several of the layers.

Data such as these can be useful for matching paint chips, for example, to identify paint left at the scene of an automobile accident. However, it is important to control the illumination and camera settings so that meaningful comparisons can be made. As pointed out in Chapter 1, with suitable calibration it is practical to match colors. However, a color camera is not capable of actually measuring colors, because the filters used on the detectors pass broad ranges of wavelengths and many different combinations of colors can produce identical red, green, and blue values, and consequently, identical intensity, saturation, and hue values.

Calibration is also required to convert measured intensity to values such as density. Figure 5.34 shows an image (acquired with a flatbed scanner using transmitted light) of a density wedge, commonly used in the darkroom to determine the optimum exposure when making photographic prints. As shown in Table 5.6 and Figure 5.35, the measured brightness values are roughly proportional to the marked exposure factors. By imaging a suitable standard, it is possible to calibrate a scanner, or a camera and light box, so that density values or other quantitative information can be obtained from samples.

The measurement procedure is the same as that shown above, except that as each pixel's brightness (on the usual 0 to 255 scale) is read, it must be converted to density (or whatever the calibrated value may be) by using a previously established calibration curve, and then these derived values must be averaged for all of the pixels within the feature. Note that it is not correct to average the brightness values and then convert the mean pixel brightness value to density, because the calibration curves are not necessarily linear. It is also important to repeat the calibration step often, because minor variations in line voltage, aging

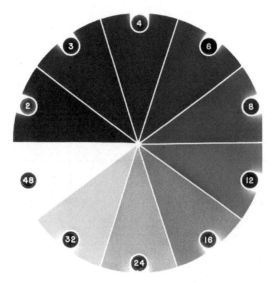

KODAK PROJECTION PRINT SCALE

**FIGURE 5.34**
Scanned image of a density wedge [steps.tif].

**TABLE 5.6**

Mean Brightness of Regions in the
Exposure Scale in Figure 5.34

| Mean Brightness | Exposure Factor |
|---|---|
| 21.71 | 2 |
| 28.32 | 3 |
| 32.65 | 4 |
| 44.25 | 6 |
| 63.97 | 8 |
| 85.98 | 12 |
| 113.01 | 16 |
| 157.18 | 24 |
| 200.49 | 32 |
| 254.99 | 48 |

or warming up of the illuminating lamp, temperature variations in the camera sensor or electronics, and a host of other variables can significantly alter the calibration. Cameras that have automatic gain and brightness circuits are practically impossible to calibrate because their characteristics change for every image.

The standard deviation of the intensity values pixels within a feature, illustrated in the example in Table 5.5, is one way to characterize the texture of an object. As noted in Chapter 2, texture is a visually recognizable characteristic that can be measured by a variety of quite different arithmetic calculations. Some of these have proved useful for applications as diverse as identifying the crops in satellite images of fields, differentiating curds and protein in microscope images of cheese, recognizing diseased cells in biopsies, and so on (Haralick, Shanmugan, and Dinstein 1973; Weszka, Dyer, and Rosenfeld 1976).

One technique that is sometimes used for evaluating the texture of the pixels in a region begins with the construction of a co-occurrence matrix. This is a table that counts the

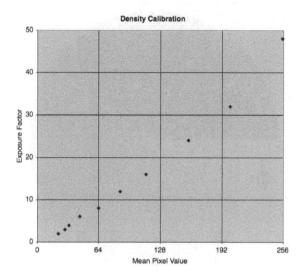

**FIGURE 5.35**
Plot of the data from Table 5.6. The highest point on the curve is shifted from a linear trend because of clipping at the maximum brightness value.

number of pairs of pixels as a function of the distance between them and the difference in their brightness levels. The former value may be measured in pixels or, if different scale images are used, may be calibrated in dimensional units. In the most general case, all possible pairs of pixels in all directions are included in the construction of the table. Sometimes, in the interests of speed or as a hangover from the days of slower computers, only pairs along the same row or column are compared, or the comparison is restricted to immediate neighbors in various directions.

The brightness difference is most often based on the 0 to 255 intensity values for the pixels, but if variations in lighting or camera settings may be encountered, the intensity can be subject to a calibration as described above. Rarely, color values (hue or saturation) may be used as well. In order to produce a table of manageable size, both the distance and difference values are usually divided down to a small number of levels. In the example shown in Figure 5.36, the pixel brightness values are represented as single digits, and pixel distances are rounded off to the nearest integer value. This small feature contains 18 pixels, producing $18 \cdot 17/2 = 153$ pixel pairs.

Because the size and shape of features may vary, the number of counts in each cell of the co-occurrence matrix must be normalized. This can be done by dividing the values in each cell of the table by the number or fraction of pixel pairs examined which are that distance apart (the totals for each distance shown in Figure 5.36b). The resulting $c_{i,j}$ values measure the probability of finding a particular brightness difference with a specified spatial separation.

These $c_{i,j}$ values are typically used to calculate various statistical properties that describe the feature's texture (Murino, Ottonello, and Pagnan 1998). Experience, or trial and error, is then used to select those properties that are useful for feature identification in particular applications. Of course, it is also possible to use the same calculation procedure for the pixels in a moving neighborhood to produce a derived image in which each pixel's value represents the local texture. This may be useful in some instance to allow thresholding to

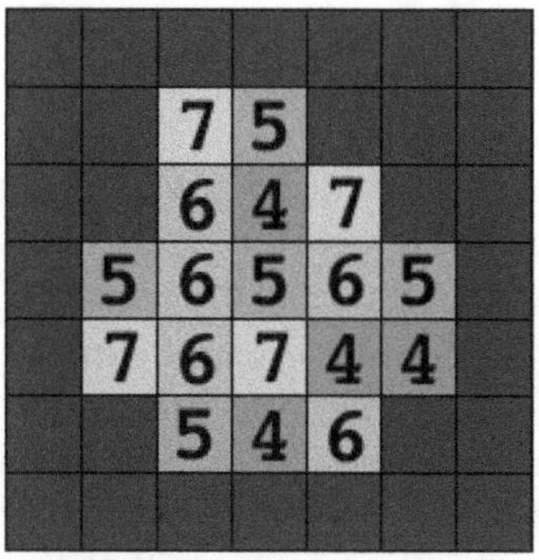

| Distance | 1 | 2 | 3 | 4 |
|----------|---|---|---|---|
| **Difference** | | | | |
| 0 | 4 | 16 | 9 | 5 |
| 1 | 24 | 16 | 18 | 6 |
| 2 | 17 | 10 | 8 | 5 |
| 3 | 4 | 5 | 3 | 3 |
| | | | | |
| totals | 49 | 47 | 38 | 19 |

(a)    (b)

**FIGURE 5.36**

Example of a co-occurrence matrix: (a) feature containing 18 pixels, each labeled with a single-digit brightness value; (b) table of the number of pixel pairs as a function of their separation distance (rounded to the nearest integer) and their brightness difference.

select textured features in an image. The fractal texture relationship shown in Chapter 2, Figure 2.49, is an example of a calculation that can be performed using the co-occurrence matrix (Peleg et al. 1984; Pentland 1983).

### 5.2.5 Problems

5.2.5.1#. Implement a program to measure the maximum caliper dimension of a feature by finding the extreme coordinates with axes rotated in steps of 22.5°. Write the results to a text file. Optionally, modify the program to use steps of 11.25° and compare the results.

5.2.5.2. Modify your program from Problem 5.2.5.1 to keep the coordinates of the extreme points and construct the bounding polygon. Verify your program by drawing the polygon onto the image. (A routine to draw a line segment between two points was introduced in Chapter 4.)

5.2.5.3. Implement a program to measure the area of each feature and report the mean value for all of the features in the image. Note that if no features intersect the image boundaries, this result should be identical to dividing the total area of all features by the number of features. Use the adjusted count to weight the individual values, and compare the result to that without correction. Features that touch (and may therefore cross) the boundaries of the image should never be measured, because their full extent cannot be known. If the image area represents a portion of the population of features, as for example in the [rice.tif] image, it is more likely that large features will intersect the image boundaries and consequently will not be measured. The adjusted count described in the text makes a statistical correction for this bias. It is not appropriate if all of the features in the population are contained within the image boundaries, as for example in the [cashews.tif] image.

5.2.5.4. Modify your program from Problem 5.2.5.3 to allow selecting other measures of feature size instead of area (e.g., length, perimeter).

5.2.5.5#. Implement a program to calculate the location of features by the centroid and the center of the bounding box, and compare the results. Mark the locations on the image (to the nearest pixel).

5.2.5.6.    Use the centroid locations of features to calculate the mean nearest-neighbor distance, and compare the result to the value expected for a random distribution based on the image area and number of features.

5.2.5.7#.   Implement a program to measure the area and perimeter of features, calculate the formfactor for each one, and write the values to a text file. Optionally, display the features with colors (hues) that represent the magnitude of the value.

5.2.5.8.    Implement a program to determine the adjusted count for each feature, as shown in Figure 5.11. This requires finding the minimum and maximum $x$- and $y$-coordinates for each feature to determine its extent. Sum the adjusted counts for all features that do not touch any image boundary. Compare this value to the number of features determined by counting features that touch the bottom and right boundaries and ignoring those that touch the top and left boundaries. This method should only be used for images that are samples of a larger field of objects, such as [rice.tif].

5.2.5.9.    Implement a program that counts the number of end points in the skeleton of each feature. Use this to label the star-shaped objects in the [stars.tif] image with unique colors distinguishing them based on the number of points in each one.

5.2.5.10.   Implement a program that counts the number of holes in each feature. Display the features with colors that represent the resulting value.

5.2.5.11.   Implement a program to measure the fractal dimension of a single feature. Write the result to a text file. Apply this to various shapes and compare the results to the visual impression of boundary roughness.

## 5.3  Classification

One of the reasons for measuring features in an image is to provide data that can be used to *__classify__* or recognize objects in images. There are two somewhat different tasks that may be encountered, both requiring feature measurement data. One is to assign a feature to a predefined class or group, typically based on a combination of measurements made on representative, known members of the class to establish the limits that define each group or class. The other, more difficult task is to separate a number of features into groups whose limits are not known, by finding clusters of similar features (indeed, sometimes even the number of classes is not known beforehand).

### 5.3.1  Multiple Criteria

Figure 5.25 showed an example in which coins and keys were distinguished by a measurement of shape. Some classification and recognition tasks can be accomplished with a single size, shape, position, or color measurement. However, it typically requires a combination of criteria to successfully distinguish between multiple classes. Determining the best combination of measurements to use can be carried out by statistical techniques in some cases. Measuring every possible parameter on a series of known objects (called the training population), and then applying techniques such as stepwise regression, principal components reduction, or other cluster-finding algorithms can result in the selection of the best combination of parameters and the discriminant values that separate the various classes.

In many situations, particularly those for which obtaining a perfectly representative set of training objects that fully cover the range of combinations of parameters is impractical, these automatic methods do not function well. However, human judgment can still be used to establish a set of rules called an expert system (but note that it is the human responsible for creating the rules, not the system that applies them, who is the expert). In general, this requires the human observer to make explicit the process by which he or she is able to recognize an object or class of objects, and to determine the corresponding computer measurement(s) that can be performed to extract similar criteria as numeric data.

Graphs showing the measurements of various parameters on example features are a powerful tool in establishing the criteria for classification. Figures 5.37 and 5.38 show an example. The image contains five classes of objects, and the graphs show two-dimensional scatterplots of several measured parameters for each object, labeled by class. It is apparent that some of the classes are easily distinguished from the others using a single parameter, whereas other classes require several criteria used in combination for unique discrimination. Some statistics programs allow the examination of scatterplots in higher dimensions, for instance by interactively rotating the $n$-dimensional point cloud about each axis on the screen.

The goal is to find a combination of measurements and corresponding limit values for each object class that can uniquely distinguish all of the classes of objects present. The measurement procedures developed previously for size, shape, and color (position is not relevant for this particular example) are adequate to accomplish the task. Implementing the various measurements and determining and performing the classification procedure is presented as Problem 5.3.2.2 below.

A popular technique for automatic object identification, which does not require a human to select the measurement parameters that are most important, nor to establish limits that define each class, is called the kNN or nearest-neighbor method. All available measurements are made on the identified objects used as training populations, and used to plot a point for each training object in a high-dimensionality space (one dimension for each measured

**FIGURE 5.37**
(See **Color insert following page 172.**) Image of features for classification [nuts.tif].

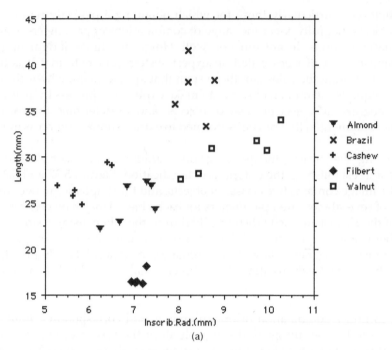

**FIGURE 5.38**
Selected scatterplots of measured parameters for the features in Figure 5.37.

parameter). Then, for each unknown object, the same measurements are performed and the location of the corresponding point is compared to those for the training populations. In the simplest implementation, the identity of the object is determined by finding the nearest point for a known object. This is shown schematically in Figure 5.39 for the case of two measurement parameters.

A somewhat more robust result can be produced by a voting method in which the identity of the majority of a small (odd) number of nearest neighbors is used. The distance to the neighbor points is calculated as a Pythagorean result (Equation 5.13). However, the apparent simplicity of this approach hides an important limitation. The various parameter axes do not have the same importance or even the same dimensions (and some parameters are redundant). Different weights $w$ shown in Equation 5.13 may be applied to each parameter to make this adjustment. Normalizing each difference by dividing by the standard deviation is often used to address this problem. It is also extremely important to use training populations that are statistically representative and equal in size to avoid biasing the result.

$$Distance = \sqrt{w_A \cdot \left(A - A_i\right)^2 + w_B \cdot \left(B - B_i\right)^2 + w_C \cdot \left(C - C_i\right)^2 + \ldots} \qquad (5.13)$$

There are many other approaches to object classification. They are usually separated into two principle categories, supervised and unsupervised, depending on whether the identities of the objects in the training populations are specified beforehand (supervised), or the software must find clusters of measurements and determine the number and identity

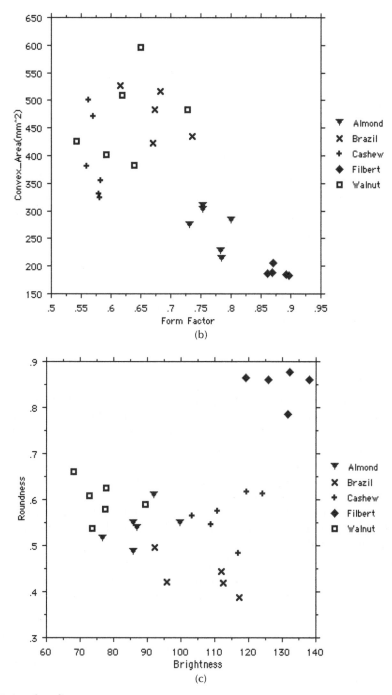

**FIGURE 5.38 (continued)**

of the different classes of objects present (unsupervised). The latter methods are generally more computationally complex. Either method can be adapted to traditional computer architectures or to parallel methods such as neural nets. Excellent texts such as Bow (1992) and Fukunaga (1990) are available covering many aspects of classification.

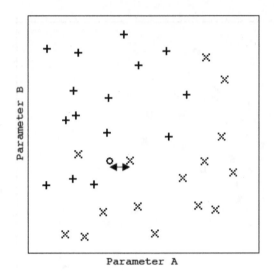

**FIGURE 5.39**
Nearest-neighbor classification. The circle represents the measurements on an unknown object. The arrow shows the distance to the nearest point corresponding to an object in the training population, in this case, a member of class X. Note, however, that if the five nearest neighbors are polled, the majority are of class +.

### 5.3.2  Problems

5.3.2.1.  Implement a program that performs the measurement of all of the various size, shape, position, and color measurements described above. Read a value from the input file to select which parameter's values are to be written out to a text file. A practical method for measuring pixel color values within features is to save a thresholded and processed binary image that delineates the features and labels them with feature identities (see Problem 5.1.5.5) as the reference image, and then perform the measurement on the original image. By reading the reference image, the pixels within features, and the feature ID, are obtained. If the pixel lies within a feature, the brightness and color values can be read from the original image.

5.3.2.2.  Execute the program from Problem 5.3.2.1 on the [nuts.tif] image to collect a set of measurement data for all of the nuts. By examining plots of the values, determine a combination of parameter measurements and limits that will characterize each class of objects.

5.3.2.3.  Modify the program from Problem 5.3.2.1 to apply the decision logic from Problem 5.3.2.2. Demonstrate your ability to select each class of objects by reading a value from the input file to choose the class to be selected, and process the image to leave just that class and erase all of the other features.

A D C <sup>A</sup> B E V
         𝓐    C 𝓐
<sup>A</sup>   B 𝓐 A   <sup>A</sup>   A   V E
   C    A D E     V E
E D   C   C   D <sup>A</sup> B
B   𝓐 V D E B   C
E B   D   B E < D
     A <sup>A</sup>       A   B
C E   C <sup>A</sup> A   C E
C   A 𝓐 > E   B   D
           C

5.3.2.4. Devise and implement a procedure to select all of the A's in the [letters.tif] image, regardless of font, size, or orientation.

5.3.2.5 Devise and implement a procedure to identify each of the features in the [letters.tif] image, regardless of font, size, or orientation.

# *Appendix: Software*

There are many ways to write image processing software. The simplest is to load all of the bytes that correspond to the image into memory, somehow apply a function or filter to the data, and write it back to the original location. There are problems with that method: handling multiple file formats, images that are too big to fit into memory (or to display), different precision levels of the pixel data, dealing with multiple color channels, data ordering, boundary conditions, etc.; but it can be very fast.

This text takes a different approach. The authors have created a fairly simple *__application program interface__* (API) to provide line-oriented random access to image data, always in an **RGBPixel** format to eliminate the problem of different formats. Pixel values are always floating point in the range of 0.0 to 255.0. The code enables a set of Photoshop plug-ins (albeit with a limited range of functionality) that is sufficient and easy to use for writing, testing, and comparing image processing and measurement algorithms. As a result, the routines can handle very large images, and the intermediate API code provides some additional boundary condition protections. The plug-ins will work with a wide variety of host programs that adhere to the Adobe Photoshop conventions for plug-in interface. The host program handles the important but essentially auxiliary chores of reading various file formats, and displaying and printing images.

This appendix discusses how to build a plug-in, how to write your own plug-in, the theory of operation both for the book's API and a portion of Adobe's API, some details of the "glue" code that implements the API, and where to get more information on developing plug-ins. Some of the information is presented more than once, in various formats and contexts, as an aid to finding and organizing it for ready reference. Examples throughout the text illustrate the use of the glue code function calls in a wide variety of routines.

Note that additional information, late additions, specific instructions for compiling plug-ins for the Macintosh, and other resources, can be found at the support Web site, **<www.intro2imaging.com>**.

## A.1  The Plug-In Source Code

Each plug-in is intended to perform a single processing or measurement function. It is generated by a separate project, with all files and resources contained in a single folder.

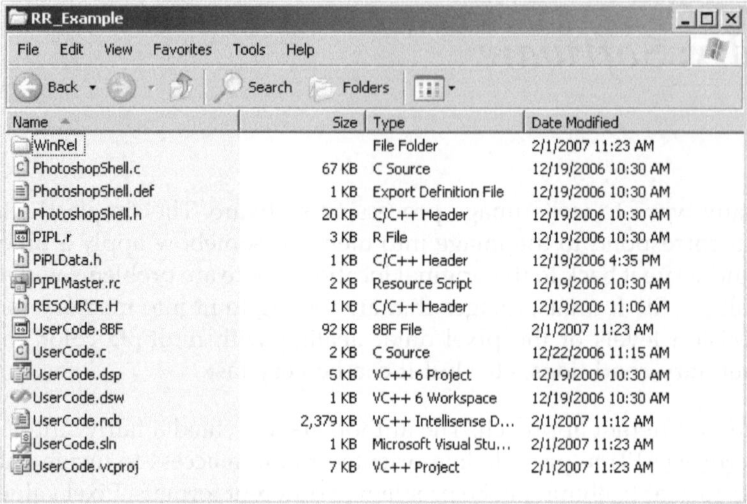

**FIGURE A.1**
A typical project folder.

## A.1.1 The Project Folder

A project folder (Figure A.1) corresponds to a single plug-in that performs an individual function on an image. Duplicating the "**Example**" folder on the companion CD is the best way to create a new project folder with everything needed to generate a new routine, as described in the Introduction. The folder contains several files that are always present, and the "**UserCode.c**" file, which should be edited to implement the desired function.

**UserCode.sln** — This is the "solution" file for Developer Studio. Double-click on it to open up the project.

**PhotoshopShell.h** — All of the structures and functions in Section A.3 are defined here, along with minimalist definitions to access Photoshop, used by the glue code.

**PhotoshopShell.c** — Contains the implementation of all of the API functions as well as internal glue code to talk to Photoshop.

**PhotoshopShell.def** — A list of the externally available functions in the DLL. Note: A Photoshop plug-in is a DLL with an extension of .8BF. For Photoshop to call the DLL, is it necessary to export a function named "**ENTRYPOINT**."

**PiPL.r** — One of the glue code files that builds a Plug-In Property List (PiPL) resource for Photoshop to recognize the plug-in.

**PiPLData.h** — This file should be edited to specify the plug-in's name in the menu, and provide a unique code used by Photoshop to enter the process into actions or history. There is also provision for a text string that is shown in the plug-in's "about" box. This may for example include a description of the plug-in function, the date it was generated, or the student's name. Details are given below.

**UserCode.c** — The user's code resides here. This is where all user-defined functions should go. The code will be called at **MainUserEntry()** as the sole entrypoint between it and Photoshop.

**UserCode.8BF** — **.8BF** files are the compiled plug-ins, which should be copied to the folder containing other Photoshop filter plug-ins (on the PC, typically located at **C:\\Program Files\Adobe\Photoshop CS3\Plug-Ins\Filters**). The name of the **.8BF** file can either be specified in the Visual Studio Properties dialog, or it is possible to leave it as **UserCode.8BF** and rename the file after it has been compiled. This location is similar for other versions of Photoshop. In all cases (Macintosh and PC) the plug-ins folder is in the same path as the Photoshop application.

**Other files** — There are quite a number of other support files for Developer Studio. For instance, the .ncb file contains the "Intellisense" data from the compiler and is quite large. It can be safely deleted. The same goes for the WinRel and the WinDebug folders. Other files are small and should be left alone.

### A.1.2   C versus C++ versus Other Languages

High-level languages can ultimately be easier to write code in, but take longer to learn initially. The goal in this book is to not show off our programming skills by crafting powerful C++ classes and subclasses, but instead to enable the student to directly see and compose his or her own algorithms. Creating a C++ class object-oriented approach is left as an exercise for the interested student, or consider using Adobe's SDK, which *is* in C++, but has a considerably higher learning curve.

As long as the plug-in is a DLL on the PC or a shared library on the Macintosh with the correct entrypoint, it does not matter in which language it was written. If you happen to be very happy with Fortran or Modula or some other compiled language, it is possible to work in that environment. It is necessary to completely replicate the Photoshop side of the API (data structures and callbacks), but how your code works internally is entirely up to you.

### A.1.3   Building a Project

The following steps will allow the student to create an image processing plug-in routine, by starting with an example project that is provided on the companion CD. The steps correspond to using Visual Studio 2005 on a Windows computer, but can be easily adapted to other compilers or platforms.

1. Duplicate the "**Example**" folder containing a minimal example project, and give the folder an appropriate name.
2. Click on the **UserCode.sln** file to open the project.
3. Edit the **PiplData.h** file. Remember to **Save** the result after editing. This file contains several items that control the appearance of the plug-in:
   a. The main menu category ("**R+R_Book**" by default) defines the name that will appear in the Photoshop Filters menu. A submenu with the individual filter plug-ins opens when this menu entry is selected. This name is a Pascal string that is always 32 characters long, so pad out a shorter entry with spaces as shown in the example and place the actual length of the string in hex at the beginning.
   b. The submenu name for the plug-in. The format for this string is identical to that for the category.
   c. A description for the "about" box. This is a C-string and may contain any printable characters including the programmer's name. Inserting a '\n' in the string will start a new line.

d. A unique signature for the plug-in that is used by Photoshop for actions and history. The format is four printable characters, at least one of which must be uppercase and one lowercase. The signature must be entered in both forward and reverse order, as shown in the example.

4. Edit the **UserCode.c** file to create the plug-in. The **#include "Photoshop-Shell.h"** statement is necessary to bring in the various interface and support subroutine calls that are described below and illustrated throughout this text.

5. Compile the plug-in. Be sure that "**Release**" is selected in the Visual Studio header bar and select **Build->Rebuild Solution**. Any errors and warnings will be reported. A warning from the linker that the filename ends in .8BF is expected and can be ignored. Remember to select **File->Save All** and **File->Close Solution** after successful compilation.

6. Copy the compiled plug-in (**.8BF** file) to the folder in which plug-ins are stored to be accessed by the host program. Rename the plug-in from **UserCode.8bf** (for example, "**PlugInName.8BF**"). The file must be of type **.8BF** for Photoshop (or other host programs that use the Photoshop plug-in interface) to recognize it.

Alternately, it is possible to put a shortcut/alias to the plug-in into the plug-ins folder, but some host programs will not accept this (Photoshop does). The host program must be told where this folder is located (in Photoshop, select **Edit->Preferences->Plug-Ins & Scratch Disks** to specify an additional folder). Some host programs do not allow specifying alternate or additional locations for plug-ins, and they must be placed only in a designated folder.

Photoshop scans the plug-ins folder and all subfolders when it is launched and recognizes the **.8BF** files there to build its Filters menu. It is OK to replace an existing plug-in file with another one having the same name while Photoshop is running, but to add a new one or to change the menu entry, the program must be terminated and relaunched.

### A.1.4  What Photoshop Looks for from a Plug-In

When it is launched, Photoshop looks at the location of the Photoshop Application, and then finds a subdirectory or folder named "**Plug-ins**." It also scans any additional folder that has been previously specified by selecting it with the command **Edit->Prefer-ences->Plug-Ins & Scratch Disks** (but some other host programs do not allow this additional location for plug-ins). It completely traverses the contents of that folder and subfolders (including all aliases or shortcuts, but some host programs do not permit subfolders or aliases) to find files that end with the extension **\*.8BF**. (There are a number of other extensions for different kinds of plug-ins, but **\*.8BF** is used for Filters.) Any file or folder whose name starts with the tilde (~) character is ignored, which can be useful for maintaining multiple sets of routines. These files are DLLs with a specific entrypoint "**ENTRYPOINT**." Other DLL entrypoints are unused by Photoshop. On the Macintosh these files are shared Libraries within a Package structure.

Once the set of **\*.8BF** filters are located, Photoshop will open the "PiPL" resource in each (see Section A.1.5) to determine the Category name (main menu entry), the Plug-In Name (name in the submenu), the modes that the plug-in will function with, and assorted other material from the property list. The bottom half of the Filter menu in Photoshop (and corresponding menus in other host programs) is built from those entries (there are some exceptions for other host programs; for example, NIH-Image does not use this information and instead lists the plug-ins by their filenames).

In many cases it is possible to replace a plug-in while Photoshop is running (but not while the plug-in itself is executing!) and have it be recognized (this is not so on the Macintosh). Alternatively, by having a shortcut or alias to the plug-in, it can be tested immediately each time it is recompiled (this works on both the PC and the Mac). (*Note: The menu name and plug-in name in Photoshop can only change when the program is relaunched.*)

Finally, if a plug-in crashes, it will ***not*** be available again until Photoshop is relaunched (more information can be found in Section A.1.6 on Debugging).

### A.1.5  PiPLs — Adobe's Plug-In Property Lists

When Adobe Photoshop is run, it scans the folder and subfolders in which the plug-ins are placed to gather information on the routines present. Most of the other host programs do this too, although some do not use the full set of information (e.g., NIH-Image) or do not perform the scan until the first time that a plug-in is used (e.g., Image-Pro Plus). The information in each plug-in, called the PiPL resource, includes the name that will appear in the menu and the types of images for which the plug-in can be used (e.g., 8 bits per channel RGB images, etc.).

The contents of the PiPL are created by the compilation process using the information in the **PiPLData.h** file in each project on the companion CD. It is important to be sure that the format requirements of the file are followed exactly. Those requirements include entering a unique four-character long ID for the plug-in, and entries for the main menu entry and submenu entry that conform to the 32-character-long format shown. The **PiPLData.h** file determines how the plug-in will show up in Photoshop. It is used when building a PiPL resource within the plug-in.

First, the file contains a unique 4-character signature for that plug-in so that Photoshop can identify it internally (this code is also used by Photoshop, but not most other host programs, to record the operations that automatic sequences called "actions" can be replayed). Both the forward order and backward (byte-swapped) order version of it are required, called **eventID** and **swappedEventID**. They must all be printable characters with at least one uppercase and one lowercase letter and, of course, they must match.

```
#define  eventID          "RdcV"     //Always 4 characters, forwards order
#define  swappedEventID    "VcdR"    //Always 4 characters, backwards order
```

Next comes the menu name and category. Both of these are ***Pascal*** style strings. They are exactly 32 characters long and begin with a byte indicating the length of the characters actually used in the string. An error in this format will prevent Photoshop from recognizing the plug-in.

```
//menu entry generation: Always 32 characters, P-string
#define pluginMenuNamePStr32 "\x193.4.3.2 Deconvolve w/ PSF          "
#define pluginCategoryPStr32 "\x08R+R BOOK                           "
```

Finally, there is a C-style string that will be displayed if "About the Plug-in" is selected within Photoshop.

```
//ABOUT BOX C-string
#define AboutBoxMessageCStr      "Example Plug-In - 3.4.3.2. Deconvolve"
```

It is legal to embed **\n** (carriage return) characters within this string, and they will be parsed correctly in the "about" box.

It is relatively simple to change both the plug-in name and menu entry to suit specific needs. However, if there are two plug-ins with the same name and category (but, of course, different file names or residing in different folders), they both will be listed in the menu, and it may be difficult to determine which is which.

### A.1.6   Debugging

Generally, the best way to get a plug-in up and running is the same as for any kind of programming:

- Get the shell up and running first. If that does not work, you're not going to get very far with the rest of the code.

- Take extra care with the boundary conditions within the code. Handle the edges of the image and values that are out-of-range or zero.

- Initialize the data structures to be large enough. A very common problem arises when code walks off either end of an array. This is especially important for local variables that are allocated on the stack, which could cause ALL of the variables to be corrupted.

- Turn on pieces of the code one at a time rather than trying to "hot-test" everything at once. Sometimes this approach can work, but to nail down where a bug is, it is often necessary to turn on/off the code for a section at a time to locate the problem by elimination. Writing back images in intermediate stages of processing can be helpful to locate and understand problems.

- Make use of the file reading/writing functions (in Section A.2.6) to log portions of code that have been reached and to export values. Even though it can be very slow when writing data from an inner loop, it can be extremely useful to find out what is actually going on inside the code.

And three Photoshop-specific debugging notes:

- If Photoshop does not recognize the plug-in at all (i.e., it does not show up in the menu), look at the **PiPLData.h** file and make sure that both of the "P-strings" are exactly 32 characters long. If they are not, the PiPL resource will be malformed and will not parse correctly. Photoshop will ignore such plug-ins.

- Photoshop looks for plug-ins *only when it launches*. Hence, if a plug-in is added to the plug-in folder while the program is running, it will not show up. It *is* possible to replace an existing plug-in with another file of the same name (PC only) while the program is running. It can be very useful to put an alias (Macintosh) or shortcut (PC) to the plug-in into the Plug-Ins folder instead; then it can be recompiled (not when the plug-in is running, of course) to provide immediate access to newer versions of the code.

- Photoshop disables a plug-in that has crashed. If selecting a plug-in appears to do nothing, it may be that it crashed, or returned an error code other than **noErr** (0) or **userCanceledErr** (**-128**). Even though you have (hopefully) corrected the errors and recompiled the plug-in, Photoshop will not call it until the program is relaunched.

## A.2 Anatomy of a Plug-In

The **Usercode.c** file for each project contains the actual image processing and/or measurement code, but it makes use of some support routines that are provided in the **PhotoshopShell.c** file. The following section examines the various parts of a typical routine in detail.

### A.2.1 The Main Loop

In the example in Code Fragment A.2.1, the most primitive of loops is used to traverse an image. First, it finds the dimensions of the image with **GetOriginalDimensions()**, then it allocates a line buffer to hold one line's worth of data with **CreateAPointer()**. Notice that there are two fields for **CreateAPointer()**. The first is the number of data elements, and the second is the size of the data element. In the example shown, the data element is an **RGBPixel**, which is defined as a **struct** in the glue code. *Note: Do not use* **malloc()** *or* **free()** *because they will make debugging more difficult.*

```
// Code Fragment A.2.1
long height, width;    //pixel dimensions of image
long x, y;             //loop counters to traverse image
RGBPixel *line;        //memory buffer

   ...

GetOriginalDimensions(&width, &height);
line = CreateAPointer(width, sizeof(RGBPixel));

   ...

for (y = 0; y < height; y++)
{
   ReadOriginalLine(y, line);
   //do some kind of processing on that line
   WriteResultLine(y, line);
}

   ...

DisposeAPointer(line);
```

Once the dimensions of the image are known, the routine can traverse the image from top to bottom (lines are numbered in the C style from **0** to **height-1** with line number **0** at the top). The heart of the main loop depends upon two calls: **ReadOriginalLine()** and **WriteResultLine()**. *Note: It is important to note that writing data to the result image, and then trying to read that line from the original, will again read the unchanged original data. There is no way to read the data back that has been written to the result image.*

Finally, **DisposeAPointer()** is used for all allocated pointers at the end of the plug-in.

*Note: It is allowed to read or write any line multiple times, just as it is allowed to traverse the image in any desired order. Just remember that the result image will consist of the last data to be written to each line.*

```
void GetOriginalDimensions(long *width, long *height);
    //Get dimensions of this image
void ReadOriginalLine(long y, RGBPixel *line);
    //read from image
void WriteResultLine(long y, RGBPixel *line);
    //write to image
```

### A.2.2  Memory Allocation

In order to allocate memory and return it to the system, the following functions should be used rather than the standard **malloc()** and **free()** C functions. They provide greater clarity and will help to avoid allocation errors by explicitly combining the length of the line (in pixels) with the size of the data type, which can vary for different purposes. Also, because they are consistently used in all the example codes and worked problems, these function names can assist in searching for occurrences of allocation and deallocation.

```
void *CreateAPointer(long length, long size);
    //allocates (length*size). NULL if it fails
void *CreateAPointerClear(long length, long size);
    //allocates and fills with zeros. NULL if it fails
void DisposeAPointer(void *pointer);
    //deallocates a pointer
```

It is important to use **DisposeAPointer()** on any pointer that has been allocated. Otherwise, it will cause a memory leak in Photoshop and will produce an eventual crash.

### A.2.3  Accessing Pixel Data

When reading or writing a line of data from the Original Image, the Result Image, or the Reference Image (see Section A.2.4), the data are always in a floating point **RGBRGBRGB** format. An **RGBPixel** structure is used on the line to hold the data:

```
typedef struct
{
    float red;
    float green;
    float blue;      //standard format for all images. USUALLY 0 to 255
} RGBPixel;
```

A pointer large enough to hold a line of data and read in a line from the Original Image can be allocated as shown in the following code fragment:

```
// Code Fragment A.2.3
long  height, width;       //pixel dimensions of image
long  x, y;                //loop counters to traverse image
RGBPixel *line;            //memory buffer
```

```
...
GetOriginalDimensions(&width, &height);
line = CreateAPointer(width, sizeof(RGBPixel));
for (y = 0; y < height; y++)
{
   ReadOriginalLine(y, line);
   //simple example to reduce color saturation by 50%
   for (x = 0; x < width; x++)
   {
      float gray = (line[x].red + line[x].green + line[x].blue) / 3.0;
      line[x].red = (line[x].red + gray) / 2.0;
      line[x].green = (line[x].green + gray) / 2.0;
      line[x].blue = (line[x].blue + gray) / 2.0;
   }
   WriteResultLine(y, line);
}
DisposeAPointer(line);
```

It is also possible to access the pixels in any order, and to read or write the red, green, and blue components independently.

If the values in the user's code are not clipped to be in the range of 0.0 to 255.0, they will be clipped in most modes by the glue code. (The exception is 32-bit mode, where the values are not clipped because Photoshop can handle −infinity to +infinity.) The glue code will convert whatever the internal format is from Photoshop into the **RGBPixel** format. Hence, in the code grayscale images will appear to be color images with equal red, green, and blue values, and 8-bit, 16-bit, and 32-bit images will all occupy the same 0.0 to 255.0 range. There is no way or need to identify what underlying format the images may be in within Photoshop.

Note: It is possible to get values out of the 0.0 to 255.0 range if Photoshop is in 32-bit mode. This can make histograms somewhat challenging.

### A.2.4 The Reference Image

Filters in Photoshop are fundamentally designed for accessing one image (or one layer) at a time. As the program has changed over the years, it is now possible to access layers other than the current one, but the SDK code required for that is dated after Adobe began requiring a signed legal document from developers. Instead, this text avoids that requirement by implementing a method of storing the current image into what is called the Reference Image. This Reference Image is stored on disk and persists after each plug-in has ended (and even when the host program is ended or the computer is turned off). There is only one Reference Image, and initializing and writing a new one will erase any previous one. It can be accessed (if there is one) with the following functions:

```
ErrType  InitializeRefImage(long width, long height);
   //initializes new Reference image (eliminates previous one)
ErrType  GetRefDimensions(long *width, long *height);
   //get Reference image dimensions, returns error if not found
```

```
void       ReadRefImageLine(long y, RGBPixel *line);
   //read and write work with replacement
void       WriteRefImageLine(long y, RGBPixel *line);
   //values that are written can be read afterwards
```

This simple set of functions allow reading and writing to the Reference Image. The format is the same as the Original and Result images: **RGBPixel**.

It is not necessary to open or close the Reference Image as that is handled automatically by the glue code.

## A.2.5   Temporary Images

Functions that require iteration on image values, and some other algorithms, need a temporary image (or sometimes more than one) of an arbitrary size and depth. **RGBPixel** is not always suitable — some may only need an **unsigned short** or others might need a complex type like **FFTelem**.

```
// Code Fragment A.2.5 — Temporary image access
typedef struct        // defined in the glue code
{
   float real;
   float imag;        //standard complex format for FFT function.
} FFTelem;
…
Image_ID fftimage; //reference to the temp image
…
fftline = CreateAPointer(fftsize, sizeof(FFTelem));
if (MakeTemporaryImage(fftsize, fftsize, sizeof(FFTelem), "FFT",
        &fftimage) != noErr)
   goto mainexit;
…
WriteTempImageLine(fftimage, y, fftline); //example of row access
…
ReadTempImageColumn(fftimage, x, fftline);//example of column access
…
```

In addition to *row*-access functions (which read a horizontal row from the image), the support code provides *column*-access functions for the temporary images. Column access is considerably slower than row access because the pixels in a column are *not* contiguous in memory, but the functions can be very powerful for rotation, performing large Gaussian functions, performing Fourier or wavelet transforms, etc.

Remember that this requires a line array whose dimension is the **height** of the image, not its **width**. When using the column functions, it is advisable to allocate the pointer to be the larger of height and width (for operations such as computing a fast Fourier transform, the two dimensions will usually be the same).

```
dimension = width;
if (height > width) dimension = height;
line2 = CreateAPointer(dimension, sizeof(RGBPixel));   //big enough
```

With this method, **line2** is large enough to handle columns and rows when read from the image.

The following functions facilitate temporary image access:

```
ErrType  MakeTemporaryImage(long width, long height, long bytesperpix,
                          char *uniqueName, Image_ID *imageID);
   //Allocate a temporary image workspace in desired byte depth
void     ReadTempImageLine(Image_ID imageID, long y, void *line);
   //read and write work with replacement
void     WriteTempImageLine(Image_ID imageID, long y, void *line);
   //just like in the Reference Image
void     ReadTempImageColumn(Image_ID imageID, long x, void *line);
   //!!must allocate line to be large enough
void     WriteTempImageColumn(Image_ID imageID, long x, void *line);
   //!!must allocate line to be large enough
```

**Image_ID** is an intentionally opaque data structure that keeps track of whether the temporary image is allocated on disk or in memory. The temporary images are *not* persistent across plug-in calls. The only image that is persistent from one plug-in call to the next is the Reference Image. However, there can only be one Reference Image, whereas it is possible to create several temporary images, as needed.

The implementation of temporary images uses memory-mapped files with unique names and IDs that are created upon demand and disposed on plug-in completion. If there is insufficient memory available, the glue code falls back on large unmapped files within the user's Temporary Items folder, and subsequent access will be extremely slow; or, there might be an error returned if there is insufficient space available. Note that it is *not* necessary to dispose of a temporary image as that is handled automatically by the glue code when the plug-in terminates.

### A.2.6 Reading and Writing Data Files

For image analysis especially, it is important to be able to output the numerical information (such as feature measurements) obtained by the plug-in. This is also useful for some image processing operations such as generating an image histogram. It is possible, of course, to write routines that plot data or perform other operations, but for the purposes of this text writing out simple text files that can be read by data analysis routines, statistical packages, and spreadsheet programs is a sufficient tool to enable further interpretation of the data.

It is also very useful when writing a plug-in program for image processing or measurement to be able to supply input information. This can include logical variables to control options in the program (e.g., to select addition or subtraction of a reference image) or provide numeric values (e.g., a kernel of values to be used for convolutions).

There are two basic modes for accessing data (text) files: reading and writing. The text file implementation is designed to be very simple. The glue code provides routines that can access simple text (*.txt) files. These files can be written or read by programs such as NotePad as well as spreadsheet programs such as Excel.

Text files can either be read from or written to, but must be opened for one kind of access or the other. It is possible to *read* multiple values from a line (e.g., if they are separated by tabs or commas), but when *writing* a new value is written on every line. It is intended to be that way for the purposes of clarity.

It is perfectly legal to have multiple text files open at one time. Remember that the call must specify which file (by using its **File_ID**) is being accessed.

### A.2.6.1   Writing to a Text File

Begin by calling **CreateTextFileToWrite()**. This has the side effect of erasing any existing file of the same name. There is *not* a mechanism to append to an existing file.

Once the target file is created and opened (which happened in the same step), data are written to it using **WriteANumber()**. This outputs floating point numbers into the text file with each subsequent call until **CloseTextFile()** is called for that file.

```
// Code Fragment A.2.6.1 — Write out an image histogram
File_ID  myOutputDataFile = 0;
…

err = CreateTextFileToWrite("c:\\histogram.txt", &myOutputDataFile);
for (i = 0; i < 255; i++)
   if (err == noErr)
      err = WriteANumber(myOutputDataFile, Histo[i]);  //output array
CloseTextFile(myOutputDataFile);
```

### A.2.6.2   Reading from a Text File

Just as a text file can be opened for writing, a text file can be opened for reading with **OpenTextFileToRead()**. Once the file is opened for reading, a floating point value is read from it (all values are converted to floating point) by using **ReadANumber()**. If there is a bad value in the file, a zero (0) is returned, hopefully with an appropriate error value.

```
// Code Fragment A.2.6.2 — Read in a 5 × 5 kernel
File_ID myInputDataFile = 0;
…

err = OpenTextFileToRead("c:\\kernel5x5.txt", &myInputDataFile);
for (i = 0; i < 5; i++)
   for (j = 0; j < 5; j++)
      if (err == noErr)
         err = ReadANumber(myInputDataFile, &(myKernel[i][j]));
CloseTextFile(myInputDataFile);
```

With the exception of **CloseTextFile()**, all of these functions return errors if there are any. For instance, when reading off the end of a text file, an **eofErr** (end-of-file error) is returned. That can be useful for counting the number of values in a file. Expect to get an

**fnfErr** (file-not-found error) if the file does not exist for reading, and so on. A table of error values is in Section A.2.8.

**File_ID** is an intentionally opaque data structure that keeps track of an opened text file so that upon plug-in termination it is possible to warn the programmer that files are left open.

The following functions are used for text file access:

```
ErrType  CreateTextFileToWrite(char *filename, File_ID *fileRef);
ErrType  OpenTextFileToRead(char *filename, File_ID *fileRef);
void     CloseTextFile(File_ID fileRef);
ErrType  ReadANumber(File_ID fileRef, float *value);
ErrType  WriteANumber(File_ID fileRef, float value);
```

### A.2.7   Escape/Cancel/Abort and the Progress Bar

The Progress Bar in Photoshop (Figure A.2) can be very useful for two purposes:

1.  Telling the user the status of processing on the image, and
2.  Affording the user an opportunity to quit processing if he or she decides that it is either taking too long or that the wrong function was chosen.

In Photoshop these are actually two separate functions — one to drive the progress bar and the other to test whether the user clicks on cancel or presses the esc key (command-period on the Macintosh). They have been combined in this text because generally they are both performed at the same time.

```
Boolean  TestAbortSetProgressBar(float fraction);
  //goes from 0 to 1.0, set the bar anyplace
  //and see if the user wants to cancel
```

This function accepts a floating point value in the range of 0 to 1. There is no requirement to go in forward order (0.0, 0.1, 0.2, …, 1.0), but that is generally accepted behavior. If this function returns a **TRUE**, the user wants to cancel. If it is **FALSE**, then the user does not want to cancel. If the user does cancel, exit the main function with the error code **userCanceledErr (–128)**.

Also, it is generally not a good idea to call this function too often as it can significantly slow down execution. Once per line is generally enough. For very large images, even less frequent calls may be desirable.

**FIGURE A.2**
The Photoshop progress bar.

```
//push progress bar while traversing the image
for (y = 0; y < height; y++)
{
  if (TestAbortSetProgressBar((float)y / (float)(height))
    goto mainexit; //go to the end and dispose memory
  ReadOriginalLine(y, line);

  ...
}
```

If this function is not used, the program may appear to be unresponsive to the operating system, causing either an hourglass (Windows) or the spinning beach ball (Macintosh), but that is not an indicator that anything has crashed. The function is vital for allowing Photoshop to accept and process events from the operating system.

Finally, even though this function is called, the progress bar may not appear, especially for very fast functions. Photoshop will wait until a short time has passed before putting up the progress bar.

### A.2.8 Error Handling

Photoshop was originally written on the Macintosh and ported by Adobe to Windows. As such, all of the errors that the program understands are based upon that same set of Macintosh errors. The errors that the glue code routines return fit into that same numeric sequence. The result will be either a **noErr (0)** or a negative value (but it is safer to assume that anything other than **noErr** is an error). Below is a short list of the most common errors:

```
typedef WORD ErrType;
#define   noErr               0
#define   userCanceledErr     (-128)
#define   readErr             (-19)
#define   writErr             (-20)
#define   openErr             (-23)
#define   dirFulErr           (-33)     // directory full
#define   dskFulErr           (-34)
#define   nsvErr              (-35)     // no such volume
#define   ioErr               (-36)
#define   bdNamErr            (-37)     // bad name
#define   fnOpnErr            (-38)     // file not open
#define   eofErr              (-39)
#define   posErr              (-40)     // cannot position before start
#define   mFulErr             (-41)     // memory full (open)
#define   tmfoErr             (-42)     // too many files open
#define   fnfErr              (-43)
#define   wPrErr              (-44)     // disc is write-protected
#define   fLckdErr            (-45)     // file is locked
#define   vLckdErr            (-46)     // volume is locked
```

```
#define   fBsyErr          (-47)      // file is busy
#define   opWrErr          (-49)      // file already open for write
#define   paramErr         (-50)
#define   rfNumErr         (-51)      // bad fRefNum
#define   wrPermErr        (-61)      // permission doesn't allow writing
#define   memFullErr       (-108)
#define   nilHandleErr     (-109)
#define   memWZErr         (-111)
```

If the plug-in completes successfully, it must return a **noErr (0)** at the end. If the user cancels it must return a **userCanceledErr (-128)**. Any other errors should be propagated back out to Photoshop to produce a meaningful error message. Also note that if anything other than **noErr** or **userCanceledErr** value is returned, that plug-in will be disabled until Photoshop is relaunched.

---

## A.3  Inside the Glue Code

The **PhotoshopShell.c, .h**, and **.def** files define several data types and provide subroutines that handle many of the specific utility tasks shown in the code examples. These files are the "glue" code that implements the API.

### A.3.1  Structures

The glue code files contain definitions for two structures that hold the pixel data in various forms.

```
struct // contained in the glue code
{
     float     red,
               green,
               blue;
} RGBPixel;
```

Data are read from and written to the host program interface in this format, which consists of three floating point values, each in the range 0 to 255, for red, green, and blue, respectively.

```
struct // contained in the glue code
{
     float     real, // real component
               imag;// imaginary component
} FFTelem;
```

This is used in the Fourier transform routines to hold complex values. There are some others that are *not* defined in the glue but which can be defined in specific routines, as appropriate.

```
struct
{
    float     magnitude,
              phase;
} FFTPhaseMag;
```

The fast Fourier transform calculations use complex numbers, which may be converted as needed from real and imaginary format to magnitude and phase.

```
struct
{
    float     L,
              A,
              B;
} LABPixel;
```

Several of the processing routines convert the RGB color coordinates to another space, such as LAB. This also requires three floating point values. A similar structure would hold hue, saturation, and intensity values, or any other color coordinates.

```
struct
{
    long      IntPxl;
} IntegerPixel;
struct
{
    float     RealPxl;
} FloatPixel;
```

Binary images used in Chapters 4 and 5 may be represented as arrays of either integers or floating point numbers as appropriate.

### A.3.2   General Routines

These are utility functions that allocate memory for reading and writing images, and access text files for input and output.

```
A.3.2.1////////////////////////////////////////////////////////////////////////
void      *CreateAPointer(long length, long size);
```

This routine allocates memory for an array that will hold a line of pixel data. The size variable defines the data type (e.g., **sizeof(RGBPixel)**). The length is the number of values in the line, so the actual memory allocated is **(length * size)** bytes. The array length is not just the number of pixels, for instance, to read a horizontal line from an image of width W pixels, because each pixel has three floating point values (red, green, and blue), the array length would be **W * 3**. Also, note that the call will match a pointer that is compatible with any data type (**void\*** is compatible with **float\***, **RGBPixel\***, etc.). Typically, the code would declare

```
    RGBPixel *Line;
```

and then allocate memory with

```
Line = CreateAPointer(width, sizeof(RGBPixel));
```

**A.3.2.2**/////////////////////////////////////////////////////////////////////////
```
void    DisposeAPointer(void *pointer);
```

At the end of each routine, dispose any pointers created to free up the memory.

**A.3.2.3**/////////////////////////////////////////////////////////////////////////
```
void    ErrorMessage(char *Message);
```

This call always takes a C-string. It puts up a dialog that can warn the user of problems (e.g., image dimensions do not match reference image, etc.). It is also used internally for the **About This Plug-in** function when invoked from Photoshop. A typical use would be:

```
ErrorMessage("You can't do that.");
```

after which the code should dispose of any declared memory pointers and return a **userCanceledErr** or other appropriate error code to the host program. The call can also be useful for determining if the code reaches a certain point in execution.

**A.3.2.4**/////////////////////////////////////////////////////////////////////////
```
ErrType  CreateTextFileToWrite(char *filename, File_ID *UniqueFile);
```

This call creates and opens a text file to write output data. It will overwrite any existing file of the same name. It takes a C-string and returns an **ErrType** equal to **noErr** if no errors were encountered. If the file name corresponds to a file that is already open (e.g., if trying to monitor the output for debugging purposes by having the file open in Note-Pad), the call will fail and the new output will not be written. It returns **File_ID** that is used for subsequent file accesses. The following code fragment shows an example of the use of this function and the related routines:

```
// Code Fragment A.3.2.4
File_ID fref = 0;
...
//create and write the file "output.txt" on the c-drive
if (CreateTextFileToWrite("C:\\output.txt", &fref) == 0)
{ // convert integer value to a float
   WriteANumber(fref, (float)IntegerCount);
   CloseTextFile(fref);
} // if the file was successfully opened
```

Note the "\\" characters in the string are used to specify a *single* backslash. The "\" character is an escape-sequence for all kinds of other embedded values such as a carriage return "\n" or a null "\0" or many others (including hex values "\x13," etc.). In order to get an actual backslash "\" character, it must appear twice in the string. Failure to do so will cause painful compiler errors and the inexplicable failure of the file to get created and thus to be written.

**A.3.2.5**/////////////////////////////////////////////////////////////////////////
```
ErrType  OpenTextFileToRead(char *filename, File_ID *UniqueFile);
```

This call opens a text file from which input data will be read. It takes a C-string and returns an **ErrType** equal to **noErr** if no errors were encountered. The **File_ID** is used for subsequent file accesses. It is not possible to read from and write to the same file at the same time. Should it become necessary to read what has been written, close the file and reopen it with the **OpenTextFileToRead()** function. To reset the file position to the beginning, say, after counting the number of entries, simply close the file and open it again.

```
// Code Fragment A.3.2.5 — Read some numbers
File_ID  fref = 0;
long     count = 0;
...
//read from the file "input.txt" on the c-drive
if (OpenTextFileToRead("C:\\input.txt", &fref) == 0)
{  // read up to 5 numbers
   for (i = 0; i < 5; i++)
      if (err == noErr)
      {
          err = ReadANumber(fref, &(myArray[i]));
          count++;//keep track of the number of values read
      }
   CloseTextFile(fref);
} // if the file was successfully opened
```

Of course, if an unexpected error is detected (usually, a file not found error), it should be reported to the host program and a graceful exit made.

**A.3.2.6**/////////////////////////////////////////////////////////////////////
```
void    CloseTextFile(File_ID UniqueFile);
```

This call closes a text file previously opened for reading or writing.

**A.3.2.7**/////////////////////////////////////////////////////////////////////
```
ErrType  ReadANumber(File_ID UniqueFile, float *value);
```

This call reads one floating point value from an opened text file. Numbers may be terminated with a space, tab, return, or comma. Alpha (non-exponential notation) characters are ignored. The file pointer is advanced to the next value in the file. The call returns a zero if no legal numeric value was present. The following code fragment shows an example of use, including related routines:

```
// Code Fragment A.3.2.7 — Read a value
if (OpenTextFileToRead("C:\\GBlur.txt", &fref) == noErr)
{  // read value from file
   err = ReadANumber(fref, &sigma);
   CloseTextFile(fref);
}
```

If attempting to count the number of values in a file, it can be useful to look at the type of error encountered. Expect the **eofErr** value when the end of file is reached. Thus, if there are 25 values in a file, on the 26th read attempt the end-of-file error will occur. Once the number of values has been determined, it is possible to close the file, reopen it, and read the data in appropriately.

**A.3.2.8**////////////////////////////////////////////////////////////////////

```
void      WriteANumber(File_ID UniqueFile, float value);
```

This call writes one floating point value to an opened text file and advances the file pointer beyond it. The number is terminated with a line feed/carriage return. Calling this function repeatedly to write a series of numbers will produce a column of values that can be read by most spreadsheet and data analysis programs, as well as text processors.

```
// Code Fragment A.3.2.8 — Write a value
if (CreateTextFileToWrite("C:\\Results.txt", &fref) == noErr)
{  // write a value to the file
   WriteANumber(fref, sigma);
   CloseTextFile(fref);
}
```

Note that there is no error returned from **WriteANumber()**. Presumably, any error would have occurred when the file was created. It is generally not a good idea to write thousands of values as it will be quite slow.

**A.3.2.9**////////////////////////////////////////////////////////////////////

```
Boolean   TestAbortSetProgressBar(float fraction);
```

This call sets the position of Photoshop's progress bar, and tests to see if the user has canceled the routine. The legal range of values for **fraction** is between 0.0 and 1.0. It returns **TRUE** if the user has pressed the esc kay (command-period on the Mac) or clicked on the cancel button. It returns **FALSE** otherwise.

The progress bar may not actually come up if the routine is fast enough. In addition to using this as a polling mechanism, it allows other events to come through to Photoshop from the operating system, so it is a good idea to use it within medium to outer loops. If it is called too often, it can slow down processing as there is a fair amount of overhead in the call, so do not call it from inner loops.

### A.3.3  Accessing the Current Image

The current image is read from and written to the host program line by line. Each pixel is read and written as three floating point numbers (red, green, and blue, respectively) regardless of the image mode or type in the host program (grayscale, color, 8-bit, 16-bit, etc.).

**A.3.3.1**////////////////////////////////////////////////////////////////////

```
void      GetOriginalDimensions(long *width, long *height);
```

This call reads the dimensions in pixels of the current image.

**A.3.3.2**////////////////////////////////////////////////////////////////////

```
void      ReadOriginalLine(long y, RGBPixel *Line);
```

This call reads one horizontal line of pixels from the current image into an allocated array. The variable **y** counts from **0** to **height−1**. Note that the pixel data can be read as many times as desired, but will not change during the operation of a procedure. In other words, the function will always read the original values, not affected by any writing done to the result image.

**A.3.3.3**/////////////////////////////////////////////////////////////////////

```
void      WriteResultLine(long y, RGBPixel *Line);
```

This call writes one horizontal line of pixels to the current image from an allocated array. Writing does not alter the current contents of the image until the plug-in is ended with no error. The **y** variable counts from **0** to **height-1**. The data are always written as three floating point values (R, G, B) per pixel and will be converted automatically to the mode or type of the image in the host program.

## A.3.4   Accessing a Semi-Permanent Reference Image

The Reference Image is a disk file holding an image used in arithmetic and other operations that require access to a second image. The file persists even after the plug-in finishes or the host program is ended. The organization of the file is simply the width and height and depth (three **uint16**'s) followed by the pixels, line by line, with three floating point numbers (structure type **RGBPixel**) for the red, green, and blue values for each pixel. The file is named **Reference Image.xzy** and is stored in the user's private space, which by default is **C:\\Documents and Settings\UserName\Local Settings\Temp\**.

**A.3.4.1**/////////////////////////////////////////////////////////////////////

```
ErrType   InitializeRefImage(long width, long height)
```

This call erases the contents of the reference image file and creates a new file with the specific dimensions of the image that will be stored there. The function returns an **ErrType** equal to **noErr**.

**A.3.4.2**/////////////////////////////////////////////////////////////////////

```
ErrType   GetRefDimensions(long *width, long *height);
```

This call returns the dimensions of the image (in pixels) in the reference image file. The image data are stored as three RGB floating point values per pixel.

**A.3.4.3**/////////////////////////////////////////////////////////////////////

```
void      ReadRefImageLine(long y, RGBPixel *Line);
```

This call reads one horizontal line of pixels from the reference image into a previously allocated array. It functions just like the **ReadOriginalLine** call described above.

**A.3.4.4**/////////////////////////////////////////////////////////////////////

```
void      WriteRefImageLine(long y, RGBPixel *Line);
```

This call writes one horizontal line of pixels to the Reference Image from a previously allocated array. It functions just like the **WriteOriginalLine** call described above. Note that it is possible to read and write to the Reference Image, and that changes are made immediately. However, the intent of the Reference Image is to create a semi-permanent storage location, much like the memory button on a calculator, and not to use it as dynamic read-write memory. The temporary images in memory (see Section A.3.5) can be accessed more quickly, and there can be multiple temporary images present.

### A.3.5 Creating and Accessing Temporary Images in Memory

Holding a temporary image (or several) in memory is useful for many purposes, including performing iterative operations, converting the image to another format (e.g., real and imaginary values for Fourier transforms, or integers for binary image measurement), or generating a grid to combine with an image. These functions create and access image arrays in memory. Note that there is no requirement that these images have the format of **RGBPixel**. They can be specified with other data types, and this affects the array dimensions.

**A.3.5.1**/////////////////////////////////////////////////////////////////////

```
ErrType  MakeTemporaryImage(long xdimension, long ydimension,
         long bytesperpixel, char *uniquename, Image_ID *uniqueFile);
```

This call allocates the memory to hold a temporary image and returns a unique **Image_ID** used to access it. There can be multiple temporary images in memory.

*Warning: The temporary images are automatically discarded when the plug-in finishes and returns to the host program.*

The dimension values are the number of pixels. Because the values associated with each pixel can be specified according to the purpose, it is necessary to define the data type. This can be either single or multiple values, e.g., **bytesperpixel** could have a value equal to **sizeof(long)** or **sizeof(float)**, or **sizeof(RGBPixel)** or **sizeof(FFTelem)**. The call returns an **ErrType** equal to **noErr** for no error. If there is insufficient memory, it is possible to get other errors. As a last resort, the glue code will attempt to create a large file on disk, which will be quite slow. This does not access Photoshop's virtual memory architecture.

**A.3.5.2**/////////////////////////////////////////////////////////////////////

```
void     ReadTempImageLine(Image_ID uniqueFile, long y, void *Line);
```

This call reads one horizontal line of values from a temporary file. The **Image_ID** specifies which temporary file to use because there may be more than one. The line pointer can be of any data type, and it is important to use **CreateAPointer()** to allocate an array of the correct size. The value of **y** can range from **0** to the value of **ydimension-1** used to create the array.

**A.3.5.3**/////////////////////////////////////////////////////////////////////

```
void     WriteTempImageLine(Image_ID uniqueFile, long y, void *Line);
```

This call writes one horizontal line of values to a temporary file. The **Image_ID** specifies which temporary file to use because there may be more than one. The line pointer can be of any data type, and it is important to use **CreateAPointer()** to allocate an array of the correct size. The value of **y** can range from **0** to the value of **ydimension-1** used to create the array.

**A.3.5.4**/////////////////////////////////////////////////////////////////////

```
void     ReadTempImageColumn(Image_ID uniqueFile, long x, void *Line);
```

This call reads one vertical line of values from a temporary file. The **Image_ID** specifies which temporary file to use because there may be more than one. The line pointer can be of any data type, and it is important to use **CreateAPointer()** to allocate an array of the correct size. The value of **x** can range from **0** to the value of **xdimension-1** used to create the array. This function is primarily intended for use with FFT and wavelet transforms of temporary images, but may also be useful in other applications such as image rotation. It will read per pixel whatever the **bytesperpixel** amount was specified in the **MakeTemporaryImage()** call.

**A.3.5.5**////////////////////////////////////////////////////////////////////////////

```
void      WriteTempImageColumn(Image_ID uniqueFile, long x, void *Line);
```

This call writes one vertical line of values to a temporary file. The **Image_ID** specifies which temporary file to use because there may be more than one. The line pointer can be of any data type, and it is important to use **CreateAPointer()** to allocate an array of the correct size. The value of **x** can range from **0** to the value of **xdimension-1** used to create the array. This function is primarily intended for use with FFT and wavelet transforms of temporary images, but may also be useful in other applications such as image rotation.

### A.3.6   Errors

The function calls listed above are reasonably protected against common programming errors. For example, trying to read or write a line from or to an image beyond the bounds of the image will not crash the program, and writing too long a line will automatically truncate the data to prevent a crash. Attempting to use a null pointer that has not been allocated will be ignored; so will attempting to read or write to a temporary image without a valid ID number.

However, such errors will produce unpredictable results, and it is still possible to write a function that will crash, including taking down the host program. Attempting to read or write beyond the length of an allocated line array is a common and definitely fatal error. It is ultimately the student's responsibility to properly protect his or her code against errors and to ensure that the programs perform correctly.

The error codes that may be returned by these calls are listed in Section A.2.8.

All errors are of type **ErrType** which is similar to the Macintosh's **OSErr** type. Because Photoshop was originally written on the Macintosh, all of the errors that it understands use those same values. Because those values are dramatically different on different OS implementations, Adobe's values are used throughout this text and code.

There is a **#define RRDEBUG 1** declaration within PhotoshopShell.c that will cause an error message to be put up if there is an incorrect parameter (at least, in many cases). It is quite easy to change that to **0** to prevent such messages from appearing.

---

## A.4   The Photoshop Interface

There have been several versions of the Adobe Photoshop plug-in API that have evolved in functionality and complexity over time. Some of the recent versions of the Photoshop

Software Development Kit (SDK) have required programmers to register with and pay fees to Adobe, but the version used by the interface code used throughout the examples in this book is based on only a subset of the freely distributed version, first released with Photoshop 3.0. The plug-ins are still fully functional with all recent versions of Photoshop up to and including the newest, Photoshop CS3 (aka Photoshop 10.0).

Many other programs for both Windows and Macintosh have adopted the same interface conventions, allowing the plug-ins to be used with them as well, ranging from free (NIH-Image, UTHSCA Image) or quite inexpensive (Adobe Photoshop Elements, Corel Paint Shop Pro, Lemke Graphic Converter), through professional graphics tools (Deneba Canvas, Corel Painter), to very expensive science-oriented packages (Media Cybernetics Image-Pro Plus). The interface code and plug-ins will work with all of these programs.

The interface defines a set of function calls and a record format that passes information to the plug-in such as the image dimensions and data format. Some of the programs mentioned above do not support the full range of formats available in Adobe Photoshop, for example, restricting image bit depth to 8 bits per channel, and may not be able to open all of the example images (or will automatically reduce their bit depth when they are opened), but the functionality of the plug-ins and the interface is not affected by this restriction, and the routines will still perform in the same way.

The data format passed by the calling program is detailed in the **PhotoshopShell.h** and **Photoshophell.c** files on the companion CD, and is summarized below. There are additional values included in the record that are not needed by or used by the interface routines or the plug-ins.

The calls from the host program to the plug-in and the callbacks from the plug-in to the host program that are used in the plug-ins are summarized in the following text. There are additional callbacks defined in the Adobe API that are not supported or used by the interface routines or the plug-ins.

### A.4.1 Adobe's Plug-In API and the Tiny Portion Used in this Text

The support code for this text uses a very small fraction of Adobe's API. There are just three callbacks (**advanceState()**, **progressProc()**, and **abortProc()**). The code uses the plug-in API as of Photoshop 3.0.4 with a small number of additions from Photoshop 4 and 5 (mostly regarding image depth and mode). The actual current SDK from Adobe allows layer access, processing suites, display functions, and so on. If you are interested in the Adobe SDK, contact Adobe at <http://www.adobe.com/devnet/>.

### A.4.2 Data Structures

**Edit_PSData** is a heavily redacted version of Adobe's **FilterRecord** — the data structure used to pass in and out data from Photoshop. Adobe has also declared a **Rect** type containing **int16** values for top, left, bottom, and right and a **Point** type containing **h** (horizontal) and **v** (vertical) **int16** values. Finally, they use an **RGBColor** structure holding **int16** values for red, green, and blue.

All of these structures are confined to use by the glue code and are not needed when accessing the API in this book.

### A.4.3  Adobe Constants

In addition to the **ErrType** values, which are really Adobe's **OSErr** values, which in turn are really a subset of Apple's **errors.h** values, a small number of other constants are used to communicate with Adobe's plug-in socket. These image modes are documented in Section A.4.6. A number of these modes are supported and are, in all cases, translated to a floating point **RGBPixel** (0 to 255) structure within the glue code.

In addition to the image mode are a small number of values that correspond to states for Adobe's plug-in state machine (see the next section).

```
// Operation selectors
#define filterSelectorAbout        0
#define filterSelectorParameters   1
#define filterSelectorPrepare      2
#define filterSelectorStart        3
#define filterSelectorContinue     4
#define filterSelectorFinish       5
```

The full Adobe plug-in SDK is much more interesting (and complex!) and allows you to saving and recalling parameters, accessing other layers, driving dialogs, etc.

### A.4.4  The Calling Sequence

Photoshop plug-ins were originally state machines. They were called first with a "**Prepare**" flag so that the plug-in could allocate memory, a "**Start**" flag to do the first part of processing, and then repeatedly the "**Continue**" flag to process a little bit at a time. Finally, the "**Finish**" flag told the plug-in to free any memory that had been allocated. There is also an "**About**" flag to tell the plug-in to put up an About Box.

In this way, Photoshop remained in charge, and the plug-ins would process the image a little at a time, but would always return to Photoshop, and they could share the same calling stack.

In Photoshop 3.0.4 Adobe introduced a callback **->advanceState()**, which provided a way under plug-in control that the plug-in could send off the data that was now processed and fetch the next portion. This completely eliminated the need for the series of **"doContinue"** flags from Photoshop and greatly streamlined plug-ins.

Fundamental to the Photoshop plug-in calling method was the convention that data would always be read and written at the same time, i.e., that the image portion just processed would be written back and the next portion to be worked on would be read, and that Photoshop would provide the buffers for data transfer. This is very efficient for some kinds of image processing functions but not for others, especially those with a large neighborhood.

In this implementation, the reading of data is decoupled from the writing of data to allow having separate memory pools rather than relying upon those provided by Photoshop. It also allows having more than one piece of image data in memory at a time. The code in the **MainUserEntry()** routine is completely isolated from the rest of the Photoshop calling sequence.

### A.4.5 Callbacks

The support code for this text uses three callbacks:

->**advanceState()** is discussed in the preceding section (Section A.4.4).

->**progressProc()** is used to push the Progress Bar (see Section A.2.7).

->**abortProc()** is used to detect if the user wishes to cancel processing (see Section A.2.7).

### A.4.6 Pixel Ranges Depend on Modes

Photoshop has a large number of image modes:

```
#define plugInModeBitmap            0
#define plugInModeGrayScale         1     //supported
#define plugInModeIndexedColor      2
#define plugInModeRGBColor          3     //supported
#define plugInModeCMYKColor         4
#define plugInModeHSLColor          5
#define plugInModeHSBColor          6
#define plugInModeMultichannel      7
#define plugInModeDuotone           8
#define plugInModeLabColor          9     //supported via transform
#define plugInModeGray16            10    //supported
#define plugInModeRGB48             11    //supported
#define plugInModeLab48             12    //supported via transform
#define plugInModeCMYK64            13
#define plugInModeDeepMultichannel 14
#define plugInModeDuotone16         15
#define plugInModeRGB96             16    //supported
#define plugInModeGray32            17    //supported
```

The API used here supports a number of these modes by translating the data into a floating point RGB image with a numeric range of 0.0 to 255.0. After processing, the data are transformed back into the original range. However, Photoshop's internal data values are not that simple.

Of the modes that are supported, both **plugInModeGrayScale** and **plugInMode-RGBColor** have 8-bit images with value ranges 0 to 255. This is normal and what most people would expect.

The **plugInModeLabColor** mode is also a trio of 8-bit values, but they are transformed to RGB prior to processing, and then transformed back to LAB with the glue code to prevent a host of nightmares. The model for the transform is not pure CIELab (see Chapter 1) and is not identical to Adobe's, but it is reversible, so many kinds of processing can be performed on LAB images with similar results to their RGB counterparts. It is described in Section A.4.7.

The two modes **plugInModeGray16** and **plugInModeRGB48** correspond to Adobe's 16-bit image mode. Normally, this would suggest a pixel value range of 0 to 65535

(0 to $2^{16} - 1$), but in this case it does not. The actual range is 0 to 32768 (0 to $2^{15}$), inclusive. Reasons for this (from the Photoshop 5 era) include: (a) having a 50% value at 16384, which is not present in 8-bit mode, and (b) a number of processing shortcuts were possible on the CPUs of the time that made this mode comparatively fast. The implications are that it really is not a full 16 bits of data per channel, and that a true 16-bit TIFF image may not be correct in the bottom bit when read in. For this text's purposes, it simply does not matter, as the data are transformed into the 0.0 to 255.0 range.

Continuing on with LAB, but in 16-bit mode: **plugInModeLab48** still uses the range 0 to 32768 (as in the grayscale and RGB cases), but the data are transformed into RGB (0.0 to 255.0) for processing and then transformed back.

Finally, the support code also supports **plugInModeRGB96** and **plugInModeGray32**. Internally, the normal data range is 0.0 to 1.0 (floating point). Because the mantissa of a floating point value is 23 bits, the range is effectively $2^{23}$ unique values. (Actually, it is more than that as there are many, many more values close to 0.0). It is possible to get values that are both above and below that 0.0 to 1.0 range because of the HDR Exposure Blending function within Photoshop CS2 (and later). The support code translates this 0.0 to 1.0 range into 0.0 to 255.0 and values above and below that are transformed accordingly. It does mean that histograms of these images can be challenging.

### A.4.7  LAB to RGB Translation

The support code uses an approximation to transform LAB into RGB and back. It has the merit of being reproducible, but it is *not* identical to Adobe's transform.

```
L,a,b -> R,G,B
L' = exp(log(L+0.05)*1.28)-0.216113344)/4.71913337521
R = (L' + a/2 + b/3 + 0.5)
G = (L' - a/2 + b/3 +0.5)
B = (L' - 2*b/3 + 0.5)
```

```
R,G,B -> L,a,b
L' = (R + G + B + 1.0) / 3.0
L = 3.032 * exp(log(L'+0.05)*0.8)-0.09102821
a = 0.333333 * (R - G) + 127.5            //or +16384 in 16-bit mode
b = 0.833333 * ((R + G) / 2.0 - B) + 127.5  //or +16384 in 16-bit mode
```

This is sufficient for many kinds of processing (sharpening, blurs, etc.), but it is very obvious that the transform is not the full CIELab if the contrast of an image is reversed in LAB mode within a plug-in and compared that to the RGB case.

### A.4.8  Unsupported Modes

The modes that are not supported include: **plugInModeBitmap**, packed bits either 0 or 1 for black and white; **plugInModeIndexedColor**, single-byte values that index into a lookup table to convert to RGB; **plugInModeCMYKColor** and **plugInModeCMYK64**, that might possibly be supported via transform but are not (the data are present in four channels instead of three, and either 8 bits or Adobe's pseudo 16 bits in the latter case); **plugInModeMultichannel** and **plugInModeDeepMultichannel**, 8-bit and pseudo

16-bit versions of data that can come in as many as 56 channels; and finally, **plugInMode-Duotone** and **plugInModeDuotone16**, which are used primarily in the printing industry.

Because the image data can come in so many formats, it is rare for a plug-in to work in all image modes. Photoshop can convert any of these modes to one of the supported ones to enable a plug-in to be applied.

### A.4.9 Lines versus Tiles

Most of Adobe's example plug-ins show tile-based access: request a rectangular portion of the image, process it, and return it to Photoshop. All of the glue code uses a line-oriented approach in which there are fewer boundary conditions to deal with (the boundaries between tiles can cause huge problems for large-area neighborhoods and iterative algorithms), and any code generated is much more appropriate for the current generation of hardware, especially considering how prefetching, L1-caching, and L2-caching work. Furthermore, tile-based FFT algorithms are a nightmare, and the authors' intent is teaching image processing, not optimization.

## A.5 Full API Summary

For convenient quick access, here is a compact summary of the contents of the API:

```
//TYPES

typedef int32 Image_ID;    //opaque structure for temp image data
typedef int32 File_ID;     //opaque structure for file data
typedef int16 ErrType;     //See list of Photoshop errors above

//STRUCTURES

typedef struct
{
   float red;
   float green;            //standard format for all images.
   float blue;             //value is USUALLY 0 to 255
} RGBPixel;

typedef struct
{
   float real;
   float imag;             //standard format for FFT function.
} FFTelem;

//FUNCTION PROTOTYPES

   //Original image (RGBPixel)-Photoshop's current layer or selection.
   //Note: you CANNOT read what you write to the image
void    GetOriginalDimensions(long *width, long *height);
   //Get dimensions of this image
```

```
void      ReadOriginalLine(long y, RGBPixel *line);
   //read from image - always get the original values
void      WriteResultLine(long y, RGBPixel *line);
   //write to image - values cannot be read afterwards

   //Reference image (RGBPixel)- archived static image for
   //two-image operations
   //Note: you CAN read what you write in the reference image
   //There is only ONE reference image
ErrType  InitializeRefImage(long width, long height);
   //initializes new Reference image
ErrType  GetRefDimensions(long *width, long *height);
   //get Reference image dimensions
void      ReadRefImageLine(long y, RGBPixel *line);
   //read from the image
void      WriteRefImageLine(long y, RGBPixel *line);
   //values that are written can be read afterwards

   //Temporary image(s) (any format) - temporary block of memory, but
   //accessed with Read/Write functions
   //Note: you CAN read what you write in any temporary image
   //There can be multiple temporary images
ErrType  MakeTemporaryImage(long width, long height, long bytesperpix,
                            char *uniqueName, Image_ID *imageID);
   //Allocate a temporary image workspace in any bytedepth
void      ReadTempImageLine(Image_ID imageID, long y, void *line);
   //read and write work with replacement
void      WriteTempImageLine(Image_ID imageID, long y, void *line);
   //just like in the Reference Image
void      ReadTempImageColumn(Image_ID imageID, long x, void *line);
   //!!must allocate line to be large enough
void      WriteTempImageColumn(Image_ID imageID, long x, void *line);
   //!!must allocate line to be large enough

   //Utility functions, memory functions are protected!
   //Don't use malloc!
void      ErrorMessage(char *message);
   //put a C-string up in a dialog
void      *CreateAPointer(long length, long size);
   //allocates (length*size). NULL if it fails
void      *CreateAPointerClear(long length, long size);
   //allocates and fills with zeros. NULL if it fails
void      DisposeAPointer(void *pointer);
   //deallocates a pointer

   //Text file access
ErrType  CreateTextFileToWrite(char *filename, File_ID *fileRef);
ErrType  OpenTextFileToRead(char *filename, File_ID *fileRef);
void      CloseTextFile(File_ID fileRef);
```

```
ErrType  ReadANumber(File_ID fileRef, float *value);
ErrType  WriteANumber(File_ID fileRef, float value);

    //Progress bar / command-period & Escape test
    //Push Photoshop's progress bar and test to see if user cancels
    //(ESC/CommandPeriod/Cancel Button)
Boolean  TestAbortSetProgressBar(float percentage);
    //goes from 0 to 1.0, set the bar anyplace
    //and, see if the user wants to cancel

    //USER ENTRY POINT
ErrType  MainUserEntry();
    //the plug-in is called. return one of the errors at the end.
    //hopefully noErr.
```

# References and Literature

## A Useful Library

There is a vast literature on image processing, some of which can be found under topic headings such as "machine vision," "image understanding," "robotics vision," "remote sensing," "morphology," "image analysis," "pattern recognition," and other similar labels. Also, many publications deal with specific applications of these methods in fields as diverse as astronomy and microscopy, and within the latter field may address topics ranging from materials and food microstructure to biology, pharmaceutical research, or medical imaging. It is not intended to present any comprehensive review of the literature, even regarding the origins or programming of the many specific techniques in this text. Rather, the emphasis is on supplying relevant references to those key papers that may be useful for further understanding a particular subject, or provide insight into key advances in the broader field.

Many of the individual procedures and algorithms presented in this text are described in several of the general texts listed here, often with variations in naming conventions and with different forms of mathematical or other descriptions and explanations. The collection of techniques presented here include those that have become more or less standard in the field, as well as some of the newer ones that represent valuable extensions of those algorithms. It is primarily the latter that are accompanied by references.

In addition to the references cited in the body of the text and listed below, there are many general books that give overviews on many of the subjects covered here. Some of them are focused primarily on the mathematical theories and bases for various operations (and some of them use specialized notation). Others use examples to show the possibilities of the various procedures, with minimal math. Still others are devoted to the actual programming of algorithms, often with minimal information on their purpose or result.

The following lists are intended to provide an overview of some of the more useful texts, in the opinion of the authors. No single book covers all of the methods presented here, and most of them include other techniques (some quite specialized or advanced) that are not covered in this text. Furthermore, the books that include software must generally include routines that acquire, display, print, and store images. This text defers those functions to a host program (such as Photoshop) to concentrate on the actual procedures for image processing and analysis.

## General Image Processing Texts

D. H. Ballard, C. M. Brown (1982). *Computer Vision*, Prentice Hall, Upper Saddle River, NJ.

K. R. Castleman (1996). *Digital Image Processing*, 2nd ed., Prentice Hall, Upper Saddle River, NJ.

R. C. Gonzalez, R. E. Woods (2002). *Digital Image Processing*, 2nd ed., Prentice Hall, Upper Saddle River, NJ.

B. Jahne (1997). *Practical Handbook on Image Processing for Scientific Applications*, CRC Press, Boca Raton, FL.

A. K. Jain (1989). *Fundamentals of Digital Image Processing*, Prentice Hall, Upper Saddle River, NJ.

W. Niblack (1986). *An Introduction to Digital Image Processing*, Prentice Hall, Upper Saddle River, NJ.

I. Pitas (2000). *Digital Image Processing Algorithms and Applications*, Wiley, New York.

W. K. Pratt, (2001). *Digital Image Processing*, 3rd ed., Wiley, New York.

R. Rosenfeld, A. C. Kak (1982). *Digital Picture Processing*, 2nd ed., Academic Press, New York.

J. C. Russ (2006). *The Image Processing Handbook*, 5th ed., CRC Press, Boca Raton, FL.

L. G. Shapiro, G. C. Stockman (2001). *Computer Vision*, Prentice Hall, Upper Saddle River, NJ.

## Programming and Examples

G. A. Baxes (1994). *Digital Image Processing: Principles and Applications*, Wiley, New York.

R. Crane (1997). *A Simplified Approach to Image Processing*, Prentice Hall, Upper Saddle River, NJ.

H. R. Myler, A. R. Weeks (1993). *Pocket Handbook of Image Processing Algorithms in C*, Prentice Hall, Englewood Cliffs, NJ.

J. R. Parker (1997). *Algorithms for Image Processing and Computer Vision*, Wiley, New York.

T. Pavlidis (1982). *Algorithms for Graphics and Image Processing*, Computer Science Press, Rockville, MD.

G. X. Ritter, J. N. Wilson (2000). *Handbook of Computer Vision Algorithms in Image Algebra*, 2nd ed., CRC Press, Boca Raton, FL.

M. Seul, L. O'Gorman, M. J. Sammon (2000). *Practical Algorithms for Image Analysis: Description, Examples and Code*, Cambridge University Press, Cambridge, U.K.

S. E. Umbaugh (1998). *Computer Vision and Image Processing: A Practical Approach Using CVIPTools*, Prentice Hall, Upper Saddle River, NJ.

## Specialized Topics

### *Morphological Processing of Binary Images*

M. Coster, J.-L. Chermant (1985). *Précis D'Analyse D'Images*, Éditions du Centre National de la Recherche Scientifique, Paris.

E. R. Dougherty, J. Astola (1999). *Nonlinear Filters for Image Processing*, SPIE, Bellingham, WA.

H. Heijmans (1994). *Morphological Image Operators*, Academic Press, New York.

J. Serra (1982). *Image Analysis and Mathematical Morphology*, Academic Press, New York.

P. Soille (1999). *Morphological Image Analysis: Principles and Applications*, Springer Verlag, New York.

### *Feature Measurement and Recognition*

S.-T. Bow (1992). *Pattern Recognition and Image Preprocessing*, Marcel Dekker, New York.

L. F. Costa, R. M. Cesar Jr. (2000). *Shape Analysis and Classification: Theory and Practice*, CRC Press, Boca Raton, FL.

K. Fukunaga (1990). *Statistical Pattern Recognition*, 2nd ed., Academic Press, San Diego, CA.

### *Color Imaging and Printing*

H. R. Kang (1999). *Digital Color Halftoning*, SPIE, Bellingham, WA.

G. Sharma, Ed. (2003). *Digital Color Imaging Handbook*, CRC Press, Boca Raton, FL.

## Computer Graphics

J. D. Foley, A. van Dam, S. K. Feiner, J. F. Hughes (1996). *Computer Graphics Principles and Practice*, 2nd ed., Addison Wesley, Boston, MA.

G. Wohlberg (1990). *Digital Image Warping*. IEEE Computer Society Press, Los Alamitos, CA.


## Stereological Measurement of Microscope Images

A. Baddeley, E. B. Vedel Jensen (2006). *Stereology for Statisticians*, CRC Press, Boca Raton, FL.

C. V. Howard, M. G. Reed (2005). *Unbiased Stereology: Three Dimensional Measurement in Microscopy*, 2nd ed., Bios Scientific, Oxford, U.K.

J. C. Russ., R. T. Dehoff (2002). *Practical Stereology*, 2nd ed., Kluwer Academic, New York.

E. R. Weibel (1979). *Stereological Methods*, Vol. 1 & 2, Academic Press, New York.


## Application-Specific Examples

E. R. Davies (1990). *Machine Vision: Theory, Algorithms, Practicalities*, Academic Press, New York.

P. Francus, Ed. (2004). *Image Analysis, Sediments and Paleoenvironments*, Kluwer, New York.

P. M. Mather (1999). *Computer Processing of Remotely Sensed Images*, Wiley, New York.

R. M. Rangayyan (2005). *Biomedical Image Analysis*, CRC Press, Boca Raton, FL.

J. C. Russ (2001). *Forensic Uses of Digital Imaging*, CRC Press, Boca Raton, FL.

J. C. Russ (2004). *Image Analysis of Food Microstructure*, CRC Press, Boca Raton, FL.

F. F. Sabins Jr. (1987). *Remote Sensing: Principles and Interpretation*, 2nd ed., Freeman, New York.

J. Sanchez, M. P. Canton (2001). *Space Image Processing*, CRC Press, Boca Raton, FL.


## Additional Specific References in the Text

J. Astola, P. Haavisto, Y. Neuvo (1990). Vector median filters. *Proc IEEE* 78: 678–668.

D. H. Ballard (1981). Generalizing the Hough Transform to detect arbitrary shapes. *Pattern Recognition* 13(2): 111–122.

V. Berzins (1984). Accuracy of Laplacian edge detectors. *Comp. Vis. Graph. Image Proc.* 27: 1955–2010.

S. Beucher, F. Meyer (1992). The morphological approach of segmentation: the watershed transformation, in *Mathematical Morphology in Image Processing* (E. R. Dougherty, Ed.), Marcel Dekker, New York.

R. N. Bracewell (2000). *The Fourier Transform and its Application*, 3rd ed., McGraw Hill, New York.

G. Braudaway (1987). A procedure for optimum choice of a small number of colors from a large color palette for color imaging. *Proc. Electr. Imaging '87*, San Francisco, CA.

J. Canny (1986). A computational approach for edge detection. *IEEE Trans. Patt. Anal. Mach. Intell.* 8(6): 679–698.

P. Campisi, K. Egiazarian (Eds.) (2007). *Blind Image Deconvolution, Theory and Applications*, CRC Press, Boca Raton, FL.

R. L. T. Cederberg (1979). Chain-link coding and segmentation for raster scan devices. *Comp. Graph. Image Proc.* 10: 224–234.

CIE (1978). *Uniform Color Spaces*, Commission Internationale de L'Eclairage Publication 15, Paris.

M. L. Comer, E. J. Delp (1999). Morphological operations for color image processing. *J. Electron. Imaging* 8: 279–289.

J. W. Cooley, J. W. Tukey (1965). An algorithm for the machine calculation of complex Fourier series. *Math. of Comput.* 19: 297–301.

A. M. Cormack (1963), Representation of a function by its line integrals with some radiological applications. *J. Appl. Phys.* 34: 2722–2727.

A. M. Cormack (1964). Representation of a function by its line integrals with some radiological applications II. *J. Appl. Phys.* 35: 2908–2913.

P. E. Danielsson (1980). Euclidean distance mapping. *Comp. Graph. Image Proc.* 14: 227–248.

I. Daubechies (1992). Ten lectures on wavelets, *Soc. Ind. Appl. Math.*, Philadelphia, PA.

J. Davidson (1991). Thinning and skeletonization: a tutorial and overview in *Digital Image Processing: Fundamentals and Applications* (E. Dougherty, Ed.), Marcel Dekker, New York.

E. R. Davies (1988). On the noise suppression and image enhancement characteristics of the median, truncated median and mode filters. *Patt. Recog. Lett.* 7:87–97.

R. O. Duda, P. E. Hart (1982). Use of the Hough Transformation to detect lines and curves in pictures. *Comm. ACM* 15(1): 11–15.

S. F. Ferson, F. J. Rohlf, R. K. Koehn (1985). Measuring shape variation of two-dimensional outlines. *Systematic Zoology* 34: 59–68.

H. Freeman (1961). On the encoding of arbitrary geometric configurations. *IEEE Trans. Elec. Comput.* EC-10: 260–268.

R. L. Graham, F. F. Yao (1983). Finding the convex hull of a simple polygon. *J. Algorithms* 4: 324–331.

N. Guil, E. L. Zapata (1997). Lower order circle and ellipse Hough transform. *Patt. Recog.* 30(10): 1729–1744.

R. M. Haralick, L. G. Shapiro (1985). Survey: image segmentation. *Comp. Vis. Graph. Image Proc.* 29: 100–132.

R. M. Haralick, R. Shanmugan, I. Dinstein (1973). Textural features for image classification. *IEEE Trans. Syst. Man Cyb.* SMC-3(6): 610–621.

P. Heckbert (1982). Color image quantization for frame buffer display. *Computer Graphics* 16(3): 297–307.

G. T. Herman (1980). *Image Reconstruction from Projections—The Fundamentals of Computerized Tomography*, Academic Press, New York.

P. V. C. Hough (1962). Methods and means for recognizing complex patterns, U. S. Patent 3,069,654.

T. S. Huang, G. T. Yang, G. Y. Tang (1979). A fast two-dimensional median filtering algorithm. *IEEE Trans. Acoust. Speech Sign. Proc.* ASSP-27: 13–18.

D. A. Huffman (1952). A method for the construction of minimum redundancy codes. *Proc IRE* 40(10): 1098–1101.

R. Hunter, A. H. Robinson (1980). International digital facsimile coding standards. *Proc. IEEE* 68(7): 854–867.

H. Hwang, R. A. Haddad (1995). Adaptive median filters: new algorithms and results. *IEEE Trans. Image Proc.* 4(4): 499–502.

A. C. Kak, M. Slaney (2001). Principles of computerized tomographic imaging, *Soc. Ind. Appl. Math.*, Philadelphia, PA.

J. N. Kanpur, P. K. Sahoo, A. K. C. Wong (1985). A new method for grey-level picture thresholding using the entropy of the histogram. *Comp. Vis. Graph. Image Proc.* 29: 273–285.

B. H. Kaye (1986). Image analysis procedures for characterizing the fractal dimension of fine particles. *Proc. Particle Technol. Conf.*, Nürnberg Germany.

S. H. Kim, J. P. Allebach (2005). Optimal unsharp mask for image sharpening and noise removal. *J. Electronic Imaging* 14(2): 023005.

J. Kittler, J. Illingworth, J. Foglein (1985). Threshold selection based on a simple image statistic. *Comp. Vis. Graph. Image Proc.* 30: 125–147.

M. Kuwahara, K. Hachimura, S. Eiho, and M. Kinoshita (1976). Processing of RI-angiocardiographic images in *Digital Processing of Biomedical Images* (K. Preston and M. Onoe, Eds.), Plenum, New York, pp. 187–202.

R. L. Lagendijk, J. Biemond (1991). *Iterative Identification and Restoration of Images*, Kluwer Academic, Boston, MA.

L. Lam, S. Lee, C. Suen (1992). Thinning methodologies—a comprehensive survey. *IEEE Trans. Patt. Anal. Mach. Intell* 14: 868–885.

S. U. Lee, S. Y. Chung, R. H. Park (1990). Comparative performance study of several global thresholding techniques for segmentation. *Comp. Vis. Graph. Image Proc.* 52(2): 171–190.

S. Levialdi (1972). On shrinking binary picture patterns. *Commun. ACM* 15(1): 7–10.

S. Loncaric (1998). A survey of shape analysis techniques. *Patt. Recog.* 31(8): 983–1010.

H. E. Lu, P. S. Wang (1986). Comment on "a fast parallel algorithm for thinning digital patterns." *Commun. ACM* 29(3): 239–242.

B. B. Mandelbrot (1967). How long is the coast of Britain? Statistical self-similarity and fractional dimension. *Science* 155: 636–638.

B. B. Mandelbrot (1982). *The Fractal Geometry of Nature*, W. H. Freeman, San Francisco, CA.

D. Marr, E. Hildreth (1980). Theory of edge detection. *Proc. Roy. Soc. Lond.* B207: 187–217.

G. A. Mastin (1985). Adaptive filters for digital image noise smoothing: an evaluation. *Comp. Vis. Graph. Image Proc.* 31: 102–121.

V. Murino, C. Ottonello, S. Pagnan (1998). Noisy texture classification: a higher order statistical approach. *Patt. Recog.* 31(4): 383–393.

K. Oistämö, Y. Neuvo (1990). Vector median operations for color image processing in nonlinear image processing, E. J. Delp. Ed., *SPIE Proc.* 1247: 2–12.

N. Otsu (1979). A threshold selection method from gray-level histograms. *IEEE Trans. Syst. Man Cyber.* SMC-9: 62–69; 377–393.

A. W. Paeth (1986). A fast algorithm for general raster rotation. *Graphics Interface '86*, 77–81.

D. P. Panda, A. Rosenfeld (1978). Image segmentation by pixel classification in (gray level, edge value) space. *IEEE Trans. Comput.* 27: 875–879.

J. R. Parker (1991). Gray level thresholding in badly illuminated Images. *IEEE Trans Patt. Anal. Mach. Intell.* 13(8): 813–819.

T. Pavlidis (1980). A thinning algorithm for discrete binary images. *Computer Graphics and Image Processing* 13: 142–157.

S. Peleg, J. Naor, R. Hartley, D. Avnir (1984). Multiple resolution texture analysis and classification. *IEEE Trans. Patt. Anal. Mach. Intell.* PAMI-6: 518.

W. B. Pennebaker, J. L. Mitchell (1992). *JPEG Still Image Data Compression Standard*, Van Nostrand Reinhold, New York.

A. P. Pentland, (1983). Fractal-based description of natural scenes. *IEEE Trans. Patt. Anal. Mach. Intell.* PAMI-6: 661.

J. C. Russ (1984). Implementing a new skeletonizing method. *J. Microscopy* 136: RP7.

J. C. Russ (1990). Processing images with a local Hurst operator to reveal textural differences. *J. Comp. Assist. Microsc.* 2(4): 249–257.

J. C. Russ (1991). Multiband thresholding of images. *J. Comp. Assist. Microsc.* 3(2): 77–96.

J. C. Russ (1993). Method and application for ANDing features in binary images, *J. Comp. Assist. Microsc.* 5(4): 265–272.

J. C. Russ (1995). Median filtering in color space. *J. Comp. Assist. Microsc.* 7(2): 83–90.

J. C. Russ, J. C. Russ (1988). Improved implementation of a convex segmentation algorithm. *Acta Stereologica* 7: 33–40.

J. C. Russ, J. C. Russ (1989). Uses of the Euclidean distance map for the measurement of features in images. *J. Comput. Assist. Microsc* 1(4): 343.

P. K. Sahoo, S. Soltani, A. Wong, Y. C. Chan (1988). Survey of thresholding techniques. *Comp. Vis. Graph. Image Proc.* 4: 233–260.

L. J. Sartor, A. R. Weeks (2001). Morphological operations on color images. *J. Electr. Imag.* 10(2): 548–559.

M. Sezgin, B. Sankur (2004). Survey over image thresholding techniques and quanitative performance evaluation. *J. Electr. Imag.* 13(1): 146–165.

C. E. Shannon (1948). A mathematical theory of communication. *Bell System Tech. J.* 27(3): 379–423.

I. Sobel (1970). Camera models and machine perception. Ph.D. thesis, Stanford University, Stanford, CA.

J. A. Stark (2000). Adaptive image contrast enhancement using generalizations of histogram equalization. *IEEE Trans. Image Proc.* 9(5): 889–896.

J. A. Stark, W. J. Fitzgerald (1996). An alternative algorithm for adaptive histogram equalization. *Comp. Vis. Graph. Image Proc.* 56(2): 180–185.

H. Schwarz, H. E. Exner (1983). The characterization of the arrangement of feature centroids in planes and volumes. *J. Microscopy* 129: 155.

P. Sussner, G. X. Ritter (1997). Decomposition of gray-scale morphological templates using the rank method. *IEEE Trans Patt. Anal. Mach. Intell.* 19(6): 649–658.

A. Tanaka, M. Kameyama, S. Kazama, O. Watanabe (1986). A rotation method for raster image using skew transformation. *Proc IEEE Comp. Vis. Patt. Recog.* 272–277.

R. J. Wall, A. Klinger, K. R. Castleman (1974). Analysis of image histograms. *Proc. 2nd Joint Int'l Conf. Patt. Recog. IEEE* 74CH-0885-4C: 341–344.

J. S. Weszka (1978). A survey of threshold selection techniques. *Comp. Graph. Image Proc.* 7: 259–265.

J. S. Weszka, C. Dyer, A. Rosenfeld (1976). A comparative study of texture measures for terrain classification. *IEEE Trans. Syst. Man Cyb.* SMC-6: 269–285.

J. B. Wilburn (1998). Developments in generalized rank-order filters. *J. Opt. Soc. Amer.-A* 15(5): 1084–1099.

G. J. Yang, T. S. Huang (1981). The effect of median filtering on edge location estimation. *Comp. Graph. Image Proc.* 15: 224–245.

T. Y. Zhang, C. Y. Suen (1984). Fast parallel algorithm for thinning digital patterns. *Commun. ACM* 27(3): 236–239.

H. Zhu, F. Chan, F. K. Lam (1999). Image contrast enhancement by constrained local histogram equalization. *Comp. Vis. Image Under.* 73(2): 281–290.

# *Index*

Entries in ***bold, italic, underline*** indicate key words.